NATURAL PRODUCTS

NATURAL PRODUCTS
A Laboratory Guide

Second Edition

Raphael Ikan

Department of Organic Chemistry
The Hebrew University of Jerusalem
Jerusalem, Israel

ACADEMIC PRESS, INC.
Harcourt Brace Jovanovich, Publishers
San Diego   New York   Boston
London   Sydney   Tokyo   Toronto

This book is printed on acid-free paper. ∞

Copyright © 1991, 1969 by ACADEMIC PRESS, INC.
All Rights Reserved.
No part of this publication may be reproduced or transmitted in any form or by any means, electronic or mechanical, including photocopy, recording, or any information storage and retrieval system, without permission in writing from the publisher.

Academic Press, Inc.
San Diego, California 92101

United Kingdom Edition published by
Academic Press Limited
24–28 Oval Road, London NW1 7DX

Library of Congress Cataloging-in-Publication Data

Ikan, Raphael, date
  Natural products : a laboratory guide / Raphael Ikan. -- 2nd ed.
    p. cm.
  Includes index.
  ISBN 0-12-370551-7
  1. Biochemistry--Laboratory manuals.   I. Title.
QD415.5.I4   1991
547'.0028--dc20   91-12738
                                                                     CIP

PRINTED IN THE UNITED STATES OF AMERICA
91 92 93 94   9 8 7 6 5 4 3 2 1

*To my wife Yael
and our four favorite natural products
Amiram, Ariel, Arnon, and Eliana*

CONTENTS

PREFACE TO THE SECOND EDITION xi
PREFACE TO THE FIRST EDITION xiii

1 ACETOGENINS

I. Flavonoids 1
   A. Introduction 1
   B. Isolation of Hesperidin from Orange Peel 9
   C. Acidic Degradation of Hesperidin 12
   D. Isolation of Naringin from Grapefruit Peel 14
   E. Synthesis of Naringin Dihydrochalcone—A Sweetening Agent 17
   F. Color Changes of Anthocyanins at Various pH Values 19
   G. Questions 22
   H. Recommended Books 22

II. Lipids 23
   A. Introduction 23
   B. Isolation of Trimyristin and Myristicin from Nutmeg 26
   C. Preparation of Azelaic Acid from Castor Oil 30
   D. Formation of n-Heptaldehyde and Undecenoic Acid from Castor Oil 32
   E. Ozonization of Methyl Oleate 34
   F. Preparation of Urea Inclusion Compounds 36
   G. Gas–Liquid Chromatography of Methyl Esters of Fatty Acids 41
   H. Thin-Layer Chromatography of Fatty Acids 44
   I. Questions 46
   J. Recommended Books 47

III. Lignans 47
   A. Introduction 47
   B. Isolation of Sesamin and Sesamolin from Sesame Oil 50
   C. Questions 53
   D. Recommended Book 54

IV. Quinones 54
   A. Introduction 54
   B. Isolation of Rhein from Rhubarb Root 59
   C. Isolation and Identification of Phenols and Quinones from Defensive Secretions of Beetles 61
   D. Questions 66
   E. Recommended Books 66

V. Phloroglucinols 66
  A. High-Performance Liquid Chromatography of Hop Bitter Substances 67

2 CARBOHYDRATES

I. Mono- and Oligosaccharides 70
  A. Introduction 70
  B. Isolation of D-Mannoheptulose from Avocado 79
  C. α-D-Glucosamine from Crustacean Shells 82
  D. Chromatographic Separation of Sugars on Charcoal 84
  E. Gas–Liquid Chromatography of Carbohydrates 86
  F. Thin-Layer Chromatography of Carbohydrates 89
  G. Separation of Carbohydrates by High-Performance Thin-Layer Chromatography 91
  H. Questions 92
  I. Recommended Books 93

II. Polysaccharides 93
  A. Introduction 93
  B. Isolation of Amylopectin and Amylose from Potato Starch 97
  C. Isolation of Wood Cellulose 101
  D. Questions 104
  E. Recommended Books 104

3 ISOPRENOIDS

I. Carotenoids 105
  A. Introduction 105
  B. Isolation of Capsanthin from Paprika 110
  C. Isolation of Lycopene from Tomatoes 113
  D. Thin-Layer Chromatography of Carotenoid Pigments in Oranges 117
  E. High-Performance Liquid Chromatography of Carotenoids in Orange Juice 120
  F. Determination of β-Carotene and Chlorophylls in Plants 123
  G. Questions 126
  H. Recommended Books 126

II. Steroids 127
  A. Introduction 127
  B. Isolation of Stigmasterol from Soybean Oil 136
  C. Degradation of Stigmasterol 139
  D. Isolation and Identification of Algal Sterols 142
  E. Isolation of Hecogenin from Agaves and Its Transformation to Tigogenin and Rockogenin 146
  F. Preparation of Δ^5-Cholesten-3-one and Δ^4-Cholesten-3-one from Cholesterol 149
  G. Preparation of Vitamin D_2 and Its Separation by Thin-Layer Chromatography 154
  H. Determination of Sterols by Digitonin 159
  I. Gas–Liquid Chromatography of Steroids 161

J. Thin-Layer Chromatography of Sterols and Stanols on Silica Impregnated with Silver Nitrate   165
K. Questions   167
L. Recommended Books   168

III. Terpenoids   168
A. Introduction   168
B. Conversion of D-Limonene to L-Carvone and Carvacrol   182
C. Isolation and Determination of Optically Pure Carvone Enantiomers from Caraway (*Carum carvi* L.) and Spearmint (*Mentha spicata* L.)   187
D. Transformation of α-Pinene to Camphor   191
E. Photopronation of Limonene to *p*-Menth-8-en-l-yl Methyl Ether   195
F. Flash Chromatography of Essential–Oil Constituents   198
G. Isolation and Determination of Grape and Wine Aroma Constituents   201
H. Qualitative and Sensory Evaluation of Aromatic Herb Constituents by Direct Headspace–Gas Chromatography Analysis   205
I. Preparation of Abietic Acid from Wood Rosin and Its Dehydrogenation to Retene   209
J. Conversion of Betulin to Allo- and Oxyallobetulin   212
K. Isolation of Cerin and Friedelin from Cork   217
L. Removal of Bitter Components from Citrus Juices with β-Cyclodextrin Polymer   220
M. Questions   224
N. Recommended Books   225

4 NITROGENOUS COMPOUNDS

I. Alkaloids   226
A. Introduction   226
B. Method of Degradation   227
C. Isolation of Caffeine from Tea   230
D. Isolation of Piperine from Black Pepper   233
E. Degradation of Piperine   236
F. Isolation of Strychnine and Brucine from the Seeds of *Strychnos nux vomica*   238
G. Synthesis of Mescaline   242
H. Gas Chromatography of Alkaloids   247
I. Thin-Layer Chromatography of Opium Alkaloids   250
J. Questions   251
K. Recommended Books   253

II. Amino Acids   253
A. Introduction   253
B. Isolation of Glutamine from Red Beet   259
C. Preparation of D-Tyrosine from L-Tyrosine   262
D. Synthesis of DL-β-Phenylalanine   265
E. Preparation of Glycylglycine by Mixed Carboxylic–Carbonic Acid Anhydride Method   269
F. Thin-Layer Chromatographic Enantiomeric Resolution of Racemic Amino Acids   273
G. Reverse-Phase High-Performance Liquid Chromatography of Amino Acids   275

  H. Gas–Liquid Chromatography of Amino Acids 277
  I. Test Tube and Glass Rod Thin-Layer Chromatography of Amino Acids 280
  J. Thin-Layer Chromatography of Carbobenzoxy–Peptides 282
  K. Questions 284
  L. Recommended Books 285

III. Nucleic Acids 285
  A. Introduction 285
  B. Isolation of Soluble Ribonucleic Acid from Baker's Yeast 290
  C. Preparation of Thymidine 3′-Phosphate 293
  D. Chemical and Enzymatic Preparation of Cyclic Nucleotides 296
  E. Ion-Exchange Thin-Layer Chromatography of Nucleotides 300
  F. Questions 303
  G. Recommended Books 303

IV. Porphyrins 304
  A. Introduction 304
  B. Isolation of Hemin from Blood 309
  C. Degradation of Hemin 311
  D. Thin-Layer Chromatography of Plant Pigments 315
  E. Isolation and Identification of Petroleum Metalloporphyrins 317
  F. Questions 320
  G. Recommended Books 321

V. Proteins 321
  A. Introduction 321
  B. Isolation of Lysozyme from Albumen 324
  C. Isolation and Separation of Proteins from Groundnuts 326
  D. Thin-Layer Chromatography of Proteins 328
  E. Questions 330
  F. Recommended Books 331

VI. Pteridines 331
  A. Introduction 331
  B. Isolation of Pteridines from the Fruit Fly, *Drosophila melanogaster* 334
  C. Synthesis of Pteridines 338
  D. Thin-Layer Chromatography of Pteridines 341
  E. Questions 343
  F. Recommended Books 343

VII. Pyrazines 343
  A. Introduction 343
  B. Isolation and Characterization of Pyrazines, the Volatile Constituents of Bell Peppers 343
  C. Questions 346
  D. Recommended Books 346

BIBLIOGRAPHY 347

SUBJECT INDEX 351

INDEX OF CHEMICAL COMPOUNDS 357

PREFACE TO THE SECOND EDITION

In the twenty-five years since the publication of the first edition, natural product research technology has advanced incredibly through the fields of chemistry, food science, geochemistry, materials science, and life sciences. Comparisons of these compounds in microorganisms, algae, animals, higher plants, and marine invertebrates are now documented. With the advent of such techniques as Raman spectroscopy, magic-angle spinning spectroscopy, high-resolution electron microscopy, x-ray crystallography, and chromatographic methods, separation of positional and regioisomers, sterioisomers, and even isotopic isomers are now possible.

This new edition has been updated to include the following: The use of biomarkers (organic compounds in the geospherical record with carbon skeletons) reflecting the upsurge in geoporphyrin research primarily due to MS, yeast RNA nucleic acid studies; reversed-phase HPLC of amino acids; brewing industry applications (HPLC evaluation of carotenoids in orange juice and of "debittered" citrus); HPTLC of carbohydrates; synthesis of a sweetening agent from citrus peels, synthesis and degradation of alkaloids and of sterols, GC/MS uses with sterols, petroleum products, and aromatic constituents of wine and grape juice, flash chromatography of essential oils, optical purity of enantiomers affecting flavors, fragrances, and pheromones, as well as studies of lattice inclusion compounds ^{1}H- and ^{13}C-NMR, MS, IR, and UV data are presented for most natural products.

Used for twenty-five years in natural products organic chemistry courses, the successful first edition has been updated and improved to meet the needs of current research.

I wish to express my thanks to the following publishers and societies for their kind permission to reproduce tables, figures, and procedures: Academic Press, Association of Official Agricultural Chemists, Elsevier Publishing Company, John Wiley and Sons, Macmillan and Company, Methuen and Company, Naturwissenschaften Editors, Pergamon Press, Perkin Elmer Company, The American Chemical Society, The American Society of Biological Chemists, The Biochemical Society (London), The

Royal Chemical Society (London), Varian Associates, Verlag Chemie, and Springer Verlag.

Sincere thanks are due to my colleagues, Drs. Bernard Crammer, Vera Weinstein and Jeana Gross (Institute of Chemistry, Hebrew University, Jerusalem), and Dr. Uzi Ravid (Department of Natural Products, Agricultural Research Organization, Newe Ya'ar).

PREFACE TO THE FIRST EDITION

The syllabus of the practical course in organic chemistry at most universities mainly covers the synthesis of aliphatic, alicyclic, aromatic, and heterocyclic compounds, as well as qualitative and quantitative analysis. For some time, the Department of Organic Chemistry at the Hebrew University, Jerusalem, has felt that it would be of value to the students, either at the undergraduate or graduate level, to take a comprehensive course in natural products, so as to become acquainted with the methods of isolating these products from plant and animal matter and of determining their structures by physical methods, degraduation and, finally, synthesis.

The study of natural products has always been the starting point of the discipline of chemistry in every country of the globe, and, in view of the importance of these organic compounds in agriculture, medicine, and industry, every student of chemistry today feels the need to acquire further knowledge in this field.

In 1961, I organized a course on the subject and prepared a manual (in Hebrew) entitled "Natural Products Laboratory Guide." In the subsequent years I have repeatedly given this course to graduate students specializing in organic and physical chemistry, biochemistry, biology, and agricultural chemistry, as well as to undergraduate students (in the last trimester of their third year of studies).

Over the seven years during which I conducted this course, I have found that it was received enthusiastically by the participants at all levels. It stimulated many of them to choose the field of natural products as the theme for their M.Sc. and Ph.D. theses.

Visiting scientists from abroad who have become acquainted with the course and its methods were favorably impressed and encouraged me to publish the manual in a form suitable for wide distribution.

The book in its present form includes the following "biogenetic" chapters:
 1. Acetogenins: flavonoids, lipids, lignans, quinones.
 2. Carbohydrates: mono-, oligo- and polyaccharides.
 3. Isoprenoids: cartenoids, steroids, terpenoids.
 4. Nitrogens compounds: alkaloids, amino acids, nucleic acids, peptides porphyrins, proteins, pteridines.

Each section of these chapters includes a general introduction and methods of isolation, degraduation, and transformation, as well as applications of chromatographic procedures. The introduction to each chapter is brief and attempts only to supply or recall knowledge in the particular field. The student, who does not always find the time to read the relevant books or reviews, will find in this introduction the required material in a "concentrated" form. Furthermore, at the end of each chapter there is a list of recommended books for additional study. Typical experiments were selected for each chapter, taking into consideration the following factors: availability of starting materials and the performance time. Most of the procedures were tailored to the modest facilities of the students' laboratory.

Each experiment is described under the following headings: Introduction, References, Recommended Reviews, Principle, Apparatus, Materials, Time, Procedure (often including spectral data). At the end of each section appears a list of questions.

A large number of experiments are described in order to give the instructor a reasonable degree of freedom.

I am deeply indebted to Professor E. D. Bergmann for his encouragement, constructive comments, and criticism. I thank also Professor J. Klein and Dr. J. Radell for their helpful guidance in the first stages of this work, and Drs. Y. Lapidot and N. de Groot for their valuable suggestions on nucleic acids.

Acknowledgments are due Mr. R. Amoils for editing the manuscript, and to Mr. Y. Lakitchewitch for drawing the formulas. Last, but certainly not least, I wish to thank my wife for her expert typing of the manuscript, and our parents for their interest and encouragement.

ACETOGENINS — 1

The compounds grouped together in this chapter are different in most chemical properties, but similar in that many of them possess an aromatic ring bearing hydroxyl substituents. Many of these compounds possess a diversity of physiological properties. All of these natural products appear to be biosynthetically related in being derived by condensation of several molecules of acetate. Speculation regarding biosynthesis of all such polyacetate compounds may be found in several biogenetics books.

I. FLAVONOIDS

A. Introduction

The flavonoid compounds can be regarded as C_6-C_3-C_6 compounds, in which each C_6 moiety is a benzene ring, the variation in the state of oxidation of the connecting C_3 moiety determining the properties and class of each such compound. The classes are shown below.

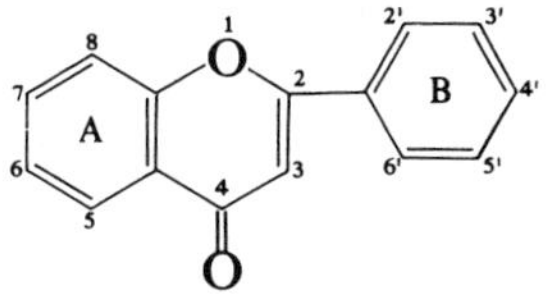

Flavone

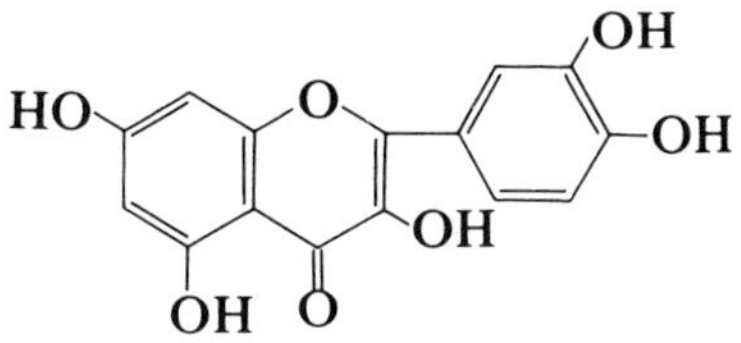

Flavonol

Flavanone

Flavanonol

Isoflavone

Chalcone

Dihydrochalcone

Aurone

Anthocyanidin

Catechin

Flavan-3,4-diol

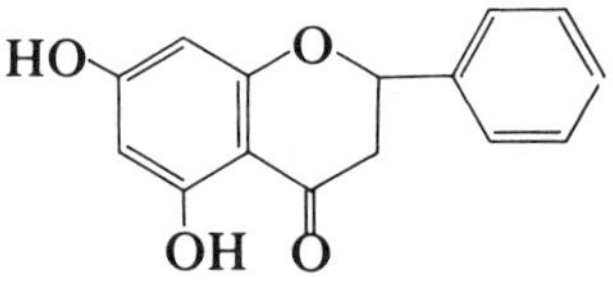

Flavonoid compounds and the related coumarins usually occur in plants as glycosides in which one or more of the phenolic hydroxyl groups are combined with sugar residues. The hydroxyl groups are nearly always found in positions 5 and 7 in ring A, while ring B commonly carries hydroxyl or alkoxyl groups at the 4′-position, or at both 3′- and 4′-positions. Glycosides of flavonoid compounds may bear the sugar on any of the available hydroxyl groups.

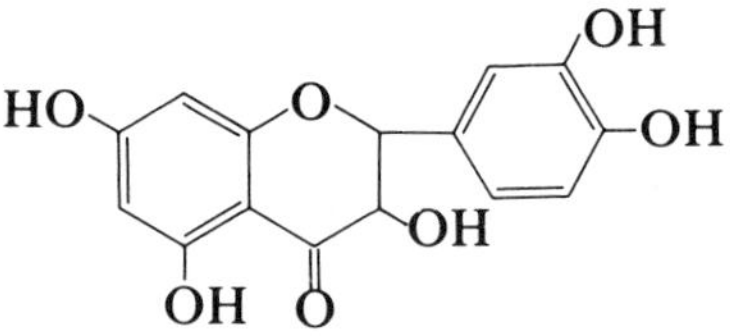

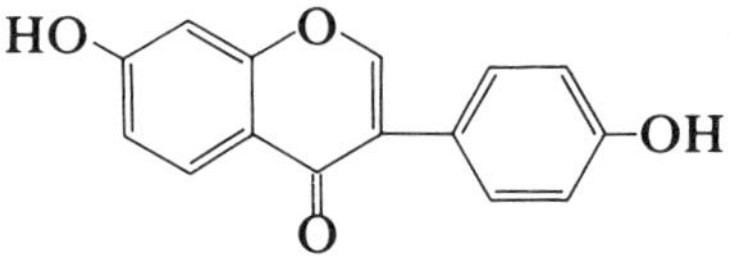

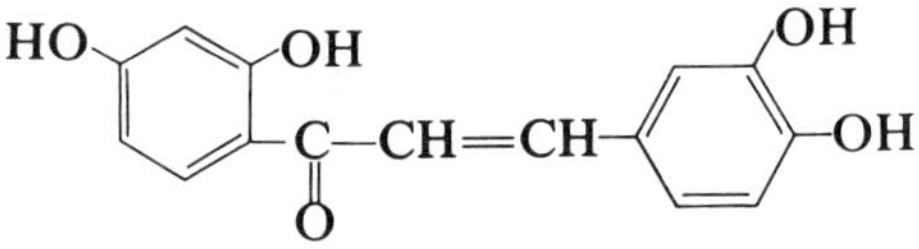

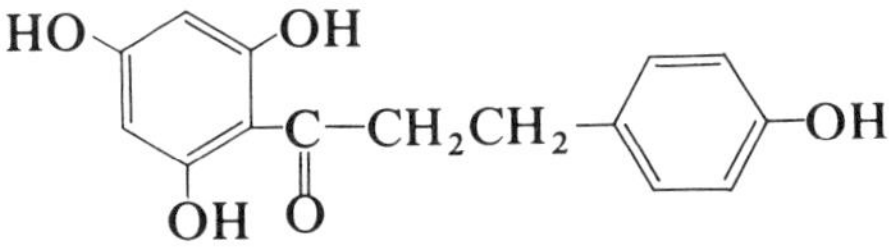

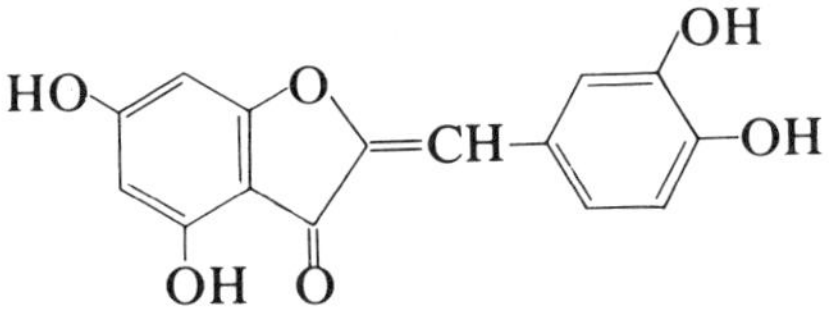

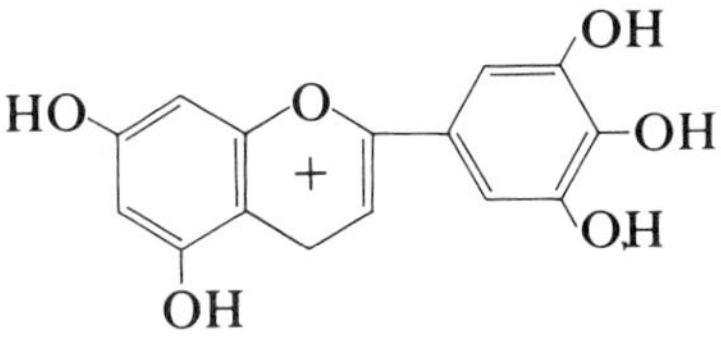

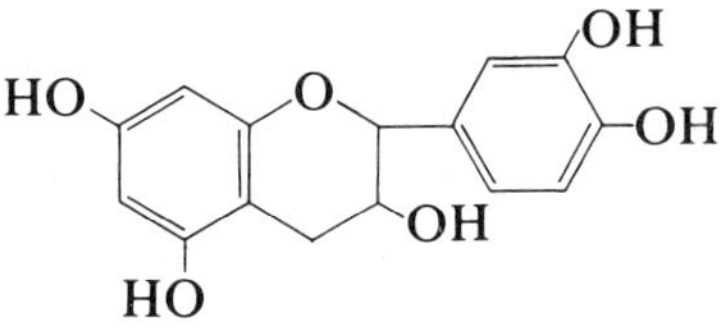

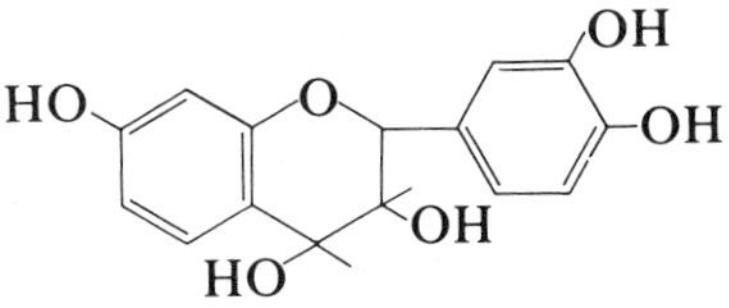

Flavonoids occur in all parts of plants, including the fruit, pollen, roots, and heartwood. Numerous physiological activities have been attributed to them. Thus, small quantities of flavones may act as cardiac stimulants; some flavones, e.g., hesperidin, appear to strengthen weak capillary blood vessels; highly hydroxylated flavones act as diuretics and as antioxidants for fats. It is also claimed that flavones behave like auxins in stimulating the germination of wheat seeds.

The possible function of this coloring matter in insect-pollinated flowers and edible fruits is to make these organs more conspicuous in order to aid seed dispersion by animals.

The fundamental method in structural studies is alkaline hydrolysis. Thus, alkaline degradation of chrysin yields phloroglucinol, acetic acid, benzoic acid, and a small amount of acetophenone.

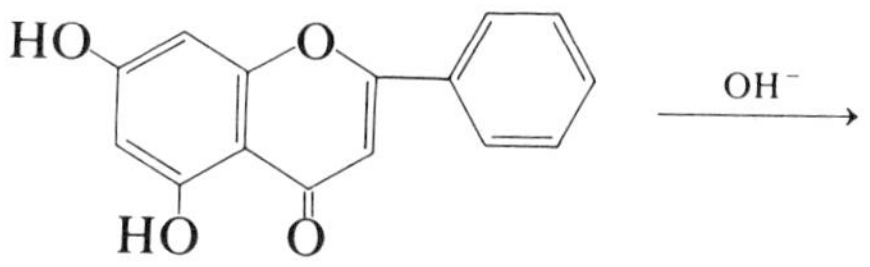

Chrysin $\xrightarrow{OH^-}$ Phloroglucinol $+ CH_3COOH + C_6H_5COOH + CH_3COC_6H_5$

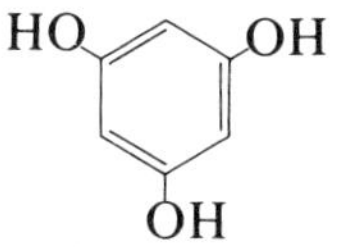

By spectroscopic measurement it is now possible to determine the structures of some flavonoid compounds on the basis of their spectra alone. Color reactions also play an important role in identification of flavonoids at the preliminary stage of analysis.

The possibility of interconversion between the various structures in this group is of considerable importance for the structural elucidation of flavonoid compounds. Thus, chalcones and flavanones are isomeric and readily undergo interconversion.

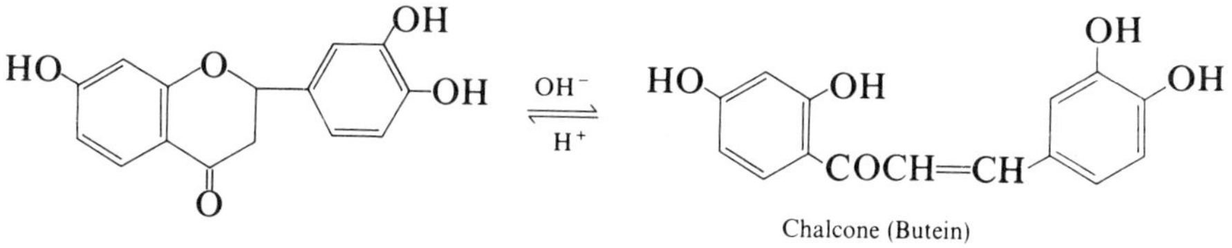

Flavanone (Butin) $\underset{H^+}{\overset{OH^-}{\rightleftarrows}}$ Chalcone (Butein)

Flavanones may be converted into flavonols and flavones.

Flavone Flavanone Flavonol

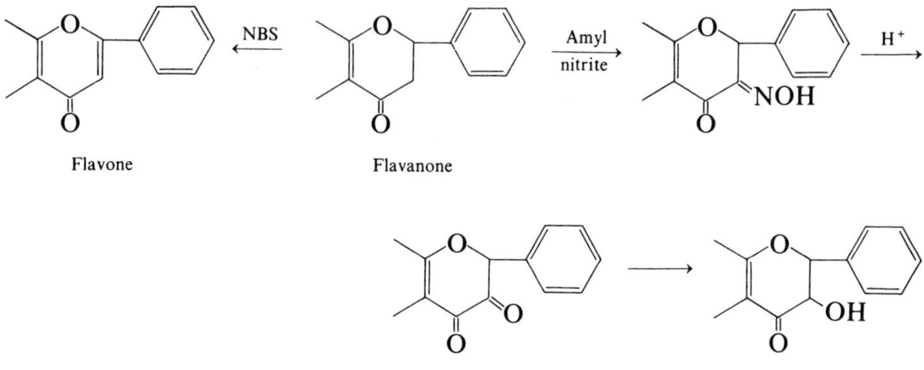

1. Flavones

In flavones, ring C is basic and forms a pyrylium salt with hydrochloric acid.

Flavone

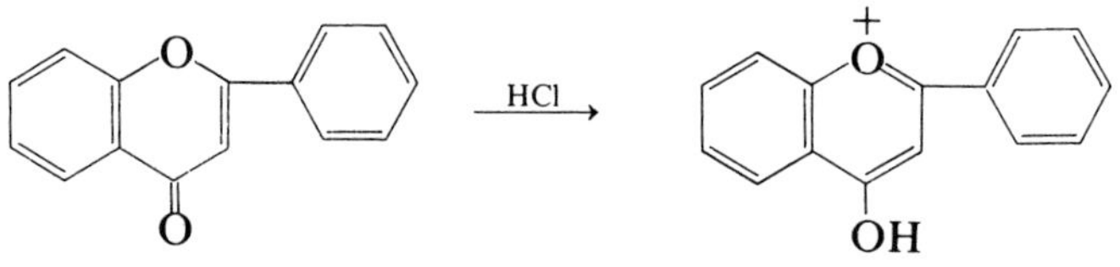

Consequently, the carbonyl group of flavone does not react normally with some carbonyl reagents such as hydroxylamine. However, it does react normally with Grignard reagents. The most widespread flavone is quercetin.

Quercetin

Primuletin

Fisetin

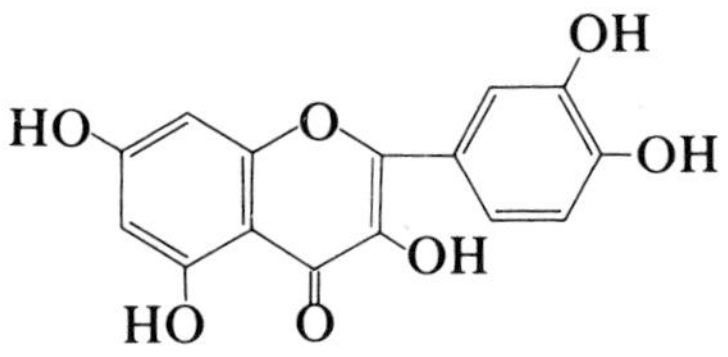

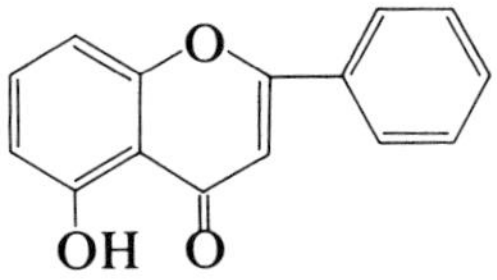

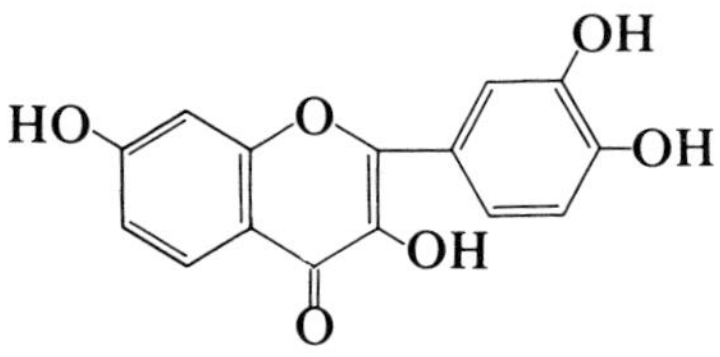

Some flavones, such as primuletin and fisetin, have only one hydroxyl group in ring A.

2. Flavanones

Flavanone has not yet been found in nature. Hydroxylated flavanones, however, do occur in nature, either in the free form or as glycosides. In plants they frequently coexist with the corresponding flavones, e.g., hesperidin and diosmin in the bark of *Zanthoxylum avicennae*, the rhoifolin and naringin in the peel of *Citrus aurantium*.

Unlike the unsaturated flavones, the saturated flavanones show reactivity of the 4-carbonyl group. The behavior of flavanones toward alkalis differs from that of flavones; the former decompose into benzaldehyde, acetic acid, and phenol under drastic conditions, whereas the latter yield phenol and cinnamic acid. Dehydrogenation of flavanones, e.g., conversion of hesperitin into diosmin, is of importance, as it makes possible the rapid identification of a new flavanone by reference to a known flavone. The following flavanones merit mentioning:

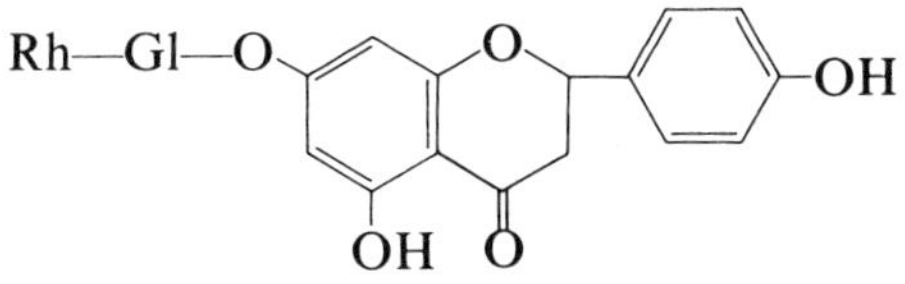

Naringin

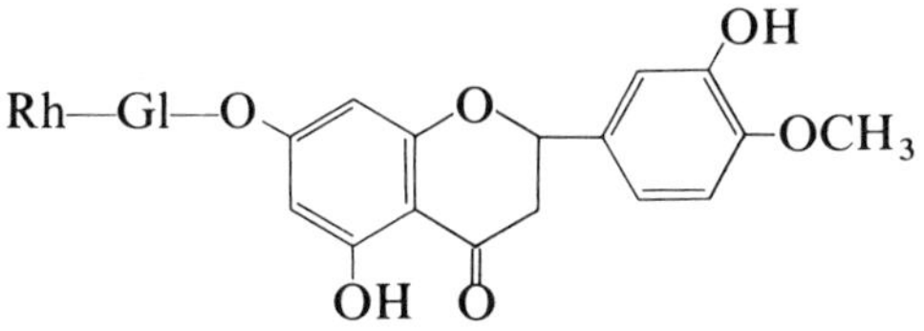

Hesperidin

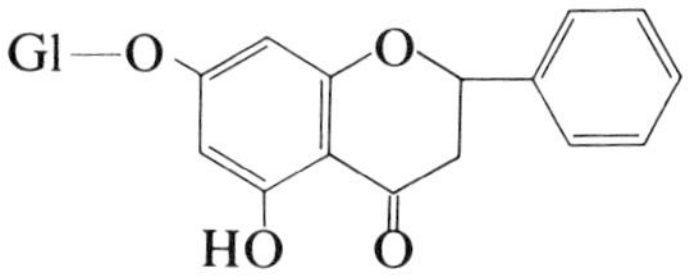

Prunin

3. Isoflavones

The isoflavones are 3-phenylchromones. At present about 35 isoflavones are known, of which the following are examples:

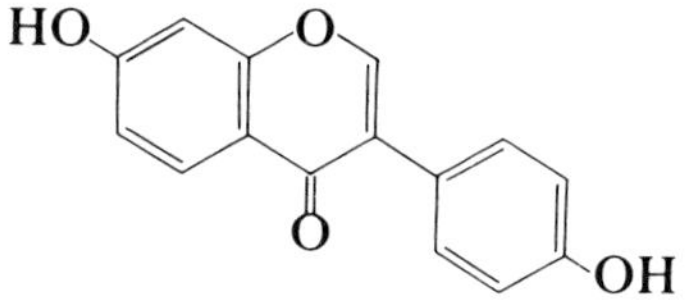

Daidzein

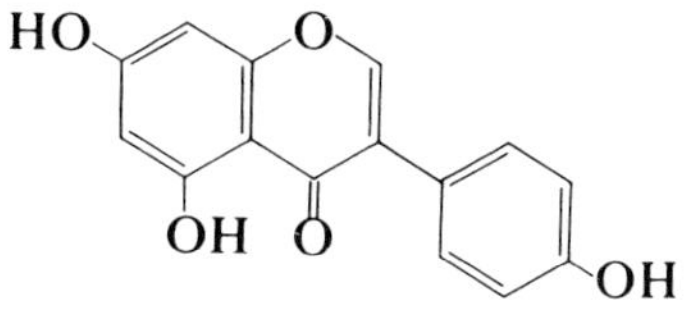

Genistein

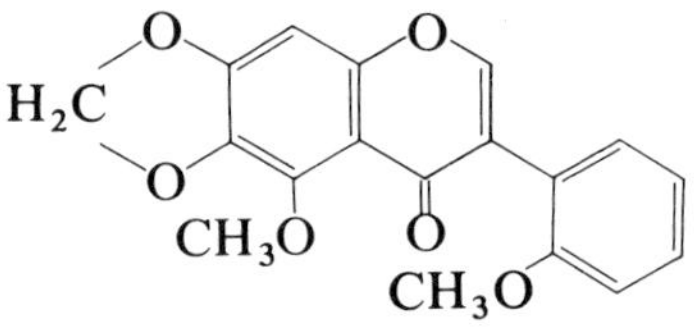

6 CHAPTER 1 Acetogenins

Tlanlancuayin

Isoflavones are degraded by alkali as follows:

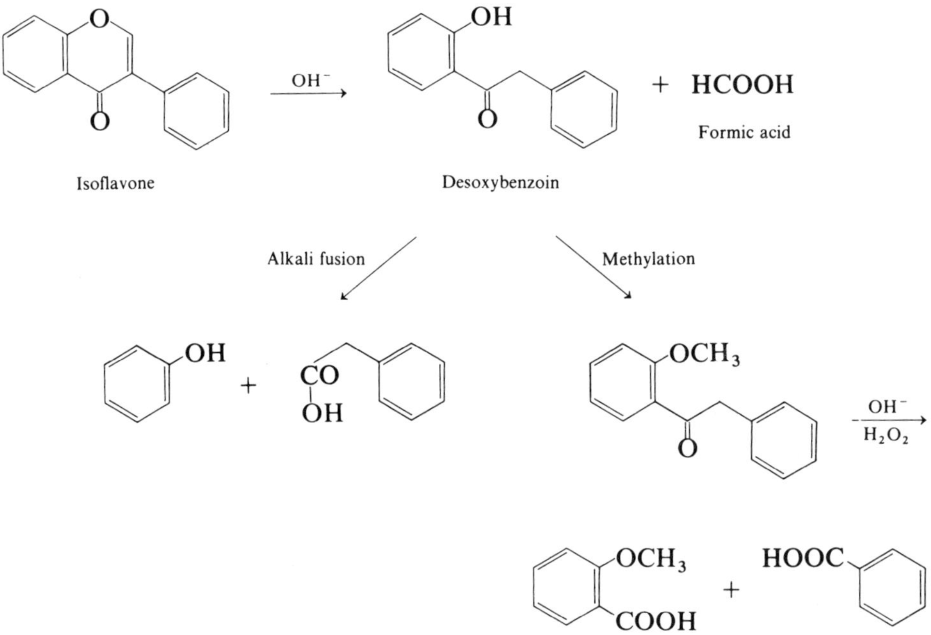

Isoflavone →(OH⁻) Desoxybenzoin + HCOOH (Formic acid)

Alkali fusion ↙   ↘ Methylation

(phenol + OH) + (CO-OH benzyl)   (OCH₃ desoxybenzoin derivative) →(OH⁻ / H₂O₂)

(o-OCH₃-COOH benzene) + HOOC-benzene

Isoflavones have shown estrogenic, insecticidal, and antifungal activity; some of them are potent fish poisons.

4. Anthocyanins

The innumerable shades of blue, purple, and violet, and nearly all the reds that appear in the cell sap of flowers, fruits, leaves, and stems of plants are due to anthocyanin pigments in the dissolved state. The sugar-free pigments are called anthocyanidins. The structure common to all anthocyanidins is the flavylium (2-phenylbenzopyrylium) ion.

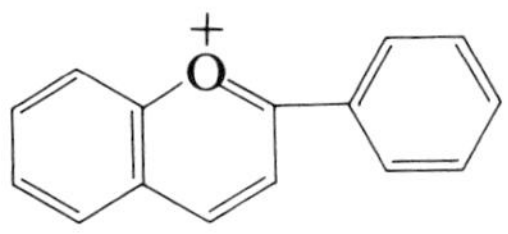

The natural anthocyanidins may be classified into three groups: pelargonidin, cyanidin, and delphinidin.

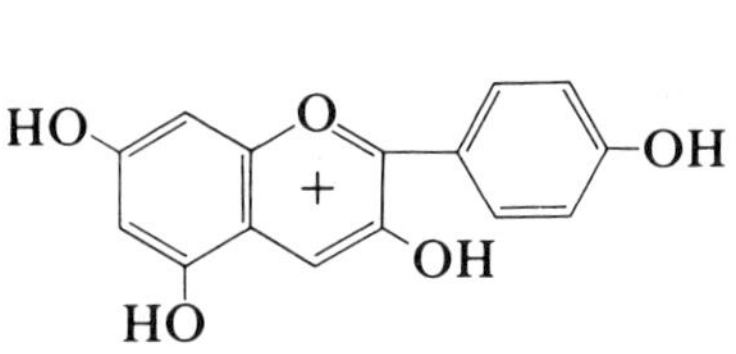

Pelargonidin

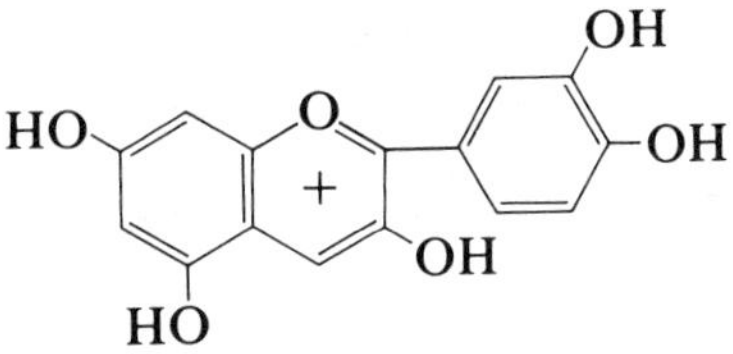

Cyanidin

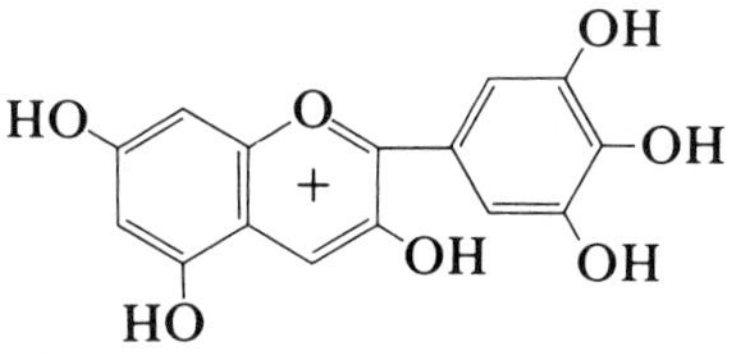

Delphinidin

Various anthocyanins can be distinguished by partition between two immiscible solvents, by their absorption spectra, and by their colors in buffer solutions of graded pH. For structure determination, anthocyanidins are treated with alkali, whereupon they form phloroglucinol and a phenolic acid:

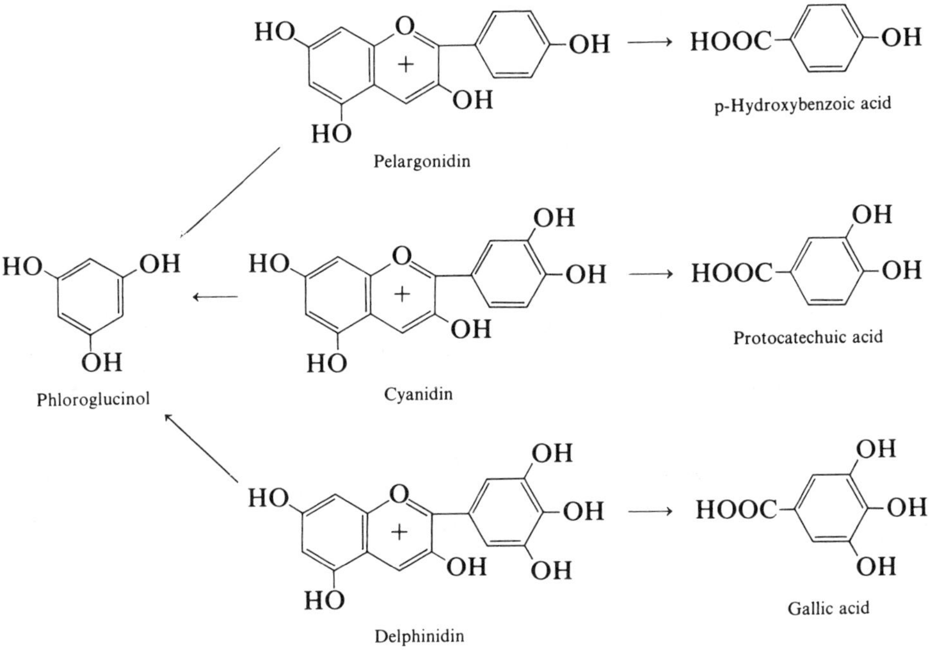

Pelargonidin → p-Hydroxybenzoic acid

Cyanidin → Protocatechuic acid

Delphinidin → Gallic acid

Phloroglucinol

5. Leucoanthocyanidins

Leucoanthocyanidins are flavan-3,4-diols. These colorless substances give red solutions with acids. Leucoanthocyanidins are widely distributed in the plant kingdom.

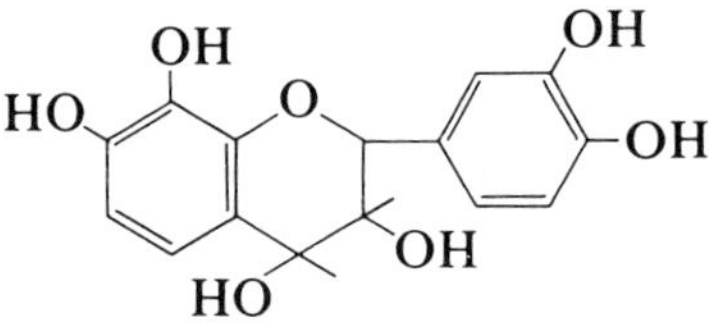

Melacacidin

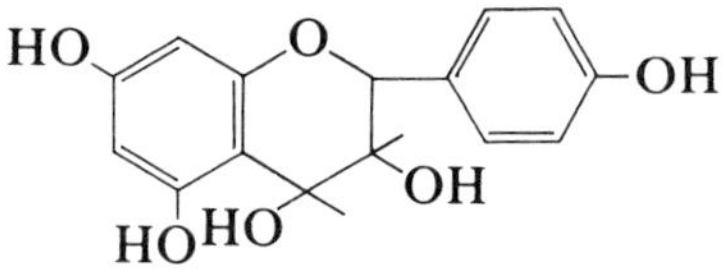

Leucopelargonidin

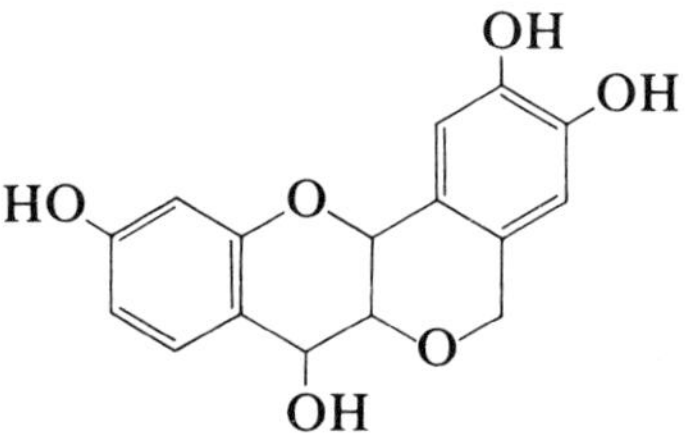

Peltogynol

Flavan-3,4-diols are sometimes obtained by reduction of flavonols of flavanonols:

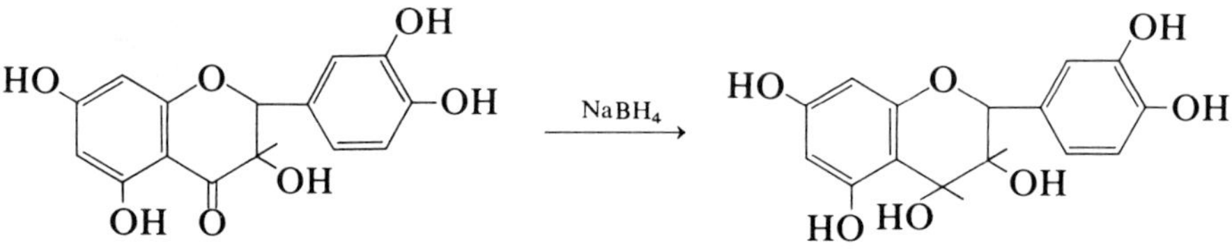

Dihydroquercetin $\xrightarrow{NaBH_4}$ Leucocyanidin

Biosynthetically, the flavonoid compounds are formed by the combination of a C_6-C_3 fragment derived from shikimic acid, e.g., *p*-hydroxycinnamic acid, with a six-carbon atom unit formed by the linear combination of three acetate units:

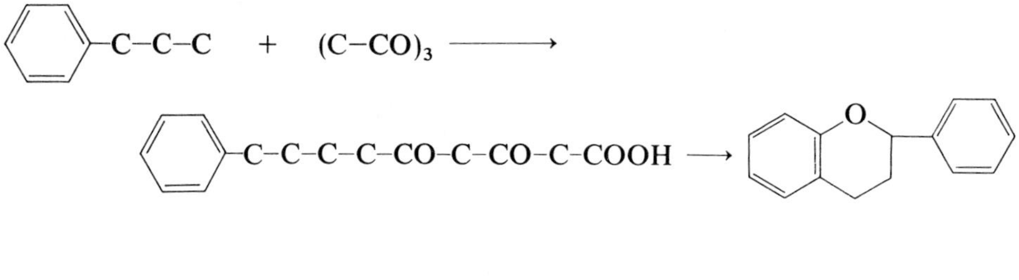

Ph–C–C–C + (C–CO)$_3$ ⟶

Ph–C–C–C–C–CO–C–CO–C–COOH ⟶ [flavonoid]

Several factors, such as the pH, complex-forming metals, and tannins, affect the colors of the anthocyanins.

The following general scheme has been proposed for the biosynthesis of compounds derived from shikimic acid: shikimic acid → prephenic acid → *p*-hydroxyphenylpyruvic acid → *p*-hydroxyphenyllactic acid → *p*-hydroxycinnamic acid → flavones.

Hydroxylation of rings A and B may occur at a stage after ring formation has been completed. Isoflavones are probably formed by phenyl migration of flavones.

B. Isolation of Hesperidin from Orange Peel

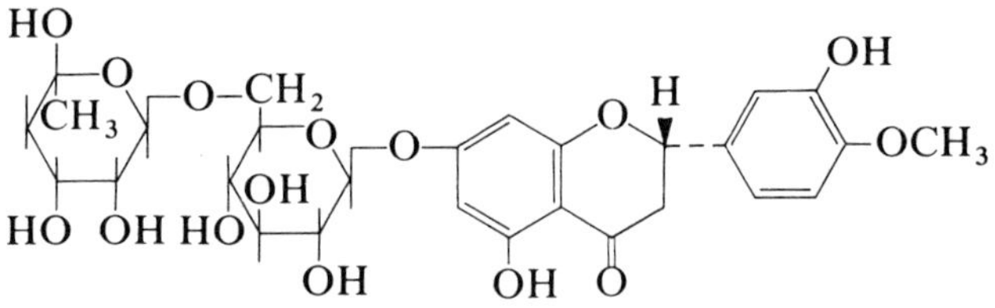

Hesperidin

1. Introduction

Hesperidin was first isolated by Leberton in 1828 from the albedo (the spongy inner portion of the peel) of oranges of the family Hesperides, and was given the name hesperidin (1). Its presence was detected in lemons by Pheffer as early as 1874 (2). Neohesperidin, an isomer of hesperidin, has been isolated together with hesperidin from unripe sour oranges cultivated in Europe (3). Horowitz and Gentili (4) isolated it from Ponderosa lemon. Hesperidin was isolated from *Citrus mitis* by Sastry and Row (5).

Neohesperidin, a bitter compound, occurs in the bitter orange, *Citrus aurantium*, while hesperidin, a nonbitter compound, is the predominant flavonoid in lemons and the ordinary sweet orange, *Citrus sinensis*.

Hesperidin decreases the fragility of blood capillaries.

2. Principle

Hesperidin can be isolated by two methods: (a) by extracting the dry peel successively with petroleum ether and methanol, the first solvent removing the essential oil and the second, the glycoside; (b) by alkaline extraction of chopped orange peel and acidification of the extract. It may be purified effectively by treatment with formamide-activated charcoal (6). Because of its highly insoluble, crystalline nature, hesperidin is one of the easiest flavonoids to isolate.

3. Apparatus

Disintegrator

4. Materials

Acetic acid	Formamide	Methanol
Calcium hydroxide	Hydrochloric acid	Orange peel
Celite	Isopropanol	Petroleum ether
Ferric chloride	Magnesium	Sodium borohydride

5. Time

Method a, 4–5 hours; method b, soaking overnight, laboratory procedure 1–2 hours.

6. Procedure

a. Method a

Sun-dried peel 200 g is powdered in a disintegrator. The powder is placed in a 2-liter round-bottomed flask attached to a reflux condenser. One liter of petroleum ether (bp 40–60°C) is added and heated on a water bath for 1 hour. The contents of the flask are filtered while hot through a Buchner funnel, and the powder is allowed to dry at room temperature. The dry powder is returned to the flask, and 1 liter of methanol is added. The contents are heated under reflux for 3 hours and then filtered hot and washed with 200 ml hot methanol. The filtrate is concentrated under reduced pressure, leaving a syrupy residue crystallized from dilute acetic acid, yielding white needles; mp 252–254°C.

b. Purification of hesperidin with formamide

A 10% solution of hesperidin in formamide, prepared by warming to about 60°C, is treated for 30 min with activated charcoal previously boiled with dilute hydrochloric acid. The formamide, when tested in a 50% aqueous solution, should be slightly acid. If it is not acid, a little glacial acetic acid or formic acid should be added. The solution is then filtered through Celite, diluted with an equal volume of water, and allowed to stand for a few hours in order to crystallize. The crystals of hesperidin are filtered off and washed, first with hot water and then with isopropanol. Two such crystallizations give a white crystalline product melting at 261 to 263°C.

c. Method b

Chopped orange peel 200 g and 750 ml 10% calcium hydroxide solution are placed in a 2-liter Erlenmeyer flask and thoroughly mixed, then left overnight at room temperature. The mixture is filtered through a large Buchner funnel containing a thin layer of Celite on the filter paper. The yellow-orange filtrate is acidified carefully to pH 4–5 with concentrated hydrochloric acid. Hesperidin separates as amorphous powder. It is collected on a Buchner funnel, washed with water, and recrystallized from aqueous formamide. If the precipitation of hesperidin on addition of hydrochloric acid is slow, it is advisable to concentrate the solution under reduced pressure.

d. Ferric chloride test
Addition of ferric chloride solution to hesperidin produces a wine red color.

e. Magnesium-hydrochloric acid reduction test
Dropwise addition of concentrated hydrochloric acid to an ethanolic solution of hesperidin containing magnesium develops a bright violet color.

f. Hesperidin

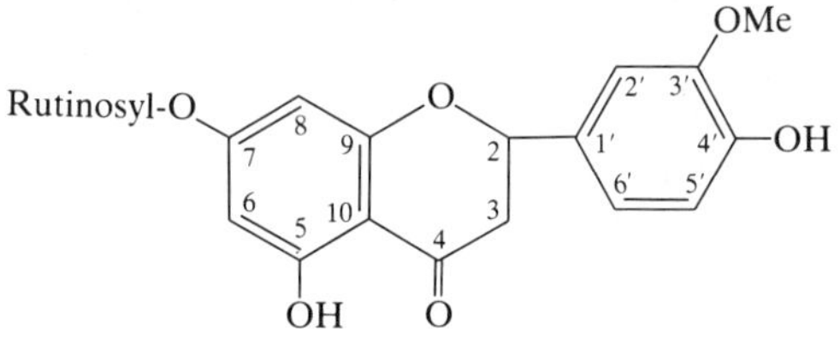

g. ^{13}C-NMR (nuclear magnetic resonance spectrometry)

C-2	δ 78.4	C-1'	131.2
C-3	42.0	C-2'	114.3
C-4	196.7	C-3'	146.7
C-5	163.0	C-4'	148.1
C-6	96.7	C-5'	112.7
C-7	165.2	C-6'	117.8
C-8	95.8	OMe	56.0
C-9	162.5		
C-10	103.5		

h. Mass spectrum
Hesperidin $C_{28}H_{34}O_{15}$, Molecular weight 610.
m/z 593(3%)$[M + H - H_2O]^+$ 303(100%)   413 (16%)
465(34%)   449(28%)

References

1. Leberton, P. (1828). *J. Pharm. Chim. Paris* **14**, 377.
2. Pheffer, W. (1874). *Bot. Ztg.* **32**, 529.
3. Karrer, W. (1949). *Helv. Chim. Acta.* **32**, 714.
4. Horowitz, R. M., and Gentili, (1960). *J. Am. Chem. Soc.* **82**, 2803.
5. Sastry, G. P., and Row, L. R. (1960). *J. Sci. Ind. Res. (India)* **19B**, 500.
6. Prichett, D. E., and Merchant, H. E. (1946). *J. Am. Chem. Soc.* **68**, 2108.

CHAPTER 1 Acetogenins

Recommended Reviews

1. Horowitz, R. M. (1961). *The citrus flavonoids*. In "*The Orange*" W. B. Sinclair, (ed.), p. 334. University of California Press, Berkeley, Calif.
2. Seshadri, T. R. (1962). *Isolation of flavonoid compounds from plant materials*. In "*The Chemistry of Flavonoid Compounds*" T. A. Geissman, (ed.), p. 6. Pergamon Press, London, England.

C. Acidic Degradation of Hesperidin

1. Introduction

In 1929, Asahina and Inubuse (1) established the molecular formula of hesperidin by showing that acid hydrolysis led to the formation of one mole each of rhamnose, glucose, and the aglycone, hesperitin. Arthur, Hui, and Ma (2) hydrolyzed hesperidin with 5% sulfuric acid in ethylene glycol and isolated the aglycone in an optically active form as the levorotatory isomer, (−)-hesperitin. They therefore concluded that hesperidin is the 7-β-rutinoside of (−)-hesperitin. This was confirmed previously by the synthesis of hesperidin by interaction of hesperitin and α-acetobromorutinose in the presence of quinoline and silver oxide (3).

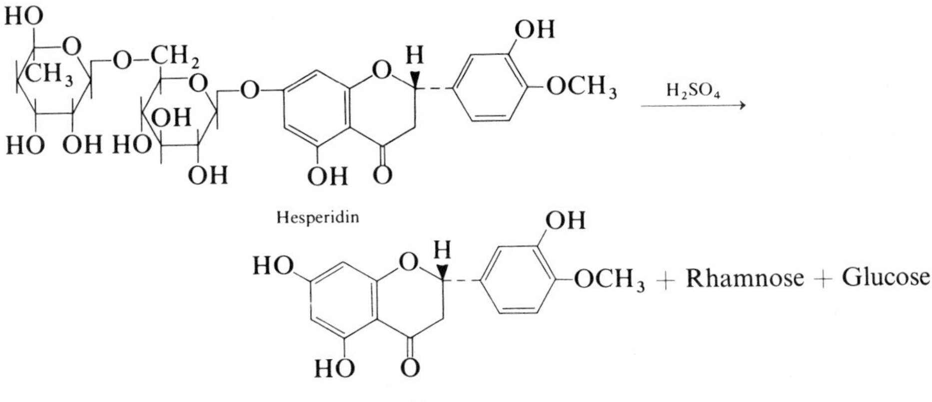

Hesperidin

Hesperitin

2. Principle

Hydrolysis of hesperidin with sulfuric acid in ethylene glycol under mild conditions furnishes the aglycone, hesperitin, and one molecule each of D-glucose and L-rhamnose (2). The carbohydrates are identified by circular paper chromatography.

3. Apparatus

   Equipment for circular paper chromatography

4. Materials

Acetic acid	n-Butanol	Hesperidin
Ammonium hydroxide	Ethanol	Silver nitrate
Anisaldehyde	Ethylene glycol	Sulfuric acid

5. Time

   2–3 hours

6. Procedure

One g hesperidin and 20 ml ethylene glycol containing 1 ml sulfuric acid are heated on a steam bath for 40 min. The clear yellow solution is poured into 50 ml water. The precipitated hesperitin is filtered on a Buchner funnel and washed with water. Crystallization from ethanol gives crystals of mp 224–226°C yield, 350 mg. The filtrate, which contains sugars, is neutralized and concentrated *in vacuo* until a thick syrup is obtained. The identification of the sugars is carried out by circular paper chromatography, the developing solvent being n-butanol–acetic acid–water, 4:1:5 (v/v). The developed chromatogram is dried to remove the solvent and then sprayed with one of the following reagents: (a) a mixture of equal volumes of aqueous $0.1N$ $AgNO_3$ and $5N$ NH_4OH; (b) a mixture of 0.5 ml anisaldehyde, 9 ml ethanol (95%), 0.5 ml concentrated sulfuric acid, and 0.1 ml acetic acid. The chromatogram is then placed in an oven at 100 to 105°C for 5 to 10 min.

		Color with Reagent	
Sugar	R_f	a	b
Glucose	0.29	brown	blue
Rhamnose	0.48	brown	green

a. UV spectrum of hesperitin

   λ_{max}^{EtOH} 289 mμ (logε 4.27)

The B ring of flavanones is not conjugated with the carbonyl group. Consequently, their absorption is strongest in the 270–290 mμ region, associated with that of the benzoyl grouping.

b. Hesperitin

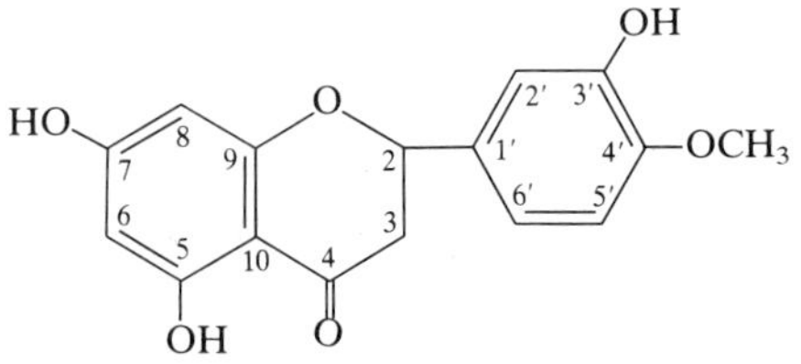

[Chemical structure of hesperitin with numbered positions]

c. ¹³C-NMR

C-2	78.5	C-1"	131.4
C-3	42.1	C-2'	114.3
C-4	196.2	C-3'	146.7
C-5	163.8	C-4'	148.1
C-6	96.2	C-5'	112.1
C-7	166.9	C-6'	118.0
C-8	95.4	CH$_3$O	55.9
C-9	163.0		
C-10	102.1		

d. Mass spectrum
Hesperitin $C_{16}H_{14}O_6$, Mol. wt. 302.
m/z 331(9%)   303(100%)   179(6%)   153(4%)
    304(18%)   287(6%)   177(1%)

References

1. Asahina, Y., and Inubuse, M. (1929). *J. Pharm. Soc. Japan* **49**, 128.
2. Arthur, H. R., Hui, W. H., and Ma, C. N. (1956). *J. Chem. Soc.* 632.
3. Zemplen, G., and Bognar, R. (1943). *Ber.* **76B**, 773.

Recommended Reviews

1. Horowitz, R. M. (1964). *Relations between the taste and structure of some phenolic glycosides.* In "*Biochemistry of Phenolic Compounds*," J. B. Harborne, (ed.), p. 545. Academic Press, New York.
2. Masami Shimokoriyama (1962). *Flavanones, chalcones, and aurones.* In "*The Chemistry of Flavanoid Compounds*," T. Geissman, (ed.), p. 286. Pergamon Press, London, England.
3. Nagy, S., and Attaway, J. A. (eds.) (1980). "*Citrus Nutrition and Quality.*" ACS Symposium Series 143.

D. Isolation of Naringin from Grapefruit Peel

1. Introduction

The albedo, the spongy inner portion of the peel, is composed chiefly of cellulose, soluble carbohydrates, pectic substances, flavonoids, amino acids, and vitamins. The flavonoid of the albedo is called naringin and was discovered in 1866 by De Vry in the flowers of grapefruit trees of Java, see (1). It was also isolated from the flowers and peel of *Citrus decumana* (2), from grapes (3), from the Japanese bitter orange (4), from leaves of *Pseudaegle trifoliata* (5), and from other sources. One of the outstanding properties of naringin is its intense bitterness; a hydrolytic enzyme (naringinase) is now added to grapefruit juice in order to hydrolyze naringin to the nonbitter naringenin (6).

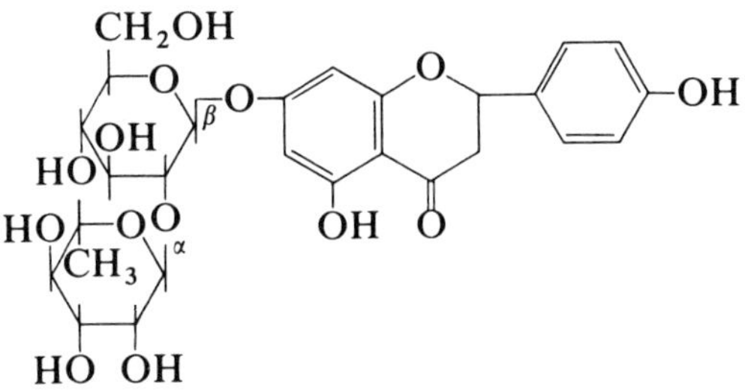

Naringin

2. Principle

In the following procedure (7, 8) naringin is extracted with hot water from the grapefruit peel along with a small quantity of pectin. Concentration of the extract to about one ninth the original volume affords naringin, as an octahydrate. Recrystallization from isopropanol gives the pure dihydrate.

3. Apparatus

Blender
Vacuum distillation assembly

4. Materials

Celite   Grapefruit peel
Ethanol   Isopropanol

5. Time

4–5 hours

6. Procedure

One part chopped grapefruit peel and four parts water are heated to 90°C, and maintained at this temperature for five min. The water extract is filtered off. Two parts water are added to the solid; the extraction is repeated at 80°C and immediately followed by filtration. The combined extracts are boiled with 1% Celite, filtered, and concentrated *in vacuo* to approximately one ninth the original volume. The concentrated extract is allowed to crystallize in a refrigerator and is then filtered as an octa-hydrate of mp 83°C (needles). Naringin (8.6 g) is dissolved in 100 ml boiling isopropanol and filtered hot. The filtrate is heated to its boiling point to initiate crystallization, then allowed to cool, filtered on a Buchner funnel, and washed with cold isopropanol; mp of dihydrate is 171°C (needles). Naringin can also be recrystallized from a small amount of hot water.

a. Naringin

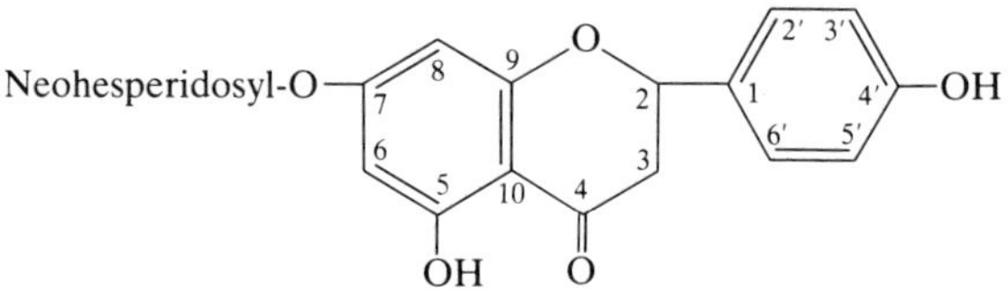

b. ^{13}C-NMR

C-2	78.6	C-1'	128.7
C-3	42.0	C-2'	128.0
C-4	196.7	C-3'	115.3
C-5	162.9[a]	C-4'	157.7
C-6	96.5	C-5'	115.3
C-7	164.9	C-6'	128.0
C-8	95.4		
C-9	162.7[a]		
C-10	103.5		

c. Mass spectrum
Naringin $C_{27}H_{32}O_{14}$, Mol. Wt. 580.
m/z 581(100%)[M + H]$^+$ 579(100%)[M − H]

References

1. Will, W. (1885). *Ber* **18**, 1311.
2. Zoller, H. F. (1918). *Chem. Zentr.* **89**, 635.
3. Willimott, S. G. and Wokes, F. (1926). *Biochem. J.* **20**, 1256.
4. Hattori, S., Shimokoriyama, M., and Kanao, M. (1952). *J. Am. Chem. Soc.* **74**, 3614.
5. Hattori, S., Hasegawa, M., Wada, E., and Matsuda, H. (1952). *Kagaku (Science)* **22**, 312.
6. Ting, S. V. (1958). *J. Agr. Food Chem.* **6**, 546.
7. Poore, H. D. (1934). *Ind. Eng. Chem.* **26**, 637.
8. Hendrickson, R., and Kesterson, J. W. (1956). *Proc Florida State Horticultural Soc.* **69**, 149.

---

[a] Values may be interchanged.

Recommended Review

1. Horowitz, R. M. (1961). *The citrus flavonoids*. In: "*The Orange.*" W. B. Sinclair, (ed.), p. 334. University of California Press, Berkeley, Calif.

E. Synthesis of Naringin Dihydrochalcone— A Sweetening Agent

1. Introduction

The peels of oranges, lemons, and grapefruits contain an array of flavonoid compounds of diverse type and structure. Two of the best known of these compounds are hesperidin, the main flavonoid constituent of oranges and lemons, and naringin, the main flavonoid constituent of grapefruit. These substances have been known for more than a century and are characterized by their abundance, accessibility, ease of isolation and intense bitterness (1).

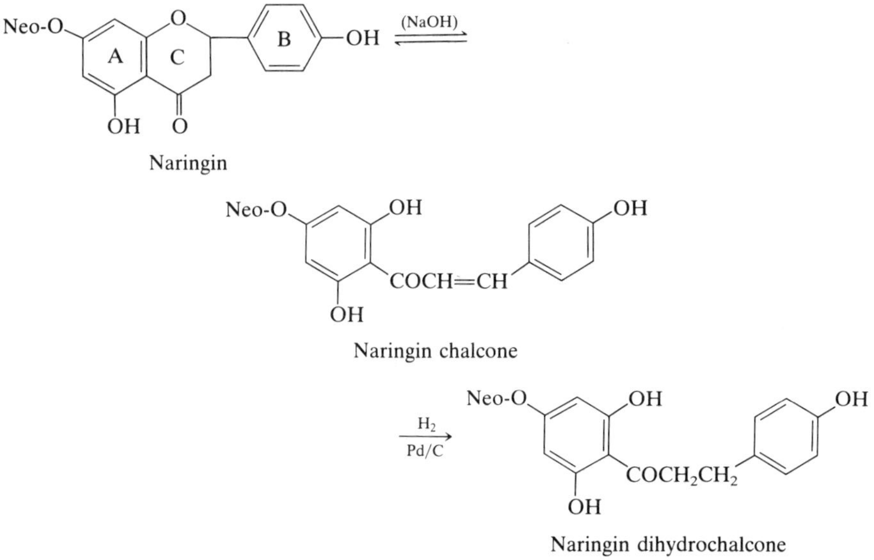

Naringin

Naringin chalcone

Naringin dihydrochalcone

The relative bitterness of these flavonoids compared to quinine is shown in the following table:

Compound	Molarity	Relative Bitterness
Neohesperidin	5×10^{-4}	2
Naringin	5×10^{-5}	20
Quinine	1×10^{-6}	100

Alkaline cleavage of ring C of naringin yields the corresponding chalcone, which is in turn hydrogenated catalytically to the corresponding intensely sweet naringin dihydrochalcone (NDHC), which is about 1000 times sweeter than sucrose.

The taste and the relative sweetness of dihydrochalcones and saccharin are summarized in the following table.

Compound	Taste	Molarity of Iso-Sweet Solution	Relative Sweetness	
			Molarity	Weight
Naringin DHC	Sweet	2×10^{-4}	1	0.4
Neohesperidin DHC	Sweet	1×10^{-5}	20	7
Saccharin (Na)	Sweet	2×10^{-4}	1	1

2. Principle

In the following procedure (2), an alkaline solution cleaves the heterocyclic C ring of naringin, forming a chalcone that is subsequently hydrogenated to an intensely sweet NDHC.

3. Apparatus

   PARR hydrogenation apparatus

4. Materials

   Naringin   Potassium hydroxide
   Pd/C (10%) catalyst   Sucrose

5. Time

   4 hours

6. Procedure

A solution of 2 g naringin in 16 g 8.5% potassium hydroxide solution is placed in a Parr pressure bottle and hydrogenated in a Parr shaker for 3 hr at an initial pressure of 50 pounds per square inch gauge (psig) in the presence of 0.29 g 10% palladium on charcoal. The catalyst is filtered and the solution, acidified to pH 6 with cold dilute hydrochloric acid. The precipitated NDHC is filtered on a Buchner funnel and washed with cold water to yield 2.3 g of the sweet product, mp 169–170°C.

The sweetness value of NDHC may be determined by panel evaluation with a minimum of five people per panel. In a typical comparison test, a

0.0045% solution of NDHC is compared with a 5% solution of sucrose. The product has a slight aftertaste.

References

1. Horowitz, R. M., and Gentili, B. (1969). *J. Agric. Food Chem.* **17**, 696.
2. Krbechek, L., Inglett, G., Holik, M., Dowling, B., Wagner, R., and Riter, R. (1968). *J. Agric. Food Chem.* **16**, 108.

Recommended Reviews

1. Crammer B., and Ikan, R. (1977). Properties and synthesis of sweetening agents. *Chem. Soc. Revs.* **6**, 431.
2. Horowitz, R. M. (1964). Relation between the taste and structure of some phenolic glycosides. In "Biochemistry of Phenolic Compounds," (J. B. Harborne, ed.), p. 545. Academic Press, New York.
3. Kinghorn, A. D. (1987). Biologically active compounds from plants with reputed medicinal and sweetening properties. *J. Nat. Products* **50**, 1009.

F. Color Changes of Anthocyanins at Various pH Values

1. Introduction

The term *anthocyanin* was proposed by Marquart in 1835 to denote the blue pigment of the cornflower. Actually, the anthocyanins encompass the entire range of red, violet, and blue pigments in plants. In general, the deepening of the visible color is brought about by the increase in the number of hydroxyl groups, as illustrated by the orange-red color of pelargonidin, the deep red of cyanidin, and the bluish red of delphinidin derivatives. Color changes ranging from violet to blue were first noted by Willstätter who observed that anthocyanins are red under acidic, violet under neutral, and blue under alkaline conditions. The other factor that controls the coloration of flowers is the coexistence of several anthocyanins, for instance malvidin and delphinidin in the berries of *Empetrum nigrum* (1) and in the flowers of *Althaea rosea* (2). A similar situation is encountered in a variety of garden plants (3).

Scott-Moncrieff (4) studied the interrelationship between color and the acidity of the cell sap. Later, Shibata, Hayashi, and Isaka (5), using petals, fruits and leaves of 200 plants, showed that all the saps are invariably acidic, irrespective of their color, the pH being around 5.5 in most cases. Robinson and Robinson (6) held the view that tannins and flavone derivatives participate in anthocyanin co-pigmentation. Other investigators suggested that the blue color is due to an insoluble organometallic complex containing magnesium or calcium, and that the violet color is produced by a mixture of blue anthocyanins with red oxonium salts (7).

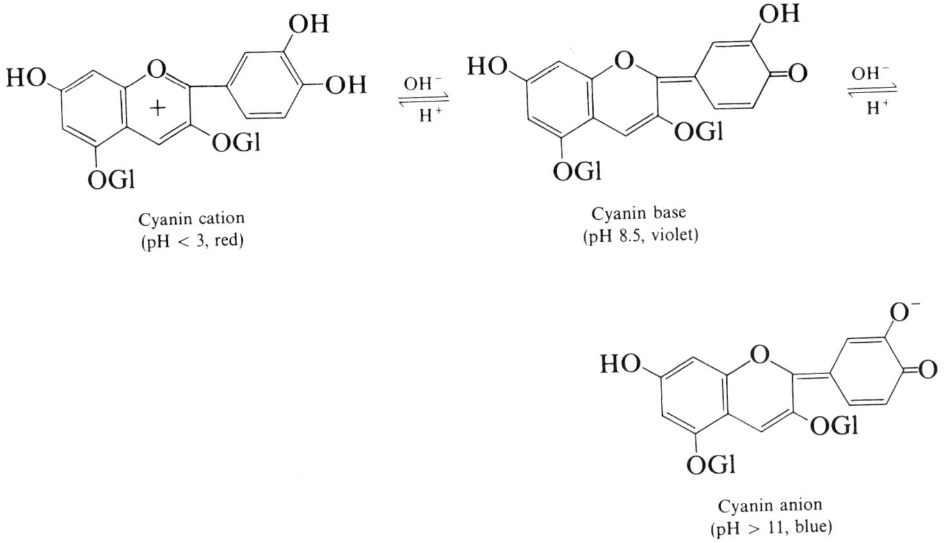

Cyanin cation
(pH < 3, red)

Cyanin base
(pH 8.5, violet)

Cyanin anion
(pH > 11, blue)

2. Principle

The following experiment demonstrates the change in color of anthocyanin as a function of pH. It is probable that the red color is due to the cationic form of anthocyanin; the violet color is characteristic of the base, while blue is the anionic form.

It is noteworthy that changes in color from red to mauve and purple, which occur during flower development, have also been ascribed to pH effects.

3. Apparatus

Burettes
Test tubes

4. Materials

Cabbage, red
Disodium phosphate (Na_2HPO_4)
Hydrochloric acid
Monopotassium phosphate (KH_2PO_4)
Pelargonium flowers
Roses
Sodium hydroxide
Tripotassium phosphate (K_3PO_4)

5. Time

   1–2 hours

6. Procedure

a. To 15 g well-washed red cabbage cut into small pieces is added 100 ml water. The pigment is extracted by boiling for 15 min and is then filtered into a measuring cylinder and made up to 50 ml with water. This solution is then introduced into a burette from which 2-ml samples are added to each of 18 test tubes containing the buffer solutions, and the resulting colors are recorded.

b. A similar procedure is adopted for pelargonium and roses.

Preparation of the Solutions of the Phosphate Buffer [a]

Test Tube	0.1N HCl	0.15M KH_2PO_4	0.15M Na_2HPO_4	0.15M K_3PO_4	pH
1	9.5	0.5	—	—	2.1
2	0.5	9.5	—	—	3.6
3	—	10.0	—	—	4.7
4	—	9.5	0.5	—	5.6
5	—	9.0	1.0	—	5.9
6	—	8.0	2.0	—	6.2
7	—	7.0	3.0	—	6.5
8	—	6.0	4.0	—	6.6
9	—	5.0	5.0	—	6.8
10	—	4.0	6.0	—	7.0
11	—	3.0	7.0	—	7.2
12	—	2.0	8.0	—	7.4
13	—	1.0	9.0	—	7.7
14	—	4.5	—	5.5	8.0
15	—	5.0	—	5.0	9.8
16	—	3.0	—	7.0	10.7
17	—	—	3.0	7.0	11.2
18	10% NaOH	—	—	—	14.0

[a] In ml.

References

1. Hayashi, K., Suzushino, G., and Ouchi, K. (1951). *Proc. Japan Acad.* **27**, 430.
2. Karrer, P., and Widmer, R. (1927). *Helv. Chim. Acta* **10**, 5.
3. Robinson, G. M., and Robinson, R. (1932). *Biochem. J.* **26**, 1647.
4. Scott-Moncrieff, R. (1939). *Ergeb. Enzymforsch.* **8**, 277.
5. Shibata, K., Hayashi, K., and Isaka, T. (1949). *Acta Phytochim. (Japan)* **15**, 17.

6. Robinson, G. M., and Robinson, R. (1931). *Biochem. J.* **25,** 1687.
7. Bate-Smith, E. C., and Westall, R. G. (1950). *Biochim. Biophys. Acta* **4,** 427.

Recommended Reviews

1. Goto, T. (1987). Structure, stability, and color variation in anthocyanins. *Prog. Chem. Org. Nat. Prod.* **52,** 113.
2. Harborne, J. B. (1968). *Flavonoids: distribution and contribution to plant color. In* "*Chemistry and Biochemistry of Plant Pigments.*" (T. W. Goodwin, ed.), p. 247, Academic Press, New York.
3. Kozo Hayashi. (1962) *The anthocyanins. In "The Chemistry of Flavonoid Compounds,"* (T. Geissman, ed.), p. 248, Pergamon Press, London, England.

G. Questions

1. Name the main groups of flavonoids.
2. What are polyphenolic compounds, and what is their function in nature?
3. Which products are obtained by alkaline treatment of flavones and anthocyanins?
4. Give examples of phenols that possess pharmacological and therapeutic properties.
5. Outline the major pathways for the biosynthesis of flavonoids.

H. Recommended Books

1. Agrawal, P. K. (1989). "*Carbon-13 NMR of Flavonoids.*" Elsevier, Amsterdam, the Netherlands.
2. Bentley, K. W. (1960). "*The Natural Pigments.*" Interscience, New York.
3. Britton, G. (1983). "*The Biochemistry of Plant Pigments.*" Cambridge University Press, New York.
4. Dean, F. M. (1963). "*Naturally Occurring Oxygen Ring Compounds.*" Butterworth, London, England.
5. Geissman, T. A. (ed.) (1962). "*The Biochemistry of Plant Pigments.*" Cambridge University Press, New York.
6. Harborne, J. B. (ed.) (1964). "*Biochemistry of Phenolic Compounds.*" Academic Press, New York.
7. Harborne, J. B., Mabry, T. J., and Mabry, H. (eds). "*The Flavonoids, Parts 1 and 2.*" (1975). Academic Press.

II. LIPIDS

A. Introduction

Lipids can be classified into three groups: fats and oils, waxes, and phospholipids. They occur in all parts of plant and animal tissue, and can be isolated by extraction with organic solvents such as ether, acetone, alcohols, chloroform, and petroleum ether.

1. Fats and Oils

The fat from a given source consists of a rather characteristic mixture of glycerides. The glycerides are of two types, simple and mixed.

$$\begin{array}{c} H_2COCOC_{17}H_{35} \\ | \\ HCOCOC_{17}H_{35} \\ | \\ H_2COCOC_{17}H_{33} \end{array}$$

Oleodistearin (mixed glyceride)

The glycerides may be readily saponified into glycerol and salts of the fatty acids.

The acids known to be present in various lipids vary in carbon content from C_{12} to C_{34} and nearly always contain an even number of carbon atoms. Many of these acids belong to the saturated aliphatic series, while others contain from one to six double bonds. Recently, fatty acids containing triple bonds and cyclic structures attached to the carbon chain have been discovered. Pure acids can be isolated from the mixtures resulting from the saponification of fats by the following procedures: fractional crystallization, fractional distillation, preparation and debromination of polybromides, adsorption chromatography, gas-liquid chromatography of methyl esters, and preparation of urea fatty-acid complexes that are stable to oxidation and are a convenient form for storing polyunsaturated acids. Urea binds fatty acids in a ratio of approximately 3:1 by weight. Desoxycholic acid is one of the compounds forming inclusion compounds with fatty acids; it forms choleic acids.

Most of the unsaturated acids, including the most important nonconjugated acids (oleic, linoleic and linolenic), occur naturally as the cis form.

A Number of the Natural Fatty Acids

Saturated acids

Caproic	$C_5H_{11}COOH$	Palmitic	$C_{15}H_{31}COOH$
Caprylic	$C_7H_{15}COOH$	Stearic	$C_{17}H_{35}COOH$
Capric	$C_9H_{19}COOH$	Arachidic	$C_{19}H_{39}COOH$
Lauric	$C_{11}H_{23}COOH$	Behenic	$C_{21}H_{43}COOH$
Myristic	$C_{13}H_{27}COOH$	Montanic	$C_{27}H_{55}COOH$

Unsaturated acids

Oleic	$C_{17}H_{33}COOH$	$CH_3(CH_2)_7CH=CH(CH_2)_7COOH$
Linoleic	$C_{17}H_{31}COOH$	$CH_3(CH_2)_4CH=CHCH_2CH=CH(CH_2)_7COOH$
Linolenic	$C_{17}H_{29}COOH$	$CH_3CH_2CH=CHCH_2CH=CHCH_2CH=CH(CH_2)_7COOH$
Arachidonic	$C_{19}H_{31}COOH$	$CH_3(CH_2)_4(CH=CHCH_2)_4(CH_2)_2COOH$
Clupanodonic	$C_{21}H_{33}COOH$	$CH_3(CH_2CH=CHCH_2)_2CH=CHCH_2$
		$(CH_2CH=CHCH_2)_2-CH_2COOH$
Nisinic	$C_{23}H_{35}COOH$	$CH_3CH_2(CH=CHCH_2)_4(CH_2CH=CHCH_2)_2CH_2COOH$

Acetylenic acids

Tariric	$C_{17}H_{31}COOH$	$CH_3(CH_2)_{10}\equiv C(CH_2)_4COOH$
Stearolic	$C_{17}H_{31}COOH$	$CH_3(CH_2)_7C\equiv C(CH_2)_7COOH$
Ximenynic	$C_{17}H_{29}COOH$	$CH_3(CH_2)_5CH=CH-C\equiv C(CH_2)_7COOH$

Hydroxy acid

Ricinoleic	$C_{17}H_{32}(OH)COOH$	$CH_3(CH_2)_5CH(OH)CH_2CH=CH(CH_2)_7COOH$

Cyclic acids

Sterulic $\quad C_{18}H_{33}COOH \quad CH_3(CH_2)_7\overset{\displaystyle CH_2}{\overset{\displaystyle /\ \backslash}{C=C}}(CH_2)_7COOH$

Chaulmoogric $\quad C_{17}H_{31}COOH \quad$ ⬠$(CH_2)_{12}COOH$

Gorlic $\quad C_{17}H_{29}COOH \quad$ ⬠$(CH_2)_6CH=CH(CH_2)_4COOH$

Although most naturally occurring fatty acids are straight-chain compounds, there are certain exceptions, such as isovaleric acid, $(CH_3)_2CHCH_2COOH$. The main reactions of unsaturated acids involve the addition of hydrogen, or halogen and oxidation.

$$CH_3(CH_2)_7CH=CH(CH_2)_7COOH \xrightarrow{H_2} CH_3(CH_2)_7CH_2CH_2(CH_2)_7COOH$$

Oleic acid $\qquad\qquad\qquad\qquad\qquad\qquad\qquad$ Stearic acid

$$CH_3(CH_2)_7\underset{Br}{\overset{|}{CH}}-\underset{Br}{\overset{|}{CH}}-(CH_2)_7COOH$$

9,10-Dibromostearic acid

Chlorine and iodine, or iodine chloride may be added in a similar manner.

Careful oxidation of oleic acid with alkaline permanganate at low temperatures gives rise to the formation of dihydroxystearic acid; at higher temperatures, the molecule is broken up into two acids of nine carbon atoms each.

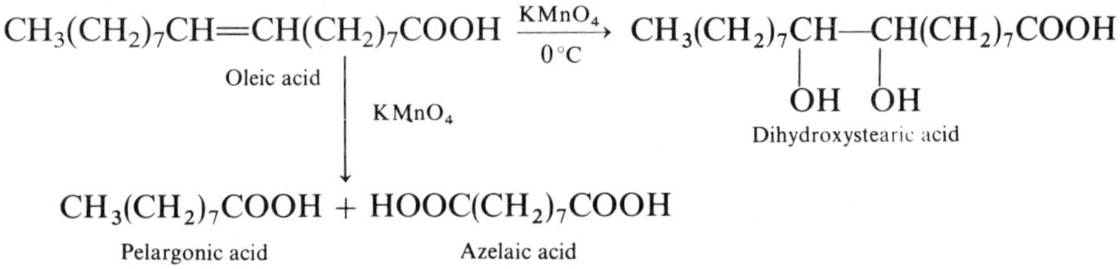

$$CH_3(CH_2)_7CH=CH(CH_2)_7COOH \xrightarrow[0°C]{KMnO_4} CH_3(CH_2)_7CH-CH(CH_2)_7COOH$$
$$\text{Oleic acid} \hspace{4cm} \underset{\text{Dihydroxystearic acid}}{OH \hspace{0.5cm} OH}$$

$$\downarrow KMnO_4$$

$$\underset{\text{Pelargonic acid}}{CH_3(CH_2)_7COOH} + \underset{\text{Azelaic acid}}{HOOC(CH_2)_7COOH}$$

Ozone adds to the double bonds of unsaturated acids, forming an ozonide, which is split into two aldehydes.

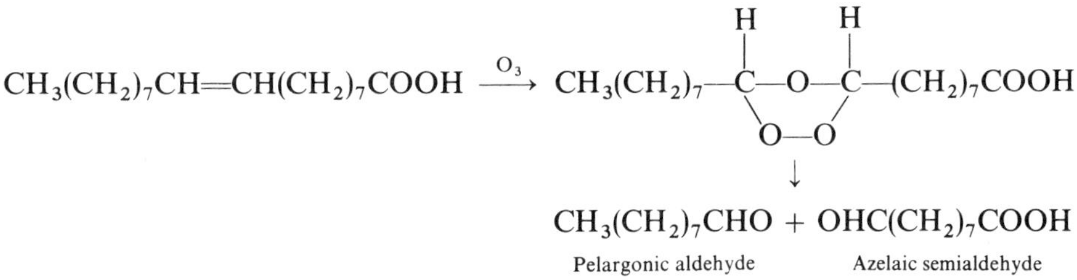

$$CH_3(CH_2)_7CH=CH(CH_2)_7COOH \xrightarrow{O_3} CH_3(CH_2)_7-\overset{H}{\underset{\underset{O-O}{\diagup}}{C}}-O-\overset{H}{\underset{}{C}}-(CH_2)_7COOH$$

$$\downarrow$$

$$\underset{\text{Pelargonic aldehyde}}{CH_3(CH_2)_7CHO} + \underset{\text{Azelaic semialdehyde}}{OHC(CH_2)_7COOH}$$

2. Waxes

The waxes consist essentially of esters of fatty acids and monohydric alcohols that contain from 24 to 36 carbon atoms. The following are examples of important natural waxes: beeswax, which contains, among other constituents, palmitic (C_{16}) and cerotic (C_{26}) acids and melissyl alcohol (C_{30}); lanolin—a waxy material obtained from wool; spermaceti—found in the head of the sperm whale and containing cetyl and oleyl alcohols and palmitic acid. The waxes are more resistant to saponification than are fats and oils.

3. Phospholipids

Phospholipids are fat-like substances containing (in addition to fatty acids such as oleic or stearic acid) phosphoric acid; a nitrogenous base such as choline, found in lecithin; or ethanolamine or serine, present in cephalins; and a polyhydroxy alcohol, either glycerol or inositol.

B. Isolation of Trimyristin and Myristicin from Nutmeg

1. Introduction

Nutmeg oil is obtained by steam distillation of the kernels of the fruit of *Myristica fragrans* Houttayn, a tree 15–20 meters high, grown in Indonesia and in the West Indies. Nutmeg oil consists of approximately 90% of terpene hydrocarbons. Major components are sabinene, and α- and β-pinene. A major oxygen-containing constitutent is terpinen-4-ol. A fraction consisting of phenols and aromatic ethers, such as safrole, myristicin, and elemicin, is responsible for the characteristic nutmeg odor.

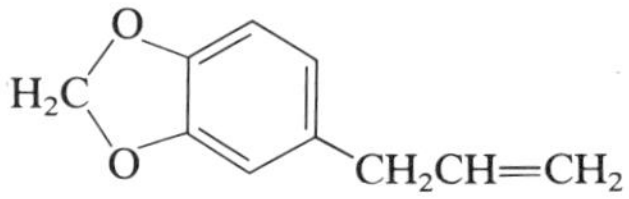

safrole

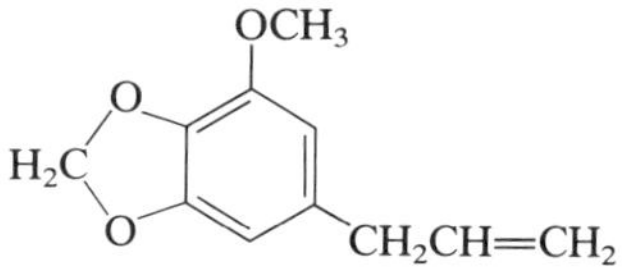

myristicin

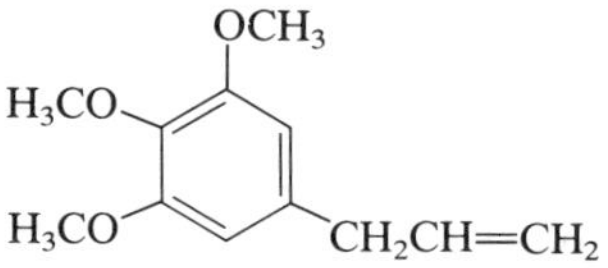

elemicin

Nutmeg oil is used mainly in food flavoring, and to a lesser extent in perfumery.

Semmler (1) isolated myristicin from the seeds of *Myristica fragrans* (nutmeg) in 1890. It was also found in French parsley (2), in the ethereal oil of the plant *Cinnamomum glanuliferum* (3), in the orthodon oil (4), and in *Ridolfia segetum* (up to 33%) (5).

Many investigators have succeeded in isolating the ester trimyristin from nutmeg by ether extraction (6, 7). Although myristicin is not a lipid, it is a byproduct of the extraction of nutmeg. Myristicin is toxic and has narcotic activity.

$$\begin{array}{c} CH_2OCOC_{13}H_{27} \\ | \\ CHOCOC_{13}H_{27} \\ | \\ CH_2OCOC_{13}H_{27} \end{array} \xrightarrow[C_2H_5OH]{KOH} \begin{array}{c} CH_2OH \\ | \\ CHOH \\ | \\ CH_2OH \end{array} + 3C_{13}H_{27}COOH$$

Trimyristin   Glycerol   Myristic acid

2. Principle

Isolation of trimyristin (ester) and myristicin (phenylpropane derivative), the two major products of nutmeg, is accomplished by extraction with chloroform. These compounds are separated by removing the solvent and then filtering.

The solid trimyristin is treated with alkali, yielding myristic acid. Myristicin is purified by column chromatography and fractional distillation. Upon bromination, it forms a solid dibromo derivative.

3. Apparatus

Reflux assembly

4. Materials

Alumina	Ether	Petroleum ether
Bromine	Hydrochloric acid	Potassium hyddroxide
Ethanol	Nutmeg	

5. Time

Isolation of trimyristin: 3 hours
Isolation of myristicin: 1–2 hours

6. Procedure

a. Isolation of trimyristin

Thirty g crushed nutmeg and 200 ml chloroform are refluxed for 90 min on a water bath, filtered through a folded filter paper, and dried on calcium chloride. The chloroform solution is then filtered, and the solvent is distilled under reduced pressure, leaving a semisolid residue, which is dissolved in 200 ml ethanol (95%). On cooling, crystalline trimyristin precipitates and is filtered off with suction and washed with cold ethanol (95%). The crystals are colorless and odorless and melt at 54 to 55°C yield, 6 g. The filtrate and washings are kept for the isolation of myristicin.

b. Trimyristin

$$\begin{array}{l} CH_2OCOC_{13}H_{27} \\ | \\ CHOCOC_{13}H_{27} \\ | \\ CH_2OCOC_{13}H_{27} \end{array}$$

c. Mass spectrum
   Trimyristin $C_{45}H_{86}O_6$, Mol. wt. 722.
   m/z 724(21%)   722(26%)   105(20%)   41(100%)
       723(55%)   495(43%)   57(46%)

d. Infrared spectrum (IR) (in CS_2)
   2700 cm^{-1}: C—H stretching vibrations
   1748: C=O stretching of saturated ester
   1363: sym. deformation vibrations of CH_3 group
   1231, 1154, 1106: asym. stretching of ester C—O—C (common to all triglycerides of long-chain fatty acids)
   1035: sym. stretching of ester C—O—C
   710: CH_2 wagging

e. Saponification of trimyristin

Five g trimyristin and 75 ml 3.5% ethanolic potassium hydroxide are refluxed for 1 hr on water bath, 150 ml water is added, and ethanol is removed under reduced pressure. After filtration, the solution is acidified with hydrochloric acid and left at room temperature until the myristic acid solidifies. The acid is filtered on a Buchner funnel and washed with water. Recrystallization from dilute ethanol gives a yield of 2 g acid, mp 51–52°C.

f. Thin-layer chromatography of nutmeg constituents

The crude extract is spotted on a thin layer plate (Silica Gel GF_{254}) and developed with dichloromethane. The dry plate is then sprayed with anisaldehyde–sulfuric acid reagent, (as described in the TLC of black pepper extract). Myristicin appears as a violet-red spot, R_f 0.72.

g. Isolation of myristicin

On concentration of the mother liquor remaining after separation of trimyristin, a residue is obtained, which is dissolved in 20 ml petroleum ether and passed through a short column containing 10 g activated alumina. Elution with 150 ml petroleum ether and evaporation of the solvent leaves an oil, which is fractionally distilled. Bp 150°C/15 mm, yield, 2.5 g.

Myristicin

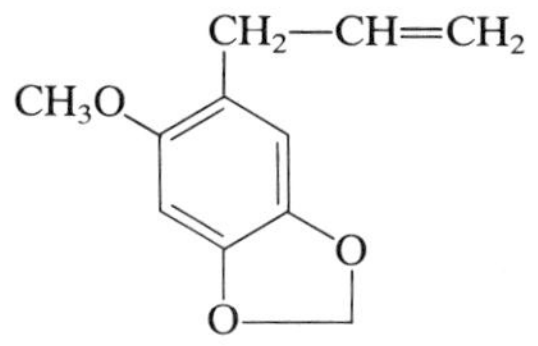

Mass spectrum
Myristicin $C_{11}H_{12}O_3$, Mol wt. 192.

m/z 192(100%)   161(14%)   119(15%)   65(16%)
     165(23%)   131(13%)   91(25%)   39(13%)

h. Preparation of dibromo myristicin

Bromine (0.4 ml) is added dropwise to 5 ml cold ether (ice bath). The resulting solution is then added slowly to 1 g myristicin in 25 ml petroleum ether until no further change in the color of bromine is observed. The dibromo derivative separates as fine needles, which are filtered and washed with petroleum ether, mp 124°C.

i. IR spectrum of myristic acid (in CS_2)

$3500-3000$ cm^{-1}: the OH of the COOH group is strongly bonded and responsible for the broad absorption band that commences at about 3500 cm^{-1} and extends into the region of the C—H stretching absorption below 3000 cm^{-1}

2650: C—H stretching vibrations

1760: a weak band due to the carbonyl group of the monomer

1708: a strong carbonyl band associated with the dimeric carboxylic acid group

1290: coupling between in-plane O—H bending and C—O stretching of dimer

940: O—H out-of-plane bending of dimer

720: CH_2 rocking

References

1. Semmler, F. W. (1890). *Ber.* **23**, 1803.
2. Thomas, H. (1903). *Ber.* **36**, 3451.
3. Pickles, S. S. (1912). *J. Chem. Soc.* **101**, 1433.
4. Huzita, I. (1940). *J. Chem. Soc. Japan* **61**, 729.
5. Gattefosse, J., and Igolen, G. (1946). *Bull. Soc. Chim. France* 361.
6. Power, F. B., and Salway, A. H. (1908). *J. Chem. Soc.* **93**, 1653.
7. Verkade, P. E., and Coops, J. (1927). *Rec. Trav. Chim.* **46**, 528.

Recommended Reviews

1. de Stevens, G., and Nord, F. F. (1955). *Natural phenylpropane derivatives.* In "*Modern Methods of Plant Analysis,*" (K. Paech, and M. V. Tracey, eds.), Vol. 3, p. 428. Springer-Verlag, New York.
2. Guenther, E. (1952). *Oils of Myristica.* In "*The Essential Oils,*" Vol. 5, p. 59. Van Nostrand, New York.

C. Preparation of Azelaic Acid from Castor Oil

1. Introduction

Azelaic acid can be prepared by oxidation of castor oil with nitric acid (1, 2), by oxidation of ricinoleic acid with nitric acid (3) and alkaline permanganate (4), by oxidation of methyl oleate with alkaline permanganate (5), by the ozonization of oleic acid and decomposition of the ozonide (6), by the ozonization of methyl ricinoleate and decomposition of the ozonide (7), by the Grignard reaction with 1,7-heptamethylenemagnesium dibromide (8), by the oxidation of dihydroxystearic acid with dichromate in sulfuric acid (9), and by other methods.

$$\text{Castor oil} \xrightarrow[C_2H_5OH]{KOH} CH_3(CH_2)_5CH(OH)CH_2CH=CH(CH_2)_7COOH$$
$$\text{Ricinoleic acid}$$
$$\downarrow KMnO_4$$
$$HOOC-(CH_2)_7-COOH$$
$$\text{Azelaic acid}$$

2. Principle

Castor oil is the glyceryl ester of ricinoleic acid. It is abundant in nature and has many industrial applications.

In the following experiment (10), castor oil is hydrolyzed with alkali, yielding crude ricinoleic acid, which is oxidized with potassium permanganate to give azelaic acid.

3. Apparatus

Stirrer

4. Materials

Castor oil	Magnesium sulfate	Potassium permanganate
Ethanol	Potassium hydroxide	Sulfuric acid

5. Time

4–5 hours

6. Procedure

Fifty g castor oil is added to a solution of 50 g potassium hydroxide in 100 ml 95% ethanol. The mixture is placed in a 500-ml flask equipped with a reflux condenser and is boiled for 3 hr. The solution is then poured into 300 ml water and acidified by addition of a solution of 10 ml concentrated sulfuric acid in 30 ml water. The acid that separates is washed twice with warm water. The yield of crude oily ricinoleic acid thus obtained is 48 g.

The 48 g ricinoleic acid is dissolved in 320 ml water containing 13 g potassium hydroxide. In a 2-liter round-bottomed flask equipped with a powerful mechanical stirrer are placed 135 g potassium permanganate and 1.5 l water at 35°C. The mixture is stirred to facilitate solution of the permanganate, and, if necessary, heat is applied to maintain the temperature at 35°C. When the permanganate has completely dissolved, the alkaline solution of ricinoleic acid is added in a single portion with vigorous stirring. The temperature rises to about 75°C. Stirring is continued for half an hour, or until a test portion added to water shows no permanganate color. To the mixture is now added a solution of 80 g concentrated sulfuric acid in 250 ml water. The acid must be added slowly and carefully to prevent too rapid evolution of carbon dioxide with consequent foaming. The mixture is heated on a steam bath for 15 min to coagulate the manganese dioxide, which is filtered while still very hot. After filtration, the manganese dioxide is placed in a 400-ml beaker and boiled with 200 ml water in order to dissolve any azelaic acid that may adhere to it. This mixture is filtered while hot, and the filtrate is added to the main portion. The combined filtrates are evaporated to a volume of about 800 ml, and this solution is cooled in ice. The crystals that separate are filtered with suction, washed once with cold water, and dried. The yield is 14–16 g material of mp 95–106°C. The crude substance is dissolved in 240 ml boiling water, filtered with suction, and allowed to cool. The crystals are filtered, washed with water, and dried; yield, 10–11 g; mp 104–6°C.

a. Azelaic acid

$$\underset{b}{CH_2}-\underset{c}{CH_2}-\underset{d}{COOH}$$
$$|\,a$$
$$(CH_2)_3$$
$$|$$
$$\underset{b}{CH_2}-\underset{c}{CH_2}-\underset{d}{COOH}$$

b. ^{1}H NMR
   a = 1.27 6H   c = 2.20 4H
   b = 1.48 4H   d = 11.93 2H

c. Mass spectrum
   Azelaic acid $C_9H_{16}O_4$, Mol. wt. 188.
   m/z 84(39%)   69(51%)   55(100%)   43(42%)
       83(41%)   60(61%)   45(46%)   41(80%)

References

1. Day, J. N. E., Kon, G. A. R., and Stevenson, A. (1920). *J. Chem. Soc.* **117**, 642.
2. Boeseken, J. and Lutgerhorst, A. G. (1932). *Rec. Trav. Chim.* **51**, 164.
3. Verkade, P. E. (1927). *Rec. Trav. Chim.* **46**, 137.
4. Maquenne, M. L. (1899). *Bull. Soc. Chim.* **21** (3), 1061.
5. Armstrong, E. F., and Hilditch, T. P. (1925). *J. Soc. Chem. Ind.* **44**, 43T.
6. Harries, C., and Tank, L. (1907). *Ber.* **40**, 4556.
7. Haller, A., and Brochet, A. (1910). *Compt. Rend.* **150**, 500.
8. von Braun, J., and Sobecki, W. (1911). *Ber.* **44**, 1926.
9. Bennet, G. M., and Gudgeon, H. (1938). *J. Chem. Soc.* 1679.
10. Hill, J. W., and McEwen, W. L. (1950). *Org. Synth. Coll.* **2**, 53.

D. Formation of *n*-Heptaldehyde and Undecenoic Acid from Castor Oil

1. Introduction

Castor oil is a readily available substance. It may undergo scission into 10-undecenoic acid and heptaldehyde, or it may be dehydrated to form 9,11-octadecadienoic and 9,12-octadecadienoic (linoleic) acids. The dehydration reaction is facilitated by the presence of dehydrating catalysts such as aluminum oxide, phosphoric anhydride, and similar substances, while the scission reaction appears to be characteristic of thermal degradation.

The pyrolysis of castor oil was first carried out by Bussy and Lecanu (1) in 1827. Some years later, Bussy identified heptaldehyde in the liquid product (2). In 1877, Krafft (3) identified the second compound as 10-undecenoic acid.

Perkins and Cruz (4), and more recently, Vernon and Ross (5), have verified the formation of 10-undecenoic acid; in the first case by distillation at 400°C (50 mm) and in the second by pyrolysis in a silica tube. Barbot (6) has proposed a mechanism for his cleavage.

$$\text{Castor oil} \xrightarrow{\Delta} \underset{\text{Heptaldehyde}}{CH_3(CH_2)_5CHO} + \underset{\text{10-Undecenoic acid}}{CH_2=CH(CH_2)_8COOH}$$

2. Principle

Pyrolysis of castor oil (which consists of about 80% triricinolein) under reduced pressure produces a mixture of heptanal and 10-undecenoic acid (7). These componds are then separated by fractional distillation. *n*-Heptaldehyde is identified as the 2,4-dinitrophenylhydrazone.

3. Apparatus

   Fractionation column
   Vacuum distillation assembly

4. Materials

   Castor oil   Colophony   2,4-Dinitrophenylhydrazine

5. Time

   2–3 hours

6. Procedure

Castor oil (150 g) and 10 g colophony are mixed in a Claisen flask fitted for vacuum distillation and heated over a bare flame. After 10 to 15 min, the distillate begins to collect in the receiving flask. When approximately 60 to 80 ml has been collected, the heating is stopped (to avoid the formation of the yellow resinous paste in the flask, which is very difficult to remove).

If two layers are formed in the distillate, the lower one (water) is removed by means of a separatory funnel, and the upper one is poured into a 500-ml flask to which is attached an efficient fractionation column; the distillate passing between 145°C and 160°C is collected (crude heptaldehyde), and when the temperature rises above 170°C, the distillation is interrupted. The crude heptaldehyde is purified by a second distillation. The yield is about 24 to 30 g. The 2,4-dinitrophenylhydrazone crystallized from ethanol forms orange needles, mp 108°C.

The residue of the fractional distillation is distilled *in vacuo*. The undecenoic acid distils between 156 and 165°C at 30 mm, giving a yield of 25 to 30 g.

a. IR (liquid film)

   3065 cm^{-1}: dimeric O—H stretching
   2915, 2850, 2650: characteristic of a dimeric carboxylic acid
   1710: C=O stretching, characteristic of a dimeric acid
   1645: C=C stretching, characteristic of the monosubstituted double bond
   1415, 1285, 1245: coupling between in-plane O—H bending and C—O stretching of dimeric acid
   990, 910: C—H out-of-plane bending of terminal methylene
   720: methylene chain rocking

b. 10-Undecenoic acid

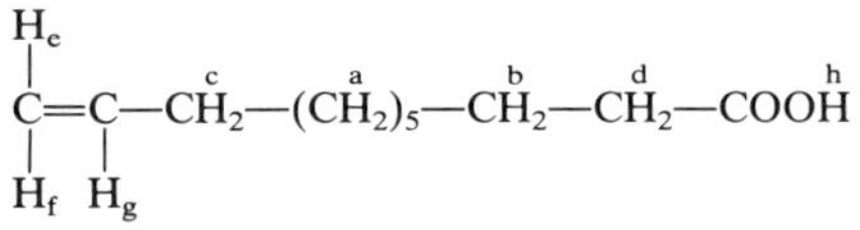

$$\underset{H_f\;H_g}{\overset{H_e}{C}=C}-\overset{c}{CH_2}-(\overset{a}{CH_2})_5-\overset{b}{CH_2}-\overset{d}{CH_2}-\overset{h}{COOH}$$

c. ^{1}H NMR

a = 1.30 (10H)   e = 4.93 (1H)
b = 1.62 (2H)   f = 4.98 (1H)
c = 2.03 (2H)   g = 5.80 (1H)
d = 2.35 (2H)   h = 10.80 (1H)

d. Mass spectrum
10-Undecenoic acid $C_{11}H_{20}O_2$, Mol. wt. 184.
m/z 96(30%)   82(30%)   69(44%)   55(100%)
    83(38%)   73(29%)   60(27%)   41(56%)

References

1. Bussy, J., and Lecanu. (1827). *J. Pharm. Chim.* **13** (2), 57.
2. Bussy, J. (1845). *J. Pharm. Chim.* **8** (3), 321.
3. Krafft, F. (1877). *Ber.* **10**, 2034.
4. Perkins, G. A., and Cruz, A. O. (1927). *J. Am. Chem. Soc.* **49**, 1070.
5. Vernon, A. A., and Ross, H. K. (1936). *J. Chem. Soc.* **58**, 2430.
6. Barbot, A. (1935). *Bull. Soc. Chim.* **2** (5), 895.
7. Dominguez, X. A., Speron, E., and Slim, J. (1952). *J. Chem. Educ.* **29**, 446.

Recommended Review

1. Sonntag, N. O. V. (1961). *Dehydration, pyrolysis, and polymerization.* In "*Fatty Acids*" (K. S., Markley, ed.), Vol. 2, p. 985. Interscience, New York.

E. Ozonization of Methyl Oleate

1. Introduction

The ozonide of oleic acid and the cleavage products obtained by treatment with water (azelaic semialdehyde, azelaic acid, pelargonaldehyde, and pelargonic acid) were first described by Harries and Thieme (1), who used no solvent for the ozonization. Subsequently, hexane (2), chloroform (3), glacial acetic acid (4), ethyl chloride (5), and ethyl acetate (6) were used as solvents. The highest yields were obtained by ozonization in glacial acetic acid, followed by reduction with zinc (4). Azelaic semi-aldehyde has been isolated, with 80% yield, as the semicarbazone by ozonization in ethyl chloride, followed by catalytic hydrogenation in methanol (5). Recent studies of the ozonization of various unsaturated compounds have shown methanol and ethanol to be superior reaction media, providing increased yields of the isolated carbonyl compounds.

Privett and Nickell (7) recently found that ozonization of methyl oleate produced nine compounds, which were identified by thin-layer chromatography.

$$CH_3(CH_2)_7CH=CH(CH_2)_7COOCH_3$$
Methyl oleate

$$\downarrow \text{1. } O_3 \quad \text{2. Zn/AcOH}$$

$$CH_3(CH_2)_7CHO \quad + \quad CH_3OOC(CH_2)_7CHO$$
Pelargonaldehyde   Methyl semiazelaaldehyde

2. Principle

In the following experiment (8), methyl oleate is ozonized in methanol. The ozonide formed is cleaved with zinc dust in acetic acid.

The resulting pelargonaldehyde and methyl semiazelaaldehyde are separated by fractional distillation.

3. Apparatus

Ozonator

4. Materials

Acetic acid
Calcium sulfate, anhydrous
Methanol
Methyl oleate

Methylene chloride
Ozone
Potassium iodide
Zinc dust

5. Time

2–3 hours

6. Procedure

Methyl oleate (24 g) is dissolved in 360 ml reagent-grade methanol in a 1-liter round-bottomed flask equipped with an inlet tube, and cooled to $-20°C$. Dry ozone is passed through the solution until the potassium iodide solution at the outlet of the reactor becomes strongly colored, indicating that ozone is no longer being absorbed. Glacial acetic acid (30 ml) is added, and the solution is warmed to 30°C. Zinc dust is added, a small pinch at a time, until a total of

12 g has been introduced, the temperature being maintained at 30 to 35°C by cooling. The mixture is then filtered and distilled on a steam bath until about one half of the methanol has been removed. Methylene chloride (100 ml) and 100 ml water are added, and the layers separated. The methylene chloride layer is washed with water until free of acid. It is then dried over anhydrous calcium sulfate. It should be noted that the washing should be carried out as soon and as thoroughly as possible; otherwise, varying amounts of the dimethyl acetal will form. Methylene chloride is removed from the solution by distillation on a steam bath. The residue is distilled *in vacuo* through a small Vigreux column, and two main fractions are obtained: 8.5 g pelargonaldehyde boiling at 37 to 47°C/0.3 mm, $n_D^{30} = 1.4193$; and 12 g methyl semiazelaaldehyde, bp 94–96°C/0.75 mm, $n_D^{30} = 1.4348$.

References

1. Harries, C., and Thieme, C. (1905). *Ann.* **343**, 318.
2. Molinari, E., and Soncini, E. (1906). *Ber.* **39**, 2735.
3. Harries, C. (1906). *Ber.* **39**, 3728.
4. Noller, C. R., and Adams, R. (1926). *J. Am. Chem. Soc.* **48**, 1074.
5. Fischer, F. G., Dull, H., and Ertel, L. (1932). *Ber.* **65B**, 1647.
6. Stoll, M., and Rouve, A. (1944). *Helv. Chim. Acta* **27**, 950.
7. Privett, O. S., and Nickell, E. C. (1964). *J. Am. Oil Chemists' Soc.* **41**, 72.
8. Pryde, E. H., Anders, D. E., Teeter, H. M., and Cowan, J. C. (1960). *J. Org. Chem.* **25**, 618.

Recommended Review

1. Kadesch, R. G. (1963). *Ozonolysis of fatty acids and their derivatives.* In "*Progress in the Chemistry of Fats and Other Lipids,*" (R. T. Holman, W. O. Lundberg, and T. Malkin, eds.), Vol.6, p. 291. Pergamon Press, London, England.

F. Preparation of Urea Inclusion Compounds

(m) Urea + (n) Guest molecule ⇌ Inclusion compound

1. Introduction

The chemistry of (inclusion) clathrate (from the Latin word *clathratus*, which means enclosed behind bars) compounds has a long history, starting at the beginning of the nineteenth century. The most important event, however, emerged from the pioneering X-ray structural work of H. M. Powell on SO_2 clathrate of hydroquinone. It was found that the size, the shape, and the chemical nature of the holes generated in an inclusion (host) lattice are the key factors in allowing the selective inclusion of the guest molecules. Chemically different species as well as constitutional isomers, positional isomers, regioisomers, stereoisomers, and even isotopic isomers are separated by this technique. In general, lattice inclusion will cause altered physical properties of the guest molecules, and toxicity may be reduced. Studying the nature of inclusion compounds requires special spectroscopic techniques for the solid state, such as Raman, magic-angle spinning spectroscopy, high-resolution electron microscopy, and X-ray crystallography.

The terms canal, channel, tube, and tunnel have been used to describe host cavities extended in one dimension without restriction on cross-sectional shape. Among the helical canal inclusion network (Fig. 1) are urea, thiourea, deoxycholic acid, tri-o-thymotide, amylase compounds, and others.

In 1940, Bengen discovered that urea formed crystalline addition compounds with a great variety of aliphatic, straight-chain molecules containing more than six carbon atoms. In the presence of straight-chain hydrocarbons,

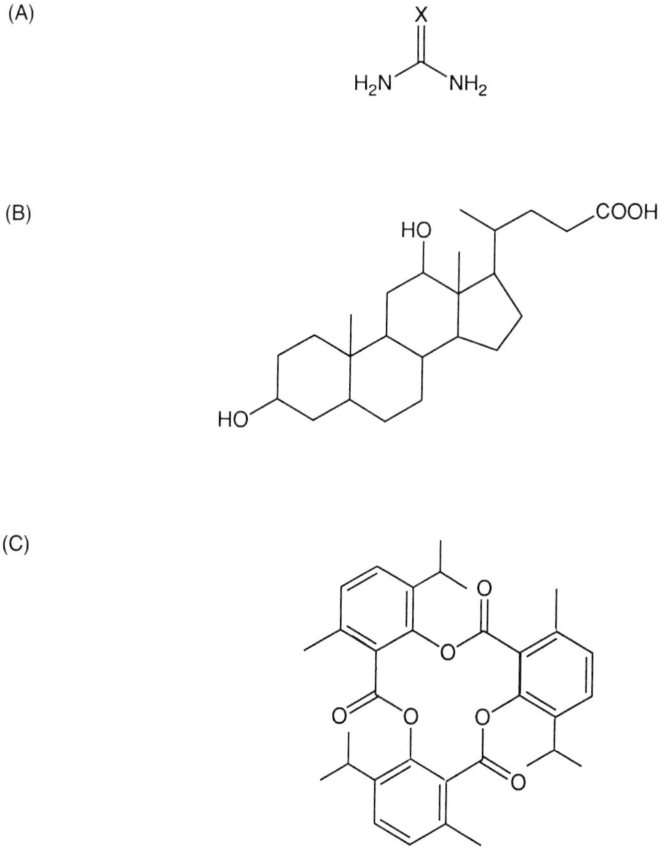

Figure 1.1 Among the helical canal inclusion network are (A) urea, when x = 0, and thiourea, when x = S; (B) deoxycholic acid; and (C) tri-o-thymotide.

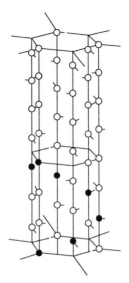

Figure 1.2 Structure of hexagonal urea. Solid circles represent urea molecules.

urea forms a hexagonal lattice with internal channels (Fig. 2). In these channels, the paraffin molecules are surrounded by urea and are thus kept firmly in position. The diameter of these channels is ca. 5 Å, thus permitting the entrance of only unbranched paraffins (1, 2) in an extended planar zigzag conformation. Branched molecules are generally excluded, unless branching occurs near the end of a long linear chain. The complexes of thiourea have a very similar structure (3), the sole difference being that in the thiourea lattice, the diameter of the channel is ca. 7 Å. Consequently, thiourea forms inclusion compounds only with larger molecules, since smaller molecules would leave too much of the channel cavity unfilled, and the structure would collapse.

Urea complexes meet almost all the requirements of an ideal derivative. For example, they form large, easily handled crystals; no special reagents or techniques are required for their preparation; the organic compound bound in the complex is readily recovered by addition of water to dissolve the urea; and the weight ratio of urea to organic compound is 3:1, thus allowing for a significant increase in the weight of material to be handled. In determining the melting point of urea complexes by the capillary method, it is observed that there is an exudation of a small amount of liquid before the melting point is reached. The temperature at which opacity occurs, the complex dissociates, and the transparent hexagonal crystals are converted to an aggregate of tetragonal microcrystals without losing their original external form is termed the dissociation temperature of a urea complex.

The preparation of urea complexes of long-chain fatty acids, esters, and alcohols has been reported by many investigators (4–6). It was found that several factors influence the yields of complexes of fatty acids. In the saturated series, the longer the chain, the higher the yield. Among the unsaturated, nonconjugated acids, increasing unsaturation lowers the yield. Conjugated isomers give higher yields of complex than do nonconjugated isomers. Trans monounsaturated acids give slightly higher yields than their cis isomers. These differences in yields of complexes from the various fatty acids suggest that the various types of acids can be fractionated by means of urea complexes. The structure of urea complexes with unsaturated acids would appear to indicate that they are excellent protectors against oxidation.

Urea inclusion compounds have been extensively applied in petrol purification, dewaxing of lubricating oil, purification of terpenes, and the separation of optical isomers.

2. Principle

Table 6 in the following description of preparations of urea complexes shows that the dissociation temperature increases uniformly with increasing chain length. Thus, the difference between the dissociation temperatures of the urea complexes of lauric and stearic acids is about 33°C, while the melting points of the commonly used derivatives of C_{12} to C_{18} fatty acids are usually close together. Therefore, this feature is of great utility for identification purposes. Another point of interest is the comparison of the dissociation temperatures of the C_{18} cis–trans isomers. The dissociation temperature of the trans isomer is significantly higher than that of the cis isomer. Since the cis compounds have some steric strain and are therefore less stable than the trans ones, they dissociate at lower temperatures. Furthermore, binary mixtures of saturated and unsaturated fatty acids and of natural oils can be separated on the basis of the difference in the rates of formation of the inclusion complexes (5).

3. Apparatus

Microscope

4. Materials

Capric acid	Linoleic acid	Palmitic acid
Caproic acid	Linolenic acid	Pelargonic acid
Caprylic acid	Linseed oil	Petroleum ether
Corn oil	Methanol	Soybean oil
Elaidic acid	Myristic acid	Stearic acid
Hydrochloric acid	Oleic acid	Tridecyclic acid
Isopropanol	Olive oil	Urea
Lauric acid		

5. Time

1–2 hours

6. Procedure

a. Urea complexes of fatty acids

The complexes of the series of pure fatty acids are prepared in the following manner: 1 g fatty acid is dissolved by warming in 30 ml methanol that has been saturated with urea at room temperature (about 16 g/100 ml). If the acid does not dissolve completely, a small quantity of isopropanol is added. Upon cooling to room temperature, the mixture deposits crystalline urea complexes, which are filtered off and dried. The products are crystallized from methanol or isopropanol, and examined under a microscope or a magnifying glass.

The dissociation temperatures of fatty acids and the molar and weight ratios of urea to fatty acid are summarized in the following table.

		Ratio of Urea to Acid	
Acid	°C	Molar	Weight
Caproic	64	—	—
Caprylic	73	6.7	2.8
Pelargonic	80.5	7.4	2.8
Capric	85	8.0	2.8
Lauric	92.5	9.7	2.9
Tridecylic	96	11.8	3.3
Myristic	103	10.6	2.8
Palmitic	114	12.0	2.8
Stearic	126	14.7	3.1
Oleic (*cis*)	110	13.6	2.9
Elaidic (*trans*)	116	13.6	2.9

b. Separation of binary mixtures of fatty acids

The following table shows series of saturated and unsaturated fatty acids that can be separated on the basis of the difference in the rates of formation of their urea inclusion complexes.

Acids	Urea (g)	Methanol (ml)	Complex Fraction	Soluble Fraction
9.9 g lauric 9.9 g stearic	30	150	10.5 g 88% stearic	8.3 g 83% lauric
3.0 g oleic 3.0 g stearic	19.3	120	3.3 g 80% stearic	2.65 g 89% oleic
1.0 g oleic 1.0 g linoleic	4.8	30	0.6 g 91.8% oleic	1.3 g 57% linoleic
2.6 g linoleic 2.6 g linolenic	13	79	1.4 g 80% linoleic	

c. Separation of saturated and unsaturated acids of natural oils

It is possible to bring about enrichment in the saturated and unsaturated acids of natural oils. The acids or esters are recovered from the complexes by warming the latter with ten times their weight of dilute hydrochloric acid. The acids are extracted with petroleum ether and dried. The complex fraction is enriched in saturated acids, while the noncomplex fraction is enriched in unsaturated acids.

Fatty Acid of	Urea (g)	Methanol (ml)	Complex (g)	Noncomplex (g)
Olive oil, 50 g	50	150	13	37.7
Corn oil, 50 g	75	150	21	26
Soybean oil, 50 g	100	100	33.5	13.7
Linseed oil, 50 g	75	150	20	29

References

1. Schlenk, W. (1949). *Ann.* **565**, 204.
2. Smith, A. E. (1950). *J. Chem. Phys.*, **18**, 150.
3. Schlenk, W. (1951). *Ann.* **573**, 142.
4. Redlich, O., Gable, C. M., Dunlop, A. K., and Millar, R. W. (1950). *J. Am. Chem. Soc.* **72**, 4153.
5. Schlenk, W., and Holman, R. T. (1950). *J. Am. Chem. Soc.* **72**, 5001.
6. Zimmerschied, W. J., Dimerstein, R. A., Wetkamp, A. W., and Marschner, R. F. (1950). *Ind. Eng. Chem.* **42**, 1300.
7. Knight, H. B., Witnauer, L. P., Coleman, J. E., and Swern, D. (1952). *Anal. Chem.* **24**, 1331.

Recommended Reviews

1. Bishop, R., and Dance, G. (1988). New types of helical canal inclusion networks. *In* "Topics in Current Chemistry, 149, Molecular Inclusion and Molecular Recognition—Clathrathes II." (E. Weber, ed.), p. 137. Springer-Verlag, New York.
2. Schlenk, H. (1954). *Urea inclusion compounds of fatty acids. In "Progress in the Chemistry of Fats and Other Lipids"* (R. T. Holman, W. O. Lundberg, and T. Malkin, eds.), Vol. 2, p. 243. Pergamon Press, London, England.
3. Swern, D. (1964). *Urea complexes. In "Fatty Acids,"* (K. S. Markley, ed.), Vol. 3, p. 2309. Interscience, New York.
4. Wosch, D., and Vogtle, F. (1987). Separation of enantiomers by clathrate formation. *In* "Topics in Current Chemistry, 140, Molecular Inclusion and Molecular Recognition—Clathrates I." (E. Weber, ed.), p. 21, Springer-Verlag, New York.

G. Gas–Liquid Chromatography of Methyl Esters of Fatty Acids

1. Introduction

Apart from the petroleum hydrocarbons, the fatty acids represent perhaps the most complex group of naturally occurring substances, and until recently no simple technique was available for their separation and identification.

Gas-liquid chromatography (GLC) was introduced by James and Martin in 1952 (1); they separated the normal saturated acids up to C_{12}. This method was extended by Cropper and Heywood (2) to the separation of C_{12}–C_{22} fatty acids in the form of their methyl esters. Many methods have been described for the preparation of methyl esters from fatty acids. The diazomethane method (3) is rapid, but should not be used for polyenoic acids, as it also adds to the double bond. Other procedures involve the use of methanol containing a catalyst such as hydrochloric acid (4); hydrochloric acid and anion exchange resin (5); and boron trifluoride (6). Orr and Callen (7) and Lipsky and Landowne (8) used polyesters as liquid phases and obtained excellent separations of saturated and unsaturated esters. A complex mixture of polyenoic acids was separated by Ackman, Burgher, and Jangaard (9). Oette and Ahrens (10) have recommended the use of the relatively less volatile β-chloroethyl esters for C_2–C_{12} acids. Similarly, Craig, Tulloch, and Musty (11) suggest the use of butyl, phenacyl, or decyl esters. The use of capillary columns in preference to packed columns was suggested by Golay (12) and used successfully by Lipsky, Lovelock, and Landowne (13) even for the resolution of cis–trans isomers.

2. Principle

The great molecular range of naturally occurring fatty acids makes it impossible to carry out a comprehensive analysis of all constituents in a single run, unless temperature or flow programming is used. The alternative approach, e.g., the one described below (14), is to separate the mixture of esters on two columns that have been adjusted to give optimal conditions for either stage. Thus two stationary phases are used: 1,4-butanediol succinate for the low-molecular esters; and diethylene glycol succinate for the high-molecular esters.

3. Apparatus

Gas chromatograph

4. Materials

1,4-Butanediol succinate
Butyric acid
Caproic acid
Caprylic acid
Chromosorb P
Dichlorodimethylsilane
Diethylene glycol succinate
Ether
Linoleic acid
Myristic acid
N-Nitrosomethylurea
Oleic acid
Palmitic acid
Potassium hydroxide
Stearic acid

5. Time

5–6 hours

6. Procedure

a. Preparation of methyl esters
Twenty to thirty mg of the following acids are methylated with diazomethane: butyric, valeric, caproic, caprylic, myristic, palmitic, stearic, oleic, and linoleic. An ethereal solution of the acids is treated with diazomethane, which is prepared as follows: in a 50-ml round-bottomed flask are placed 6 ml 50% aqueous potassium hydroxide solution and 20 ml ether. The mixture is cooled to 5°C, and 2 g nitrosomethylurea is added with shaking. The flask is fitted with a condenser set for distillation. The lower end of the condenser carries an adapter dipping below the surface of 5 ml ether contained in a 30-ml Erlenmeyer flask and cooled in an ice–salt mixture. The reaction flask is placed in a water bath at 50°C and brought to the boiling point of the ether with occasional shaking. The ether is distilled until it comes over colorless, which is usually the case after two thirds of the ether has been distilled. Under no circumstances should all the ether be distilled. The ether solution in the receiving flask contains about 0.5 g diazomethane. If a dry solution of diazomethane is required, the ether solution of diazomethane is allowed to stand for 3 hr over pellets of potassium hydroxide.

Warning! Diazomethane is not only toxic, but also potentially explosive. Hence, one should wear heavy gloves and goggles and work behind a safety screen or a hood door with safety glass. It is also recommended that ground joints and sharp surfaces should not be used. Diazomethane solution should not be exposed to direct sunlight or placed near a strong source of artificial light.

b. Preparation of columns
U-shaped or coiled glass columns, 5 mm in diameter and 6 ft in length, are used. The columns are treated with 5% dichlorodimethylsilane in toluene, washed with toluene and methanol, and dried. The partitioning medium used for the esters of lower molecular weight is 1,4-butanediol succinate (BDS) in a ratio of 1:5 (w/w) on Chromosorb P, 60 to 80 mesh. For the esters of higher molecular weight, diethylene glycol succinate (DEGS) is used in a ratio of 1:5 (w/w) on Celite 545 (80 to 100 mesh) siliconized by treatment with dichlorodimethylsilane. Columns are packed by gradual addition of the coated support accompanied by repeated tapping. After packing, the columns are preconditioned by baking for several hr at 200°C in a stream of nitrogen with a flow rate of 50 ml per min.

c. Operating parameters

For the BDS column, the conditions are as follows: column temperature, 88°C; flash heater, 206°C; argon pressure, 17 psi; flow rate, 30 ml per min. The conditions for the DEGS column are: column temperature, 180°C; flash heater, 285°C; argon pressure, 41 psi; flow rate, 60 ml per min.

References

1. James, A. T. and Martin, A. J. P. (1952). *Biochem. J.* **50**, 679.
2. Cropper, F. R., and Heywood, A. (1953). *Nature*, **132**, 1101.
3. Poper, R., and Ma, T. S. (1957). *Microchem. J.* **1**, 245.
4. Stoffel, W., Chu, F., and Ahrens, E. H. (1959). *Anal. Chem.* **31**, 307.
5. Hornstein, I., Alford, J. A., Elliot, L. E., and Crowe, P. E. (1960). *Anal. Chem.* **32**, 540.
6. Metcalfe, L. D., and Schnitz, A. A. (1961). *Anal. Chem.* **33**, 363.
7. Orr, C. H., and Callen, J. E. (1958). *J. Am. Chem. Soc.* **80**, 249.
8. Lipsky, S. R., and Landowne, R. A. (1958). *Biochim. Biophys. Acta* **27**, 666.
9. Ackman, R. G., Burgher, R. D., and Jangaard, P. M. (1963). *Can. J. Biochem. Physiol.* **41**, 1627.
10. Oette, K., and Ahrens, E. H. (1961). *Anal. Chem.* **33**, 1847.
11. Craig, B. M., Tulloch, A. P., and Murty, N. L. (1963). *J. Am. Oil Chemists' Soc.* **40**, 61.
12. Golay, M. J. E. (1958). *In:* "*Gas Chromatography*" (V. J. Coates, H. J. Noebels, and I. S. Fagerson, eds.), p. 1. Academic Press, New York.
13. Lipsky, S. R., Lovelock, J. E., and Landowne, R. A. (1959). *J. Am. Chem. Soc.* **81**, 1010.
14. Vorbeck, M. L., Mattick, L. R., Lee, F. A., and Pederson, C. S. (1961). *Anal. Chem.* **33**, 1512.

Recommended Reviews

1. James, A. T. (1960). *Qualitative and quantitative determination of the fatty acids by gas-liquid chromatography.* In "*Methods of Biochemical Analysis*" (D. Glick ed.), Vol. 8, p. 1. Interscience, New York.
2. Marcel, S. F., and Lie Ken Jie. (1980). The characterization of long-chain fatty acids and their derivatives by gas-liquid chromatography. *Adv. Chromatogr.* **18**, 3.
3. Woodford, F. P. (1964). *Gas-liquid chromatography of fatty acids.* In "*Fatty Acids*" (K. S. Markley, ed.), Vol. 3, p. 2249. Interscience, New York.

H. Thin-Layer Chromatography of Fatty Acids

1. Introduction

In a mixture of naturally-occurring fatty acids, differences in both chain length and degree of unsaturation are usually encountered, and certain *critical pairs* of acids, e.g., oleic and palmitic, or linoleic and myristic, are inseparable on thin layers of various adsorbents. Such pairs were separated by low-temperature thin-layer chromatography, by reversed-phase partition chromatography on siliconized plates (1), and by bromination or hydrogenation of double bonds before chromatography.

Methyl esters of fatty acids may be separated, according to their degree of unsaturation, on silica gel plates impregnated with silver nitrate, by forming a loose complex between silver ions and the electrons of double bonds (2,3). An alternative method (outlined below) is the formation of adducts with mercuric acetate (4). It was found that cis compounds react with mercuric acetate 10 times quicker than trans compounds (5). When the adducts are separated by thin-layer chromatography and decomposed with hydrochloric acid, the esters can be fractionated according to their chain lengths by GLC.

In comparison with the method involving the use of mercuric acetate, thin-layer chromatography on layers impregnated with silver nitrate has the advantage of eliminating the stages of adduct-formation and the dissociation of the separated adducts. Both methods are of great value when used in conjugation with GLC.

2. Materials

Acetic acid	Mercuric acetate	Petroleum ether
Chloroform	Methanol	n-Propanol
Ethanol	Methyl elaidate	S-Diphenylcarbazone
Ether	Methyl oleate	Silica gel G
Iodine	Methyl stearate	Silver nitrate

3. Procedure

a. Preparation of plates

1. The suspension for five plates (20 × 20 cm) is prepared by shaking 30 g silica gel G and 60 ml water for 30 sec and applied uniformly to a thickness of 0.25 mm with an applicator. After 30 min at room temperature, the plates are heated in an oven at 125 to 130°C for 45 min.
2. After cooling, two plates are sprayed with concentrated aqueous methanolic silver nitrate solution, 5% relative to silica gel, and then activated again at 120° for 30 sec. This method permits impregnation of only part of the plate, which can thus be used for comparative chromatography. The plates are used immediately after cooling.

b. Preparation of mercury addition compounds of unsaturated esters

The reagent consists of a solution of 14 g mercuric acetate in 250 ml methanol, 2.5 ml water, and 1 ml acetic acid. Of this solution 25 ml is added to 1 g methyl ester and allowed to react in a stoppered flask stored in the dark at room temperature for 24 hr. Methanol is evaporated at 30°C under reduced pressure or in a stream of nitrogen, and the dry residue is dissolved in 50 ml chloroform. The chloroform solution is washed with 5 × 25 ml portions of water to remove excess mercuric acetate, and dried on sodium sulfate.

c. Development

The samples are applied in chloroform solution along a line 2 cm above the rim of the plate. To obtain optimal resolution, the solvent systems are applied consecutively. First the saturated methyl esters ($R_f = 0.9$) are separated from the acetoxy-mercurimethoxy derivatives of all unsaturated methyl esters ($R_f = 0.1$) by means of a mixture of petroleum ether–ether (4:1), within 1.5 to 2 hr. Then a mixture of n-propanol and acetic acid (100:1) is used to separate, in the same direction, the acetoxymercurimethoxy derivatives of unsaturated esters. This solvent is allowed to travel 3–4 hr.

d. Detection

After spraying with s-diphenylcarbazone in 95% ethanol, the acetoxymercurimethoxy compounds appear as purple spots on a light rose background. The esters of saturated acids are located as faint brown spots between the two solvent fronts by exposing the chromatoplate to iodine vapors.

e. Plates impregnated with silver nitrate

Development of the esters is accomplished with ether–petroleum ether (5:95). The spots are located by spraying with 50% sulfuric acid and charring.

References

1. Malins, D. C., and Mangold, H. K. (1960). *J. Am. Oil Chemists' Soc.* **37**, 576.
2. Morris, L. J. (1962). *Chem. Ind. (London)*, 1238.
3. Bergelson, L. D., Dyatlovitskaya, E. V., and Voronkova, V. V. (1964). *J. Chromatogr.* **15**, 191.
4. Mangold, H. K., and Kammereck, R. (1961). *Chem. Ind. (London)* 1032.
5. Jantzen, E., and Andreas, H. (1959). *Ber.* **92**, 1427.

Recommended Reviews

1. Mangold, H. K. (1965). *Aliphatic lipids.* In Stahl, E. "*Thin-layer Chromatography*," p. 132, Academic Press, New York.
2. Marcel, S. F., and Lie Ken Jie. (1980). The characterization of long-chain fatty acids and their derivatives by thin-layer chromatography. *Adv. Chromatogr.* **18**, 35.
3. Nichols, B. W. (1964). *The separation of lipids by thin-layer chromatography. Lab. Practice* **13**, 299.

I. Questions

1. What is the relationship between the degree of unsaturation of an oil or fat and the melting point? Why must vegetable oils have lower melting points than animal fats?
2. How are oils hydrogenated? What chemical and physical changes take place?

3. Discuss the factors operative in making a fat rancid.
4. What are the main applications of urea and thiourea complexes?
5. Why do most fatty acids contain an even number of carbon atoms?
6. Which products are obtained industrially from castor oil?
7. Outline some methods used for isolation and characterization of natural lipids.
8. Why do only methyl groups appear as side chains in the natural "branched" fatty acids?

J. Recommended Books

1. Deuel, H. J. (1957). *"The Lipids."* Interscience, New York.
2. Gunstone, F. D. (1958) *"An Introduction to the Chemistry of Fats and Fatty Acids."* Chapman and Hall, London, England.
3. Hanahan, D. J. (1960). *"Lipid Chemistry."* John Wiley, New York.
4. Hilditch, T. P. (1956). *"The Chemical Constituents of Natural Fats."* John Wiley, New York.
5. Markley, K. S. (1947). *"Fatty Acids."* Parts 1–4. Interscience, New York.
6. Ralston, A. W. (1948). *"Fatty Acids and Their Derivatives."* John Wiley, New York.

III. LIGNANS

A. Introduction

Compounds containing the following structural unit are abundant in nature.

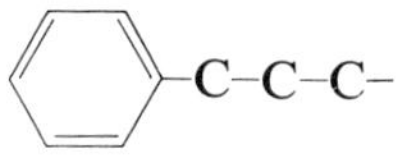

$$\text{C}_6\text{H}_5\text{-C-C-C-}$$

Thus lignin is probably formed by the polymerization of a phenolic phenyl-propane derivative to a complex cross-linked macromolecule. It has been suggested that the key role is lignin formation is played to coniferin, the glucoside of coniferyl alcohol.

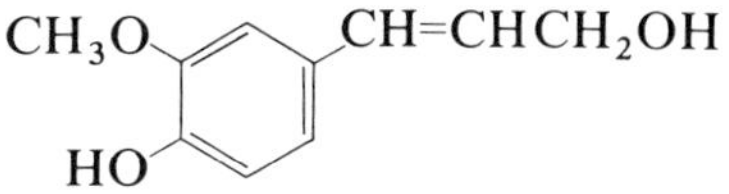

$$\underset{\text{Coniferyl alcohol}}{\text{CH}_3\text{O-C}_6\text{H}_3(\text{OH})\text{-CH=CHCH}_2\text{OH}}$$

Natural phenolic resins (lignans) are closely related to lignin since they are dimers of C_6—C_3 units linked at the β-position. A variety of forms is known, including diarylbutane derivatives such as guaiaretic acid,

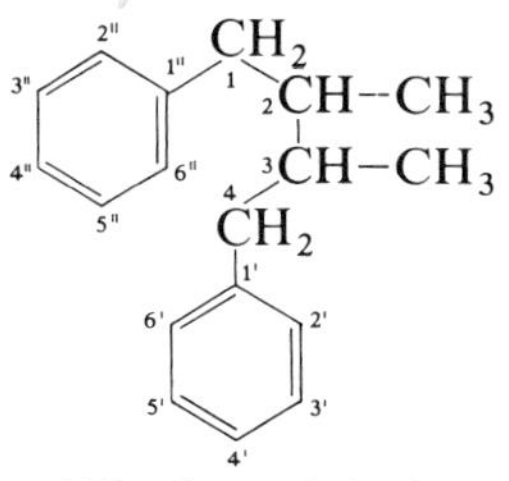

1,4-Diarylbutane derivative

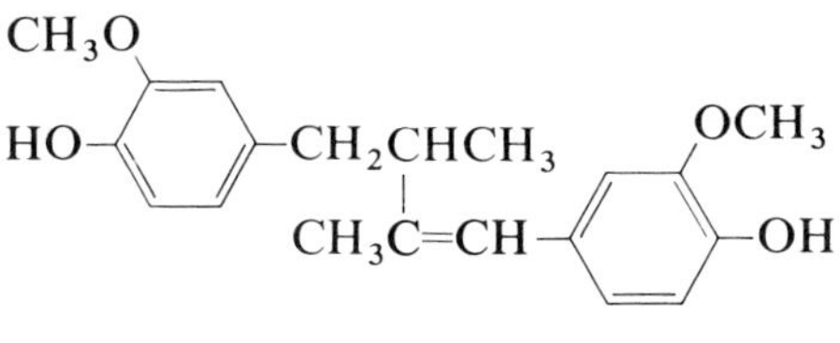

Guaiaretic acid

and tetrahydrofuran derivatives such as olivil, lariciresinol, and sesamin, which can be represented by the following structures:

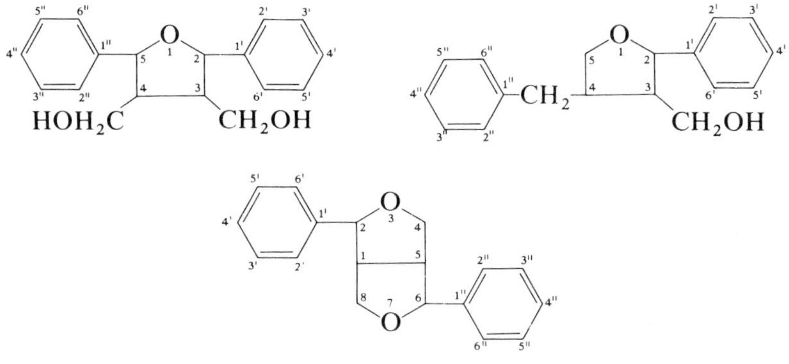

Tetrahydrofuran derivatives

The majority of natural lignans possess the 4-hydroxy-3-methoxyphenyl grouping seen in guaiaretic acid, olivil, and coniferyl alcohol.

Some lignans are sensitive to acids. Thus, olivil is converted into the isomeric isoolivil, and lariciresinol, into isolariciresinol.

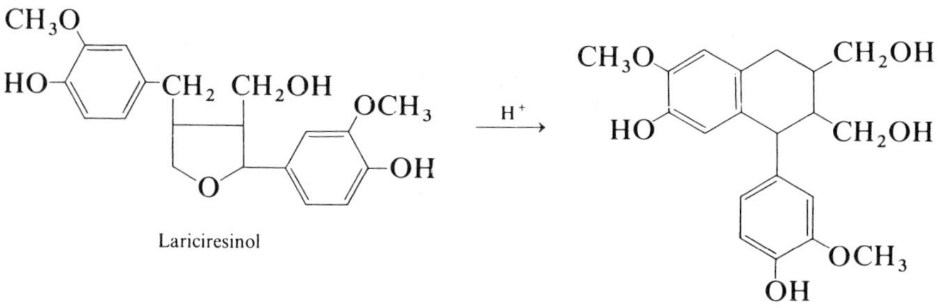

Lariciresinol

$\xrightarrow{H^+}$

Isolariciresinol

III. Lignans

Methylation and dehydrogenation reactions have made possible several correlations within the lignan series. Thus methylation of sesamin converts it into the dimethyl ether of pinoresinol.

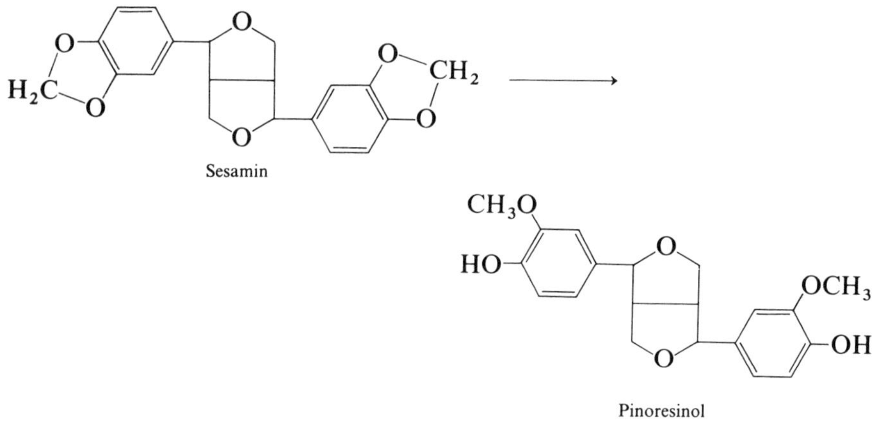

Isoolivil can be dehydrogenated and converted into a lactone that can also be obtained by dehydrogenation and subsequent methylation of conidendrin.

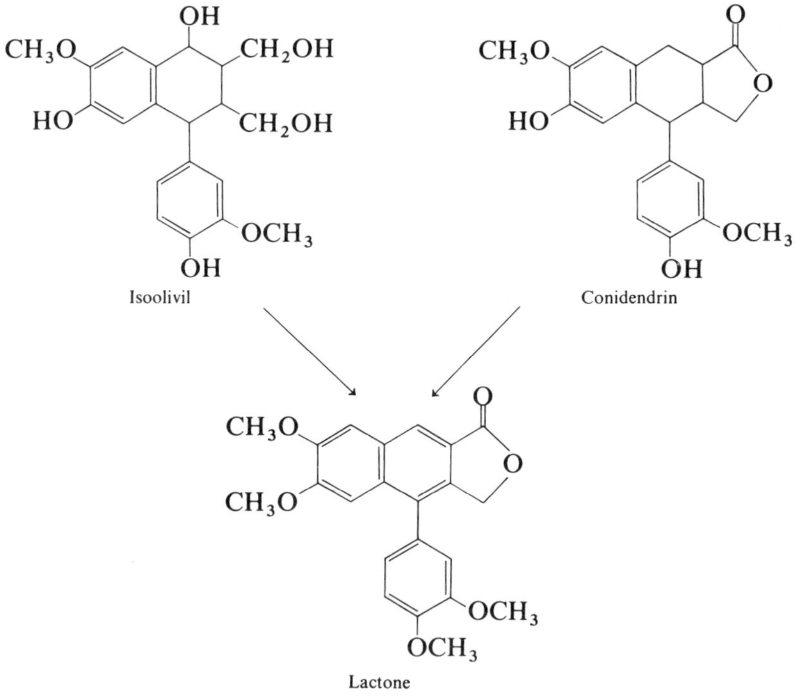

Hydrogenolysis of galgravin gives rise to the formation of the dimethyl ether of dihydroguaiaretic acid, while treatment with perchloric acid produces a dihydronaphthalene derivative.

Galgravin

Pd/H$_2$ HClO$_4$

Dimethyl ether of dihydroguaiaretic acid

Dihydronaphthalene derivative

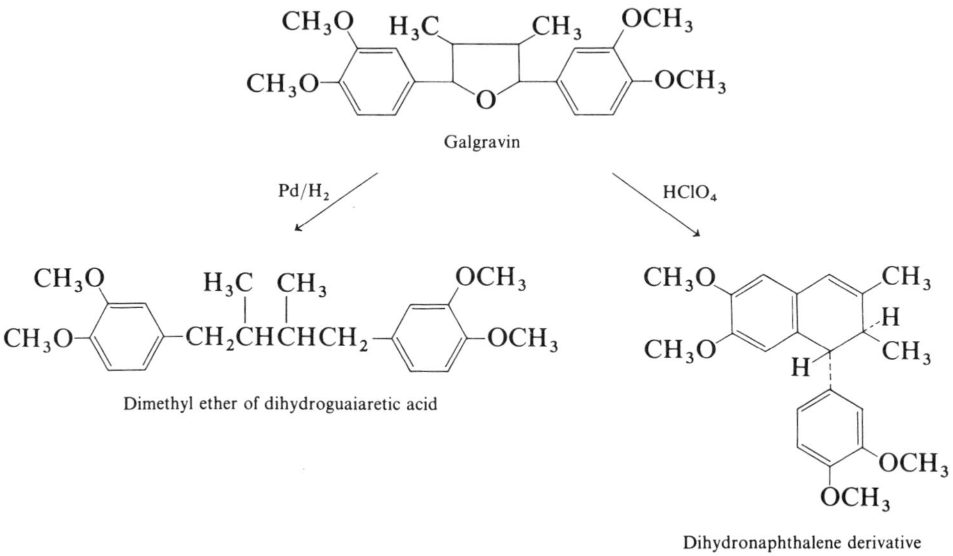

Many lignans are used as antioxidants, and others, as insecticides. The essential feature of the biosynthesis of lignans is the oxidative coupling of 4-propenyl-phenols, such as coniferyl alcohol, at the β-position of the C$_3$ chain.

B. Isolation of Sesamin and Sesamolin from Sesame Oil

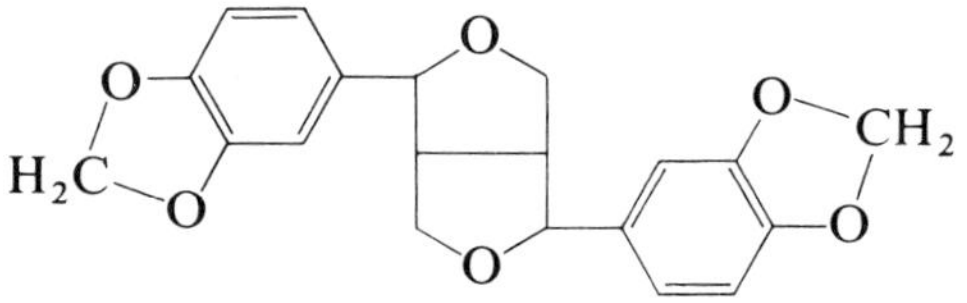

Sesamin

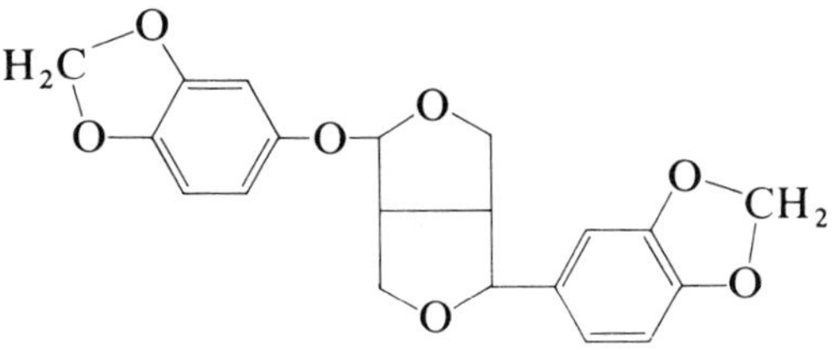

Sesamolin

1. Introduction

The seed oil of *Sesamum indicum*, commonly known as sesame oil, has two minor constituents, sesamin and sesamolin. Three stereoisomers of sesamin are known: sesamin, asarinin, and epiasarinin.

Sesamin was first isolated by Tocher from sesame oil in 1893 (1) and later by Boeseken and Cohen (2). Sesamin has also been isolated from the bark of various *Fagaro* species (3), from *Chamaecyparis obtusa* (4), from *Ocotea usambarensis* (5), from the bark of *Flindersia pubescens* (6), and from the fruit of *Piper guineense* (7). Unlike sesamin, sesamolin has not been noted to occur in any genus other than *Sesamum*. Sesamin was isolated by molecular distillation of sesame oil (8). Details on how to obtain pure sesamin and sesamolin by extraction from sesame oil have been reported by Halsam and Haworth (9) and Beroza (10).

2. Principle

Sesame oil, which contains a large amount of glycerides, is passed through an alumina column and eluted with petroleum ether (9). The fractions obtained are examined by Badouin's color test for detection of lignans. Appropriate fractions are collected, extracted with ether, and treated with alkali. Sesamin and sesamolin are separated by fractional crystallization.

3. Apparatus

Column for chromatography
Soxhlet extractor

4. Materials

Acetic acid	Ethanol	Nitric acid
Acetic anhydride	Ether	Petroleum ether
Alumina	Furfuraldehyde	Potassium hydroxide
Chloroform	Hydrochloric acid	Sesame oil
Dioxane	Methanol	

5. Time

4–5 hours

6. Procedure

A glass column 10 cm in length and 3.5 cm in diameter is filled with 75 g alumina in petroleum ether. 100 ml sesame oil is passed down the column and eluted with petroleum ether, and the fractions collected are examined by Badouins' test, which develops a deep red color in the aqueous phase when the oil is shaken with concentrated hydrochloric acid and 2% furfuraldehyde solution. The portion immediately below the strong yellow band in the column is removed and continuously extracted with ether in a Soxhlet apparatus for 3 hr. Removal of the solvents gives a yellow oil, which is saponified with 5% alcoholic potassium hydroxide for 1 hr. Water (100 ml) is added, and the soap solution is thrice extracted with 30-ml portions of ether. Removal of the ether gives about 2 g yellow resin, which is dissolved in 10 ml ether and left overnight, whereupon 0.5 g crystalline sesamin is precipitated. Crystallization from ethanol yields rod-like needles of mp 122°C. The filtrate, after removal of the ether, is dissolved in 1 ml chloroform, and petroleum ether (80–100°C) is added until the onset of cloudiness. Sesamolin separates as a white solid, which is crystallized from ethanol as white plates, mp 93–94°C; yield, 0.1–0.15 g.

a. Nitrosesamin

A solution of 0.25 ml nitric acid (d = 1.4) in 3 ml acetic acid–acetic anhydride (2:1) is added over a period of 10 min to a solution of 1 g sesamin in 8 ml of the same solvent, and the temperature is kept at 15 to 20°C. Dilution with 15 ml water gives a yellow gum that crystallizes from dioxane–methanol as yellow needles of mononitrosesamin, mp 139–140°C, $[\alpha]_D - 70°$; yield, 0.4 g.

b. Sesamolin

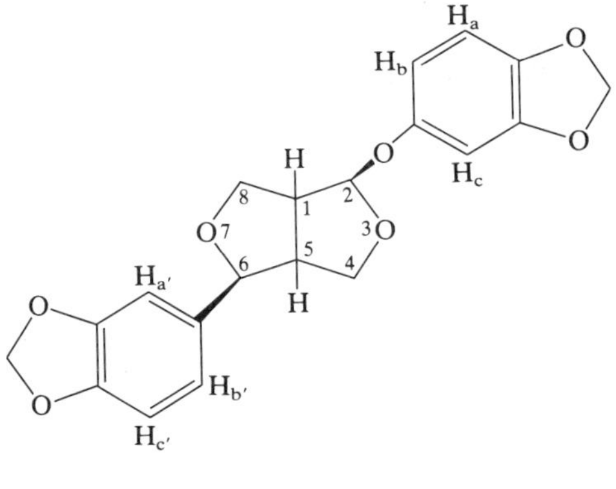

c. ^{1}H NMR [100 MHz]

δ 3.2(q, J 8.5), H-1

5.43(s), H-2

3.88(dd, J 2.2 and 8.5), H_a-4

4.07(dd, J 6.0 and 8.5) H_e-4

2.85(m), H-5

4.28(d, J 5.5), H-6

3.55(dd, J 8.0 and 8.5), H_a-8

4.37(t, J 8.5) H_e-8

6.65(d, J 8.8), H_a

6.44(dd, J 3.8 and 8.8), H_b

6.56(d, J 3.8) H_c

6.85 – 6.75($H_{a'}$, $H_{b'}$, $H_{c'}$)

d. Mass spectrum

Sesamolin $C_{20}H_{18}O_7$, Mol. wt. 370.
m/z 370(M^+)
233
203
138

e. IR spectrum of (+) sesamin (in KBr)
2850 cm^{-1}: sym. stretching of CH_2
1605, 1500, 1445: phenyl ring
1250: asym. stretching of =C—O—C
1200: in-plane bending of 1,2,4-substituted phenyl
1120, 1058: asym. stretching of cyclic ether C—O—C
930: most characteristic of methylenedioxy groups related to C—O stretching
860, 800: out-of-plane C—H bending, two adjacent hydrogen atoms

References

1. Tocher, J. F. (1893). *Ber.* **26,** R591.
2. Boeseken, J., and Cohen, W. D. (1928). *Biochem. Z.* **201,** 454.
3. Carnmalm, B., Erdtman, H., and Pelchowicz, Z. (1955). *Acta Chem. Scand.* **9,** 1111.
4. Masumura, M. (1955). *Nippon Kagaku Zasshi* **76,** 1318.
5. Carnmalm, B. (1956). *Acta Chem. Scand.* **10,** 134.
6. Hollis, A. F., Prager, R. H., Ritchie, E., and Taylor, W. C. (1961). *Australian J. Chem.* **14,** 100.
7. Hansel, R., and Zander, D. (1961). *Arch. Pharm.* **294,** 699.
8. Bhat, S. G., Kane, J. G., and Sreenivasan, A. (1956). *J. Am. Oil Chemists' Soc.* **33,** 197.
9. Halsam, E., and Haworth, R. D. (1955). *J. Chem. Soc.* 827.
10. Beroza, M. (1954). *J. Am. Oil Chemists' Soc.* **31,** 302.

Recommended Reviews

1. Budowski, P. (1964). *Recent research on sesamin, sesamolin and related compounds. J. Am. Oil Chemists' Soc.* **41,** 280.
2. Erdtman, H. (1955). Lignans. In *"Modern Methods of Plant Analysis,"* (K. Paech, and M. V. Tracey, (eds.), Vol. 3, p. 428. Springer-Verlag, New York.
3. Hearon, W. M., and MacGregor, W. S. (1955). *The naturally occurring lignans. Chem. Revs.* **55,** 957.

C. Questions

1. Define the terms lignan and lignin.
2. Classify lignans into their principal groups.
3. Describe the occurrence and structure of olivil and its transformation to isoolivil.
4. What are the commercial uses of lignans?
5. Outline the biosynthesis of lignans.

IV. QUINONES

A. Introduction

The quinones form a large group of natural pigments and are found mainly in plants; many of them have also been isolated from microorganisms such as fungi and lichens, and also from marine animals and certain insects. In general, the quinones are yellow, red, or brown in color, but when present as salts of hydroxyquinones, their colors are purple, blue, or green. The natural quinones play a part in the oxidation–reduction processes of living matter, and some of them have antibiotic properties. The naturally occurring quinones are divided into the following groups:

1. Benzoquinones

These occur mainly in fungi and insects.

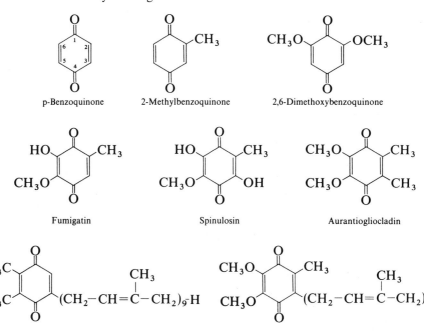

The most important benzoquinones produced by higher plants are the ubiquinones (coenzyme Q) and plastoquinones.

2. 2,5-Dihydroxybenzoquinones

These occur principally in the higher fungi. Some mold products are diquinones related to fumigatin and spinulosin.

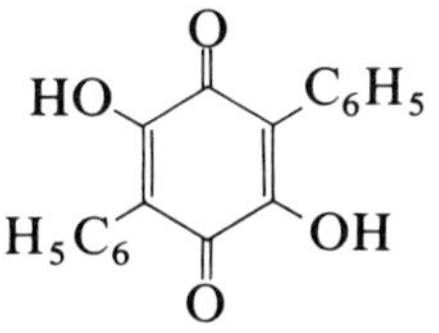

Polyporic acid

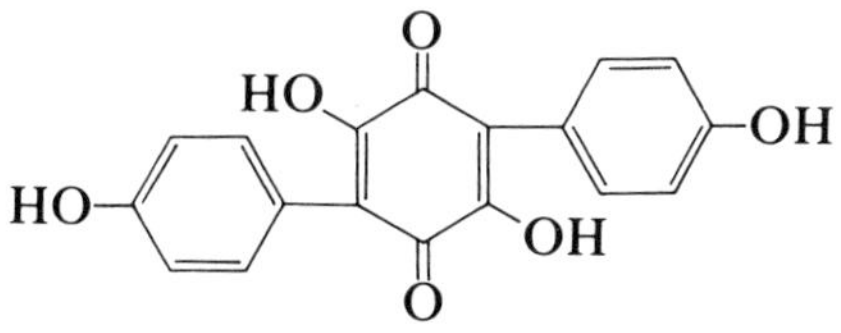

Atromentin

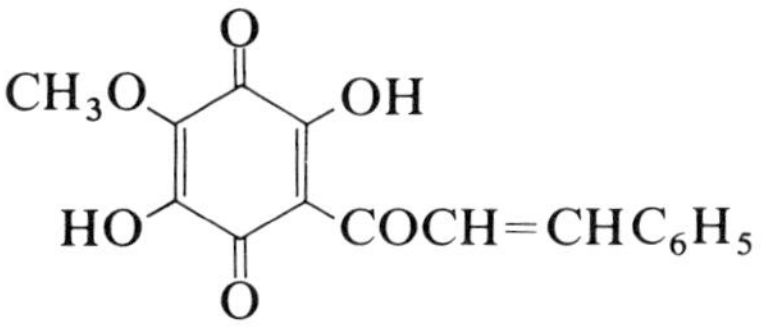

Pedicinin

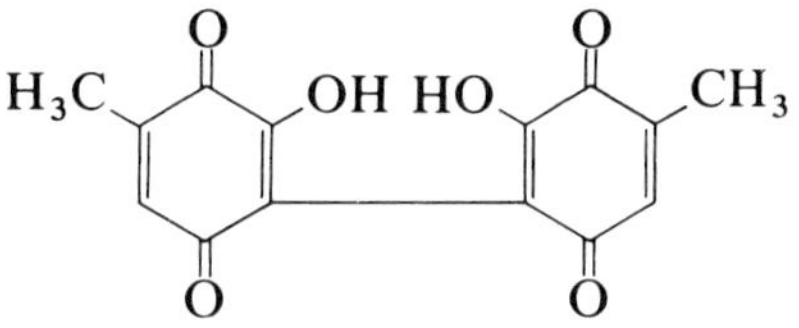

Phoenicin

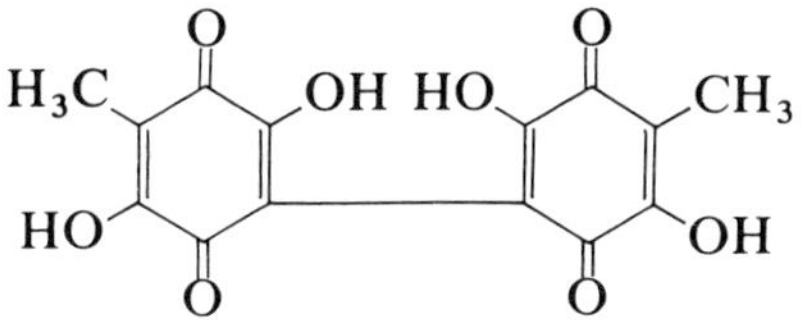

Oosporein

3. Naphthoquinones

These are mostly isolated from plants.

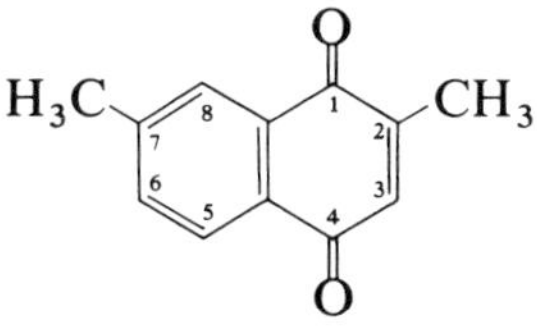

Chimaphilin

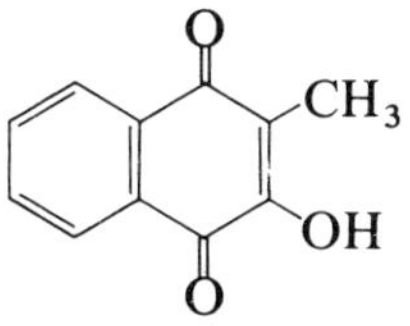

Phthiocol

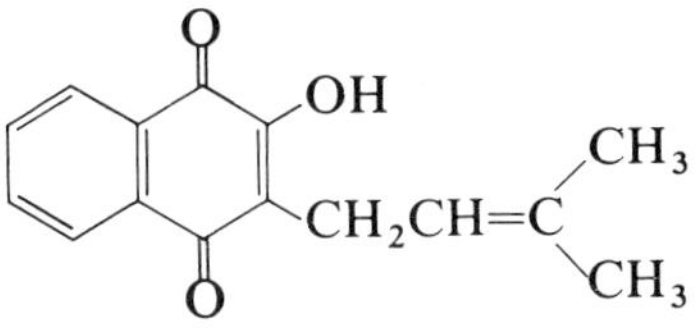

Lapachol, 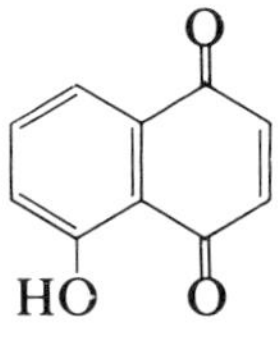 **Juglone**, 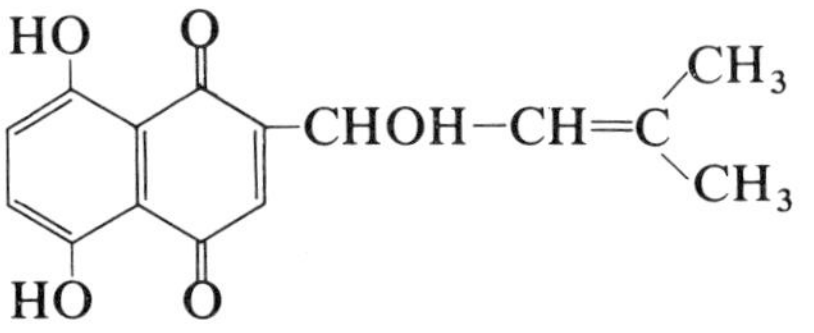 **Alkannin**, 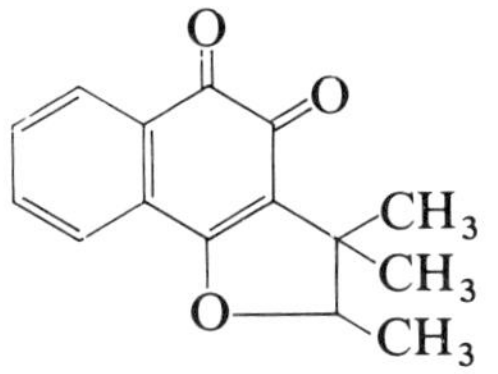 **Dunnione**

The following are animal pigments:

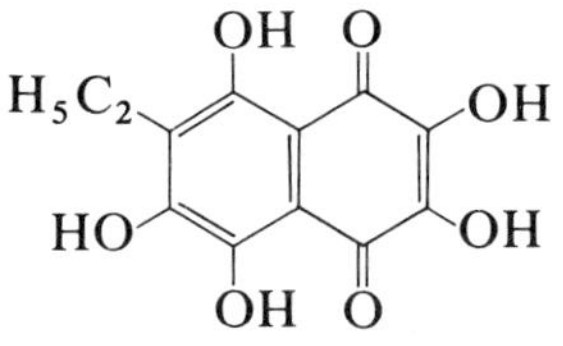

Echinochrome A, 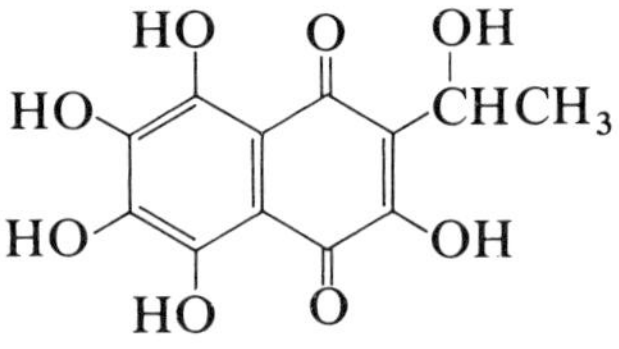 **Spinachrome A**

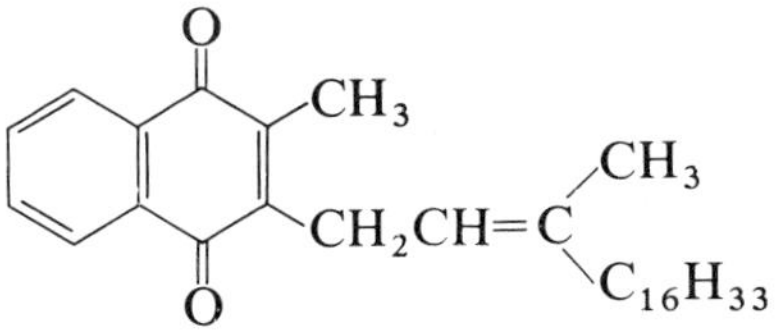

Vitamin K$_1$

4. Anthraquinones

This group is the largest. Many of these pigments occur in the Rubiaceae, Polygonaceae, Rhamnaceae, and Leguminosae.

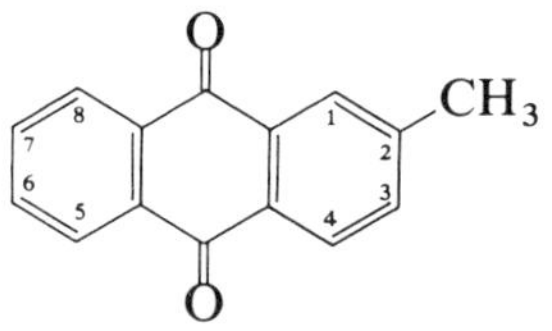

IV. Quinones

Tectoquinone

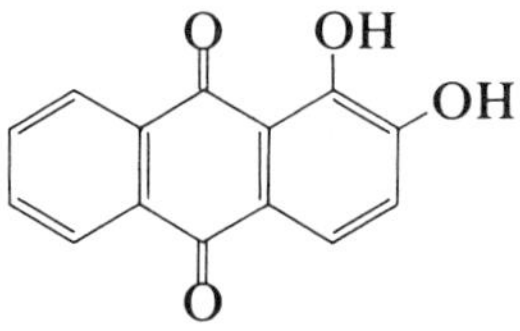

Alizarin

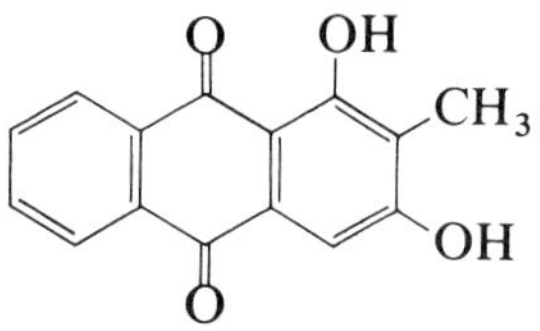

Rubiadin

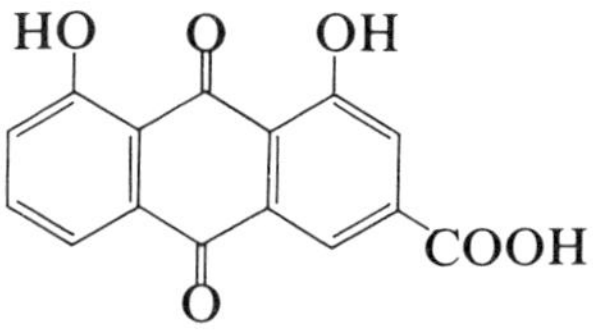

Rhein

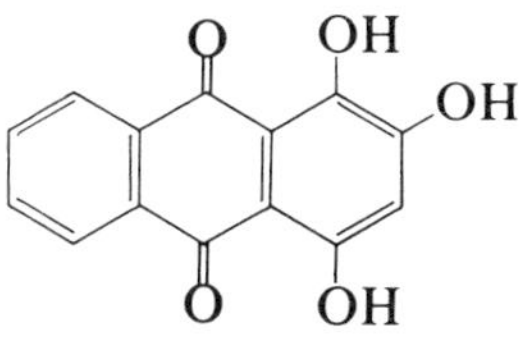

Purpurin

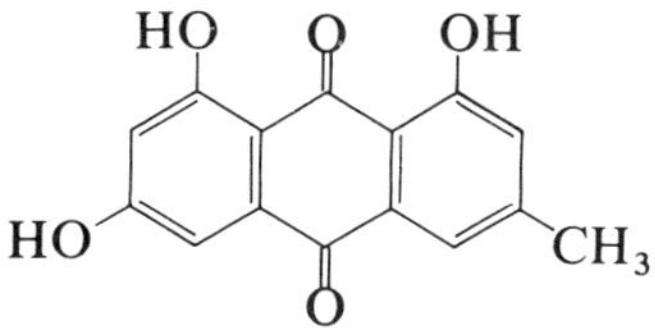

Emodin

The best known insect pigment is carminic acid.

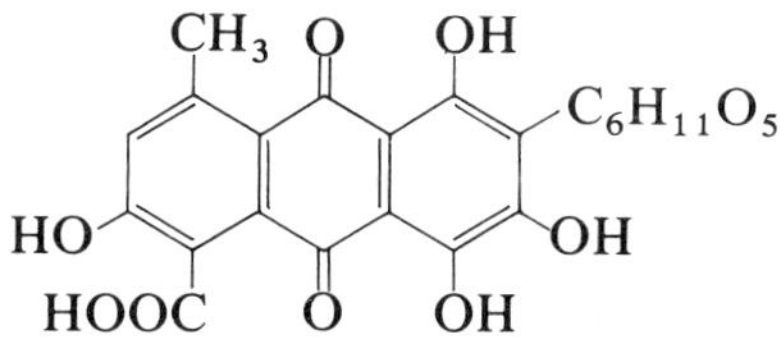

Carminic acid

A more complex, extended anthraquinone, closely related to emodin, is hypericin, which occurs in *Hypericum* species. It is formed by stepwise intramolecular coupling.

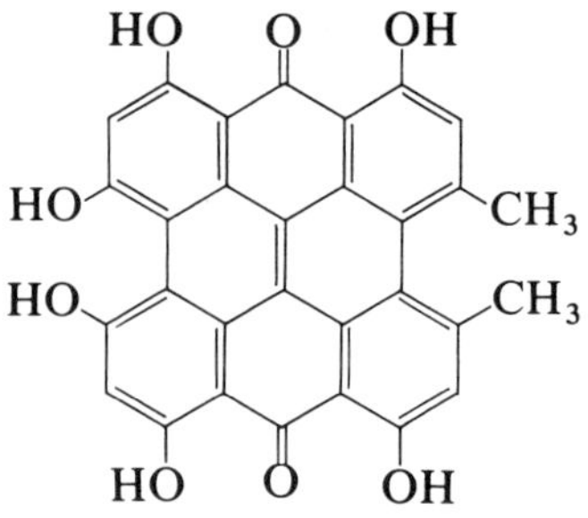

Hypericin

Among the dianthraquinones, skyrin—a mold product—merits mentioning.

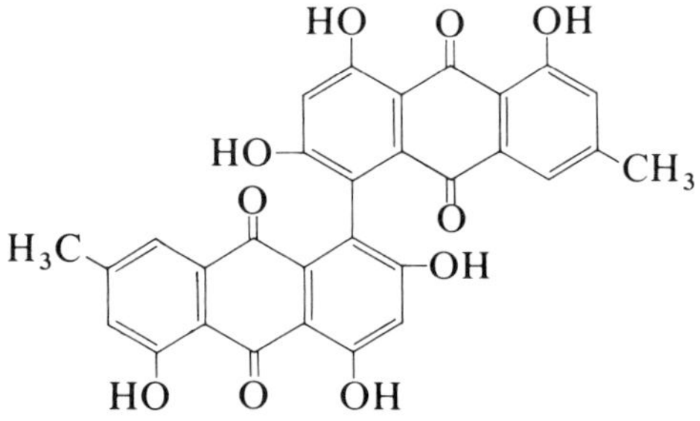

Skyrin

Although natural products related to phenanthrene are numerous, quinones of this series are rare.

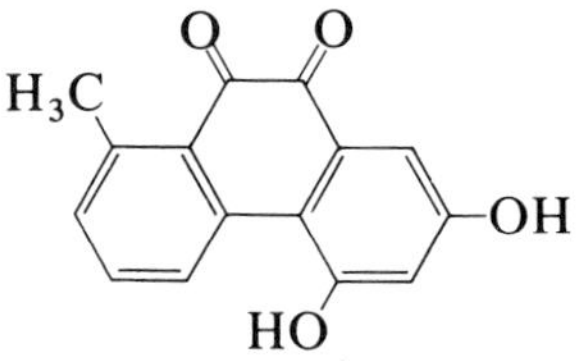

Denticulatol

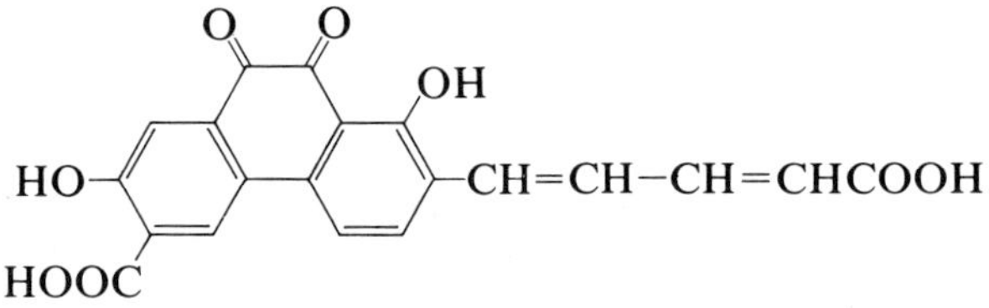

Thelephoric acid

It is interesting to note that the quinones produced by insects are either very simple or very complex, e.g., the aphin pigments. Thus, *Aphis fabae* contains a yellow pigment that is converted by enzymes into an orange and then a red compound, erythroaphin-fb. Zinc dust distillation of this compound forms derivatives of perylene, benzoperylene, and coronene.

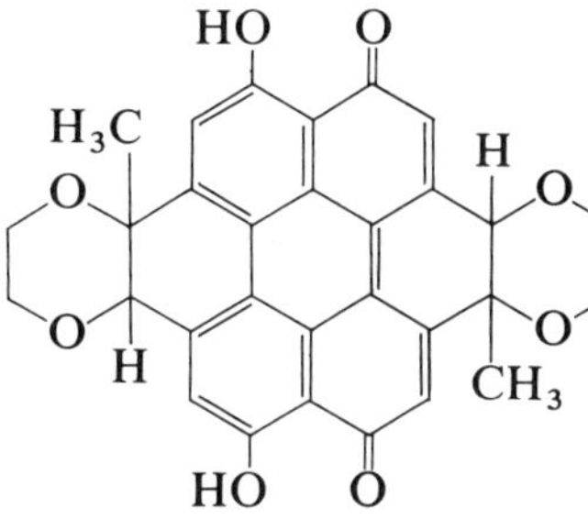

Erythroaphin-fb

Most of the natural quinones arise by the acetate-malonate pathway, while the shikimic acid route is seldom utilized for quinone biosynthesis.

B. Isolation of Rhein from Rhubarb Root

1. Introduction

Rhein was first isolated from Chinese rhubarb in 1895 (1). It is present in various *Rheum* species, partly in the free state and partly as a glycoside. It has been isolated from the roots of *Rheum palmatum* (2), from senna leaves (*Cassia angustifolia*) (3), and from *Cassia fistula* (4). The antibiotic substance cassic acid found in *Cassia reticulata* (5) has been identified with rhein. Other natural sources of rhein are the roots of *Rumex andreaeanum* (6) and the Brazilian species of *Cassia alata*, where it occurs mainly in a reduced glycosidic form (7).

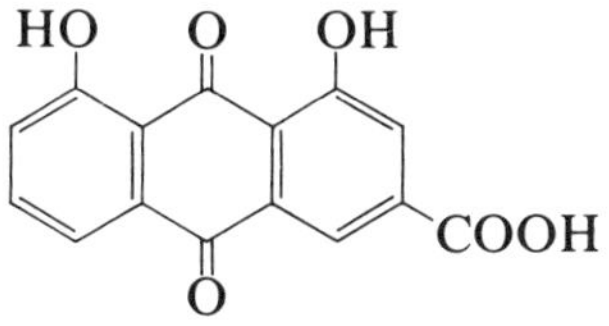

Rhein

2. Principle

In the following experiment (5), rhubarb root is extracted thoroughly with water, concentrated *in vacuo*, and then extracted with methyl isobutyl ketone. Rhein is recovered from the latter with sodium bicarbonate followed by acidification.

3. Apparatus

Extractor

60   CHAPTER 1   Acetogenins

4. Materials

Acetic acid
Acetic anhydride
Acetone
Hydrochloric acid

Methyl isobutyl ketone
Rhubarb roots
Sodium acetate
Sodium hydrogen carbonate

5. Time

6–7 hours

6. Procedure

A batch of 100 g coarsely ground rhubarb root is placed loosely in a fluted filter paper over a wad of glass wool in an extractor, and the active material is extracted for three periods (1, 1.5, and 2.5 hr), using 750 ml water for each extraction. The combined solution is concentrated under reduced pressure to about 100 ml, and the syrupy concentrate is extracted with methyl isobutyl ketone in a continuous extractor until the extract is almost colorless. The organic solution is then shaken in a separatory funnel with small portions (10–25 ml) of 5% sodium hydrogen carbonate solution until the typical reddish color ceases to appear in the extracts. The aqueous alkaline extract is cooled in ice and acidified to about pH 2 with cold, dilute hydrochloric acid, and the tan amorphous precipitate is centrifuged, washed with water, and dried *in vacuo*. The yield of crude rhein is about 200 mg. Previous removal of dark pigment with acetone, followed by cold acetic acid, is necessary for subsequent successful crystallization. The substance is crystallized from acetic acid as pale yellow needles, mp 326–329°C.

a. Thin-layer chromatography of rhein

A drop of rhein solution (in chloroform) is spotted on a thin-layer plate (Silica Gel GF_{254}) and developed with ethyl acetate–methanol–water 8:1:1.

b. Detection UV_{254} (after exposing the plate to ammonia vapors)

It reveals rhein as a red-violet spot, R_f 0.32, and, UV_{365} as a pale orange spot.

c. UV spectrum of rhein

λ_{max}^{MeOH} 230 mμ (logε 3.7); 260 (3.3); 430 (3.2)

Introduction of substituents into the anthraquinone nucleus, whose absorptions are 252 mμ (logε 4.68), 262 mμ (logε 4.30), 272 mμ (logε 4.26), 323 mμ (logε 3.26), and 410 mμ (logε 1.78), shifts the last band to longer wavelengths.

d. Preparation of rhein diacetate

A solution of 100 mg rhein and 150 mg dry sodium acetate in 30 ml acetic anhydride is refluxed for 15 min and poured into 175 ml ice water. The pale yellow material is filtered and crystallized from acetic acid. Mp 250–251°C; yield, 80 mg.

References

1. Hesse, O. (1895). *Pharm. J.* **1**, 325.
2. Tschirch, A., and Eijken, P. A. A. F. (1905). *Chem. Zentr.* **76**, 144.
3. Tutin, F. (1913). *J. Chem. Soc.* **103**, 2006.
4. Modi, F. K., and Khorana, M. L. (1952). *Indian J. Pharm.* **14**, 61.
5. Anchell, M. (1949). *J. Biol. Chem.* **177**, 169.
6. Tsukida, K., Yoneshige, M., and Tsujioka, J. (1954). *J. Pharm. Soc. Japan* **74**, 382.
7. Hauptmann, H., and Nazario, L. L. (1950). *J. Am. Chem. Soc.* **72**, 1492.

Recommended Reviews

1. Sijnsma, R., and Verpoorte, R. (1986). Anthraquinones in the Rubiaceae. *Prog. Chem. Org. Nat. Prod.* **49**, 79.
2. Thomson, R. H. (1957). Anthraquinones. In *"Naturally Occurring Quinones,"* p. 158. Butterworth, London, England.
3. Thomson, R. H. (1965). Quinones: Distribution and biosynthesis. In *"Chemistry and Biochemistry of Plant Pigments,"* (T. W. Goodwin, ed.), p. 309. Academic Press, New York.

C. Isolation and Identification of Phenols and Quinones from Defensive Secretions of Beetles

1. Introduction

Among the arthropods, chemical defense substances play an important role in the relationship of the organism to its predators.

Many tenebrionid beetles manufacture and store large quantities of p-benzoquinones within their defensive glands (1–3). The first mention of tenebrionid defensive secretions are those of Gissler (4) and Williston (5), who observed that seven species of Eleodes were able, upon being disturbed, to eject a "pungent, vile-smelling liquid." Certain species of tenebrionids have entered the folk literature as "circus bugs" from their conspicuous habit of standing on their heads when disturbed.

In 1943 Alexander and Barton (6) reported the presence of ethylquinone in the secretions of the flour beetle *Tribolium castaneum*. Tolu- and ethylquinone were subsequently found in other tenebrionids: *Diaperis maculata* (7), *Eleodes hispilabris* (8) and *Eleodes longicollis* (9). All three of these reports indicated that hydrocarbons were also present in the defensive secretions. They were identified from *E. longicollis* (10) as 1-nonene, 1-undecene and 1-tridecene.

Quinones have also been found in the secretions of several species of *Blaps* (11), *Tenebrio molitor* (12), *Alphitobius diaperinus* (13), *Zophobas rugipes* (14) and *Blaps sulcata* and *Blaps wiedemanni* (15).

It has been suggested that the quinones are generated from phenol glucosides by hydrolysis and subsequent oxidation (16).

The Central American tenebrionid beetle, *Zophobas rugipes*, has a pair of prothoracic defensive glands secreting phenols and a pair of abdominal defensive glands secreting quinones (17). It has been assumed that the phenol precursors are probably oxidized to quinones.

An interesting mechanism (see Fig. 3) has been proposed for the generation of quinones from phenols in the defensive organ of the bombardier beetle.

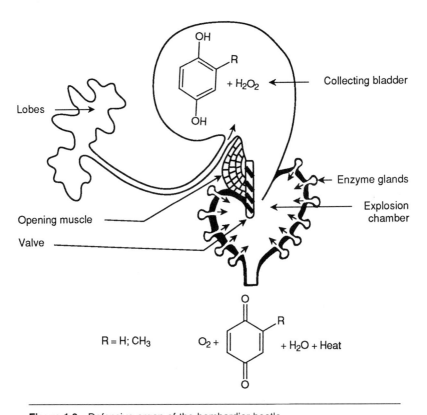

Figure 1.3 Defensive organ of the bombardier beetle.

IV. Quinones

2. Principle

Blaps, flour beetles, or any other tenebrionids can be obtained from entomological laboratories. The pungent secretion of the quinones is obtained by slightly pressing the body of the insect. It is collected with fine capillaries and analyzed by chromatographic (TLC) and spectroscopic (NMR, IR, MS) methods. **Care should be taken when handling these lachrymatory and skin irritating compounds!**

The following procedure is based on (15).

3. Apparatus

Infrared spectrophotometer
Mass spectrometer
Nuclear magnetic resonance spectrometer

4. Materials

Blaps
p-Benzoquinone
Carbon disulfide
2,4-Dinitrophenylhydrazine
Flour beetles
Glass capillaries
Hydroquinone
α-Naphthol

5. Time

3 hours

6. Procedure

a. Collection of secretion

Pressing the body of Blaps causes the discharge of pungent, yellowish-brown liquid, which is collected with the help of fine capillaries and stored in a deep-freeze refrigerator. The secretion consists of two phases, aqueous and hydrophobic. The nonaqueous phase is soluble in carbon disulfide.

The CS_2 solution is subjected to various chromatographic and spectroscopic studies, whereas the aqueous phase serves for the identification of glucose.

b. TLC of the secretion

The crude secretion (or its CS_2 solution) is spotted on a silica gel GF_{254} plate (0.25 mm thick) and developed with petrol–ether–ether, 7:3. The quinones and hydroquinones are detected with the help of UV light (254 nm) and by spraying with 2,4-dinitrophenylhydrazine solution and chloranil reagent, (prepared by dissolving 0.5 g chloranil in 10 ml methanol).

Separation and Identification of Quinones and Hydroquinones by TLC

Zone on Chromatogram	Compound[a]	R_f	Visible	2,4-DNP	Chloranil Reagent
A	Hydroquinone	0.12	—	—	Blue
B	2-Methyl hydroquinone	0.14	—	—	Blue
C	2-Ethyl hydroquinone	0.16	—	—	Blue
D	p-Benzoquinone	0.48	Yellow	Orange-brown	—
E	2-Methyl benzoquinone	0.53	Yellow	Orange-brown	—
F	2-Ethyl benzoquinone	0.66	Yellow	Orange	—

[a] All compounds absorb at UV (254 nm) light.

c. Quinones and hydroquinones

Infrared Spectra of Benzoquinones (in CS_2)

Absorption (u)	Characteristic of
6.05	Carbonyl group
9.2 and 12.35	2-methyl-1,4-benzoquinone
11.02 and 12.0	2-ethyl-1,4-benzoquinone

Ultraviolet Spectra

Compound	λ_{max}^{EtOH}(nm)
1,4-Benzoquinone	242
2-Methyl-1,4-benzoquinone	245
2-Ethyl-1,4-benzoquinone	247
Hydroquinone	292
2-Methyl hydroquinone	293
2-Ethyl hydroquinone	294

d. Mass spectra

The following molecular ions are 136, 122 and 108, and correspond to ethyl-, methyl- and benzoquinone, respectively.

The following is a fragmentation scheme of benzoquinone:

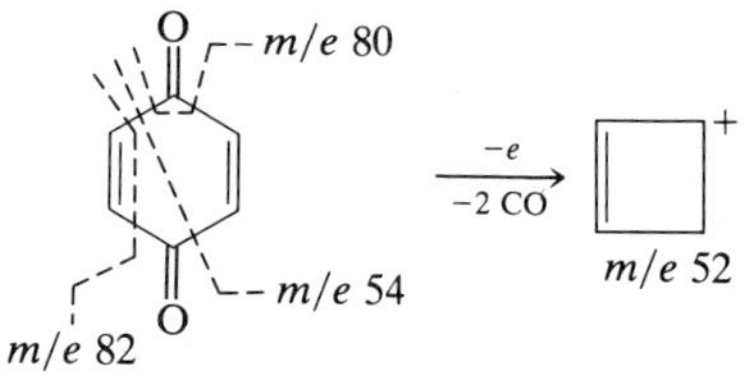

^{1}H-NMR

Quinone	Structural Unit	δ^a (ppm)
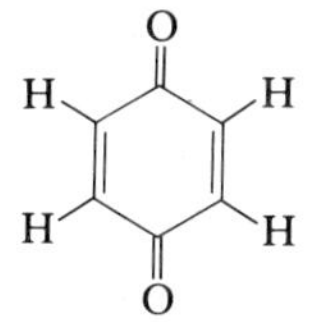 1,4-benzoquinone (H at all four positions)	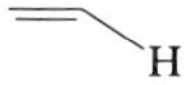 =CH–H	6.75 (s)
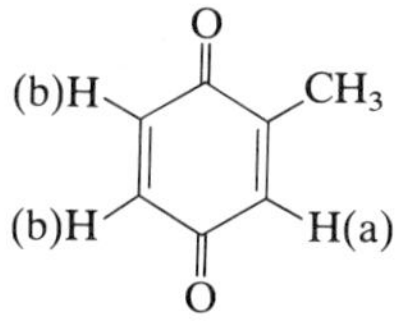 2-methyl-1,4-benzoquinone ((b)H, (b)H, H(a), CH$_3$)	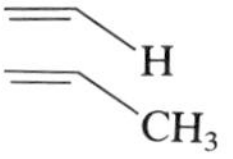 =CH–H =C–CH$_3$	(a) 6.58 (m) (b) 6.70 (s) 2.01 (d)
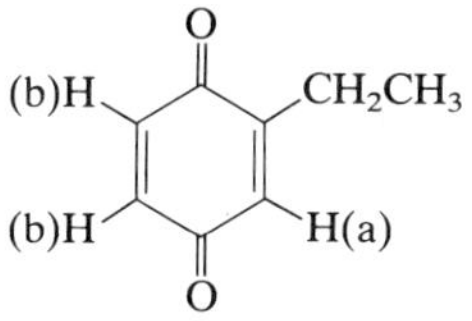 2-ethyl-1,4-benzoquinone ((b)H, (b)H, H(a), CH$_2$CH$_3$)	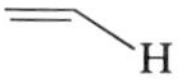 =CH–H —CH$_2$— —CH$_3$	(a) 6.58 (m) (b) 6.70 (s) 2.45 (q) 1.16 (t)

^{a} Values (in CS$_2$) refer to tetramethylsilane ($\delta = 0$). Letters in parentheses refer to doublet (d), multiplet (m), quartet (q), singlet (s), and triplet (t).

References

1. Schildknecht, H. (1963). *Angew. Chem.* **75**, 762.
2. Eisner, T., and Meinwald, J. (1966). *Science* **153**, 1341.
3. Weatherston, J. (1967). *Quart. Rev. Chem. Soc. Lond.* **21**, 267.
4. Gissler, C. G. (1879). *Bull. Brooklyn Ent. Soc.* **2**, 7.
5. Willistton, S. W. (1884). *Psyche, Camb.* **4**, 168.
6. Alexander, P., and Bartton, D. H. R. (1943). *Biochem. J.* **37**, 463.
7. Roth, L. M., and Stay, B. (1958). *J. Insect Physiol.* **1**, 305.
8. Blum, M. S., and Crain, R. D. (1961). *Ann. Ent. Soc. Am.* **54**, 474.
9. Chadha, M. S., Eisner, T., and Meinwald, J. (1961). *J. Insect Physiol.* **7**, 46.
10. Hurst, J. J., Meinwald, J., and Eisner, T. (1964). *Ann. Ent. Soc. Am.* **57**, 44.
11. Schildknecht, H., and Weis, K. H. (1960). *Z. Naturf.* **15b**, 757.
12. Schildknecht, H., and Kramer, H. (1962). *Z. Naturf.* **17b**, 701.
13. Tseng, Y. L., Davidson, J. A., and Menzer, R. E. (1971). *Ann. Ent. Soc. Am.* **64**, 425.
14. Tschinkel, W. R. (1969). *J. Insect Physiol.* **15**, 191.
15. Ikan, R.; Cohen, E., and Shulov, A. (1970). *J. Insect Physiol.* **16**, 2201.
16. Eisner, T., McHenry, F., and Salpeter, M. M. (1964). *J. Morph.* **115**, 355.
17. Tschinkel, W. R. (1969). *J. Insect Physiol.* **15**, 197.

Recommended Reviews

1. Weatherston, J. (1967). The chemistry of arthropod defensive substances. *Quart. Rev. Chem. Soc.* **21**, 287–313.
2. Weatherston, J., and Percy, J. E. (1970). Arthropod defensive secretions. *In* "Chemicals Controlling Insect Behavior." (M. Beroza, ed.), 287–313 Academic Press, New York.

D. Questions

1. What are the functions of quinones in nature?
2. Outline the biosynthesis of quinones.
3. Describe methods of isolation and identification of lapachol and skyrin.

E. Recommended Books

1. Morton, R. A. (ed.) (1965). *"Biochemistry of Quinones."* Academic Press, New York.
2. Thomson, R. H. (1957). *"Naturally Occurring Quinones."* Butterworth, London, England.
3. Patai, S., and Rappoport, Z. (eds.) (1988). *"The Chemistry of Quinonoid Compounds."* Wiley, New York.

V. PHLOROGLUCINOLS

A variety of natural products contain the phloroglucinol nucleus in their structure. It can be regarded as derived from three acetate units.

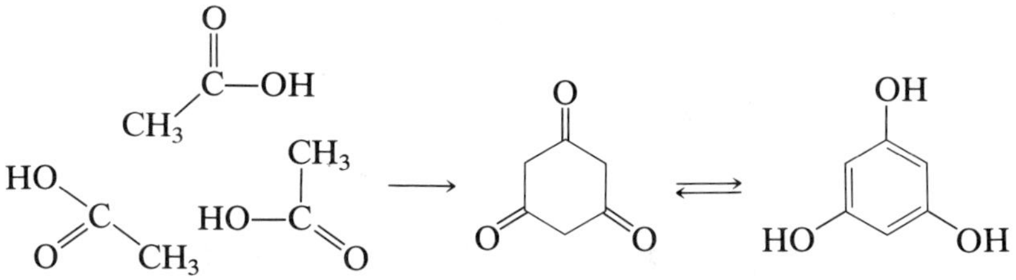

As indicated above, phloroglucinol can be represented as a tautomeric triketone, and many reactions of phloroglucinol derivatives are best understood by reference to this ketonic structure.

The bitter substances found in hops (*Humulus lupulus*) have isoprenoid side chains attached to the phloroglucinol nucleus or to a cyclopentatrione ring. In view of the well-known anionoid activity of the phloroglucinol nu-

V. Phloroglucinols

cleus, it is considered that the introduction of the acyl side chains such as $COCH_2CH(CH_3)_2$ of humulone may occur through the isopentenyl cation $(CH_3)_2C{=}CH{\cdot}C^+H_2$. Plants containing compounds possessing a phloroglucinol moiety have been used as antihelminic drugs, and many of them possess insecticidal and bacteriocidal properties.

A. High-Performance Liquid Chromatography of Hop Bitter Substances

1. Introduction

The flavor of beer, like that of many foods and beverages, is composed of many volatile and nonvolatile compounds present in a definite blend.

Principal classes of volatiles of beer are acetals, acids, alcohols, aldehydes, amides, amines, esters, furans, hydrocarbons, ketones, lactones, phenols, pyrazines, pyridines.

Modern separation and identification techniques place the number (of volatile and nonvolatile compounds in beer) close to 800. Only a small number of these compounds are "flavor active."

The hop plant (*Humulus hupulus*) is a climbing herbaceous plant belonging to the natural family Moraceae and the natural order of Urticales. The lupulin glands secrete the bitter resin and essential oil at the base of the female flowers.

The resin fraction of hops consists of two major groups of compounds, the α-acids (humulenes) and β-acids (lupulones), which together normally form 10–15% of the weight of the dried hops. These compounds are of mixed biogenetic origin and may be regarded as phloracylphenones, modified by introduction of isopentenyl side-chains and derived biogenetically from the desoxy α acids.

Both the α and β acid fractions of hops consist of a mixture of homologs in which the structure of the acyl side-chain varies. The proportionate composition of the α and β acid fractions varies widely with the variety of hop. Thus European hops generally contain large quantities of humulone in relation to the other homologs, whereas American and Japanese varieties frequently contain almost as much cohumulone as humulone. These differences may be quite significant, since cohumulone is normally better utilized in the brewing process and may give rise to less desirable flavor characteristics than humulone.

In the brewing of beer, hops are boiled with malt and other cereal extracts (wort) in the brew kettle for a given period. During this period, the three major hops α-acids (I) humulone, cohumulone and adhumulone and the minor β-acids (II) undergo heat-induced isomerization into iso-α-acids (III) (each α-acid yields a cis-trans pair of stereoisomers), and hulupones (IV).

Beer therefore contains six major bitter iso-α-acids. Quantitation of these compounds in isomerized hop extracts, worts, and beers is very important in the brewing industry. Various HPLC methods have been used (1–3) for separating the bitter principles of hops.

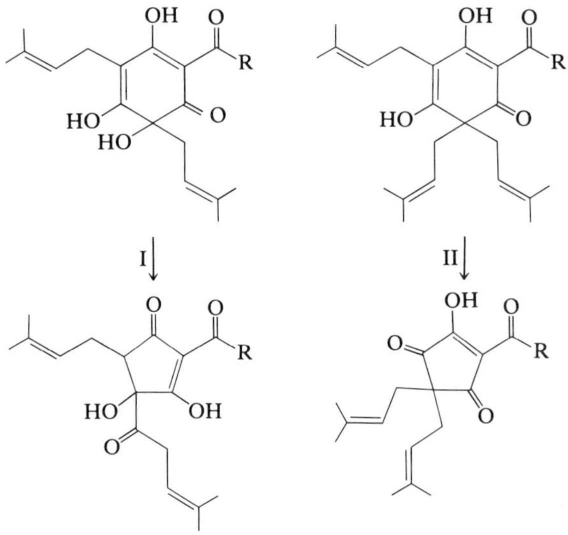

III   IV

R = CH$_2$CH(CH$_3$)$_2$: humulone (I) and lupulone (II)
R = CH(CH$_3$)$_2$: cohumulone and colupulone
R = CH(CH$_3$)CH$_2$CH$_3$: adhumulone and adlupulone

2. Principle

In the following procedure (4), the beer extract is analyzed by the HPLC technique using sodium acetate–acetic acid–buffer–methanol–water eluant. Addition of BHT (butylhydroxy toluene), a well-established antioxidant, to the eluting solvent and to the extract improves the results.

3. Apparatus

HPLC apparatus
Lichroma tubing packed with C$_{18}$ silica gel
Syringe

4. Materials

Acetic acid   Isooctane
Beer   Methanol
BHT   Sodium acetate

5. Procedure

Beer (20 g) is weighed into a beaker and then transferred with minimal foaming into a 100 ml separation funnel, acidified with 2 ml 3N HCl and extracted with 2 portions (25 ml each) isooctane. Twenty-five ml of the isooctane layer is completely evaporated in a rotavapor. HPLC of the hop acids is performed on a polar octadecyl (C_{18}) phase, which was treated with silica gel (10 μm), treated with octadecyltrichlorosilane and containing ca. 17% bonded organic material. The column is 25 × 0.46 cm Lichroma tubing. Elution is carried out with the following elution system: methanol–water–acetic acid (85:15:1). Addition of BHT to the sample solution (1%) and to the eluting solvent (1%) improves the results.

The order of elution of the hop acids is cohumulene, adhumulene, humulone, colupulene, and lupulone.

References

1. Verzele, M., and Dewaele, C. (1981). *J. Chromatogr.* **217,** 399.
2. Anderson, B. (1983). *J. Chromatogr.* **262,** 448.
3. Knudson, E., and Siebert, K. (1983). *J. Am. Soc. Brew. Chem.* **41,** 51.
4. Verzele, M., and de Potter, M. (1978). *J. Chromatogr.* **166,** 320.

Recommended Reviews

1. Birch, A. J. (1957). Biosynthetic relations of some natural phenolic and enolic compounds. *Prog. Chem. Org. Nat. Prod.* **14,** 186.
2. Ashurst, P. R. (1967). The chemistry of hop resins. *Prog. Chem. Org. Nat. Prod.* **25,** 63.
3. Charalambous, G. (1981). Volatile constituents of beer. *In* "Brewing Science, 2." (J. R. A. Pollock, ed.), p. 167. Academic Press, New York.
4. American Society of Brewing Chemists. (1976). "Methods of Analysis." St. Paul, New York.
5. "Brewing Science." Verzele, M., and Pollock, J., eds. (1979). Academic Press, New York.
6. Meilgaard, M. C., and Peppard, T. L. (1986). The flavour of beer. *In* "Food Flavours, Part B," p. 99. Elsevier, Amsterdam.

CARBOHYDRATES — 2

I. MONO- AND OLIGOSACCHARIDES

A. Introduction

The carbohydrates are among the most abundant constituents of plants and animals. In plants they constitute the membranes of the cell walls. They can be divided into the following classes: monosaccharides, e.g., arabinose, $C_5H_{10}O_5$, glucose, $C_6H_{12}O_6$; oligosaccharides, e.g., sucrose, $C_{12}H_{22}O_{11}$, raffinose, $C_{18}H_{32}O_{16}$; and polysaccharides, e.g., starch, cellulose $[(C_6H_{12}O_6)_n - (n-1)H_2O]$.

1. Monosaccharides

The monosaccharides are polyhydroxyaldehydes or polyhydroxyketones and are classified according to the length of the carbon chain and the nature of the carbonyl group. The following are examples of tetroses, pentoses, and hexoses (both "aldoses" and "ketoses").

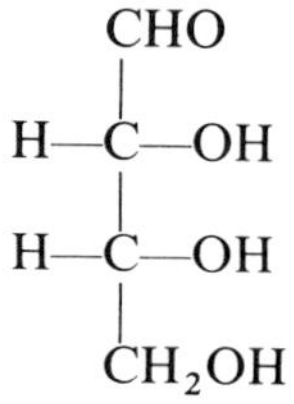

```
      CHO                CH₂OH              CHO
       |                   |                 |
   H—C—OH               C=O              H—C—OH
       |                   |                 |
   H—C—OH              H—C—OH            H—C—OH
       |                   |                 |
     CH₂OH               CH₂OH           H—C—OH
                                             |
                                           CH₂OH
   D-Erythrose         D-Erythrulose       D-Ribose
```

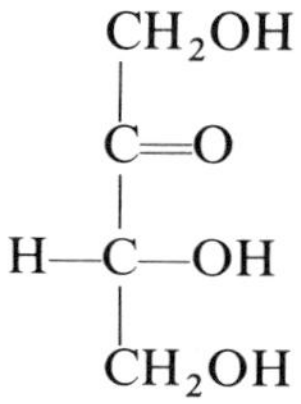

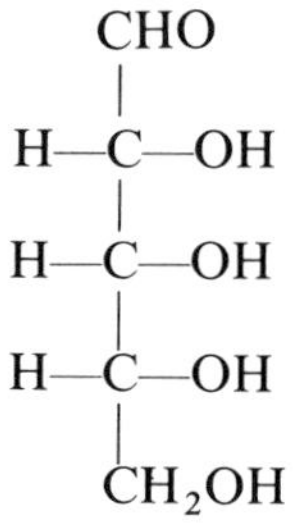

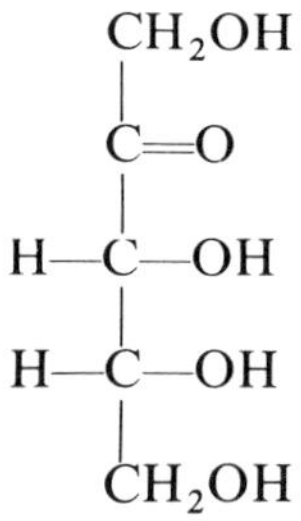

```
   CH₂OH              CHO              CH₂OH
    |                  |                |
    C=O           H—C—OH              C=O
    |                  |                |
  H—C—OH          HO—C—H            HO—C—H
    |                  |                |
  H—C—OH           H—C—OH            H—C—OH
    |                  |                |
   CH₂OH            H—C—OH            H—C—OH
                       |                |
  D-Ribulose         CH₂OH            CH₂OH
                    D-Glucose         Fructose
```

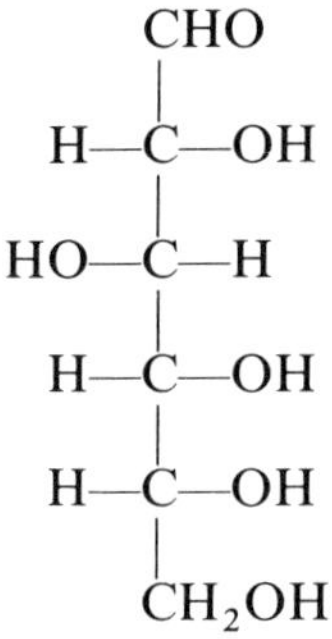

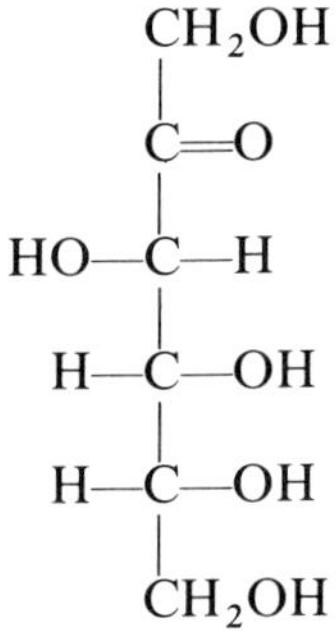

Heptoses are found in certain plants and bacteria. For example, L-glycero-D-mannoheptose has been isolated from the cell walls of *Escherichia coli*, and sedoheptulose from the avocado pear.

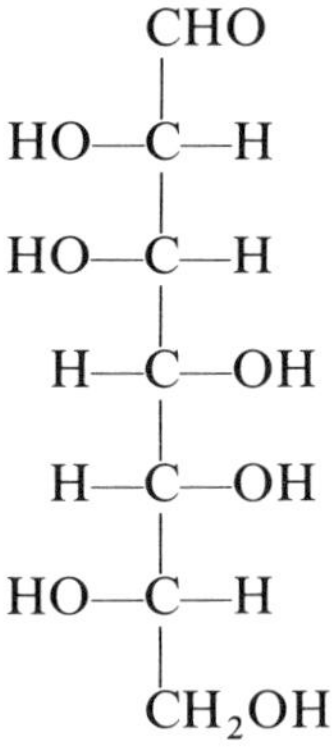

```
        CHO                      CH₂OH
         |                        |
      HO—C—H                      C=O
         |                        |
      HO—C—H                   HO—C—H
         |                        |
       H—C—OH                   H—C—OH
         |                        |
       H—C—OH                   H—C—OH
         |                        |
      HO—C—H                    H—C—OH
         |                        |
        CH₂OH                    CH₂OH
  L-Glycero-D-mannoheptose      Sedoheptulose
```

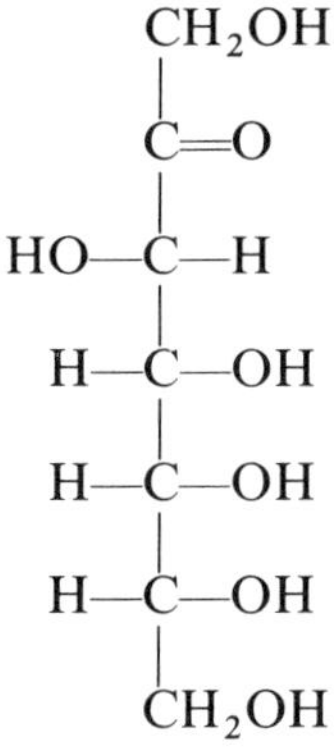

Their importance in the photosynthetic carbohydrates was recently demonstrated by Calvin and his coworkers.

The consensus is that the steric structure of a sugar is related to the two antipodes of glyceraldehyde.

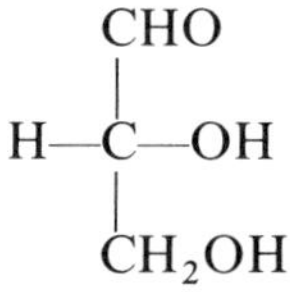

```
        CHO                   CHO
         |                     |
      H—C—OH               HO—C—H
         |                     |
        CH₂OH                 CH₂OH
   D-Glyceraldehyde       L-Glyceraldehyde
```

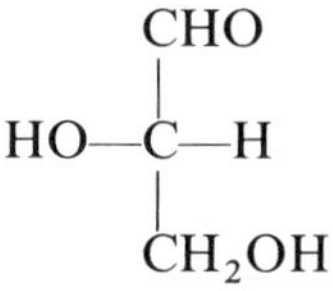

Thus the configuration of the H and OH groups on the lowest asymmetric carbon atom (next to the —CH₂OH group) of a sugar indicates whether the sugar is derived from D- or L-glyceraldehyde.

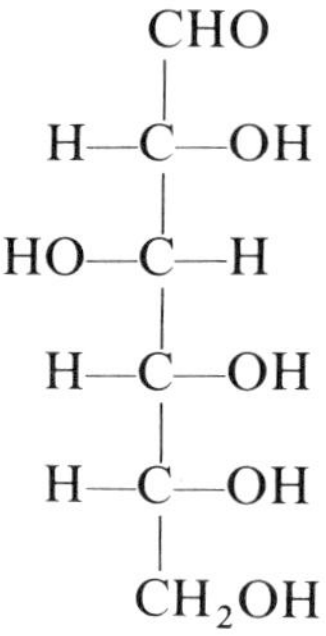

CHAPTER 2 Carbohydrates

```
    CHO                CHO
H—C—OH            HO—C—H
HO—C—H             H—C—OH
H—C—OH            HO—C—H
H—C—OH            HO—C—H
   CH₂OH              CH₂OH
  D-Glucose           L-Glucose
```

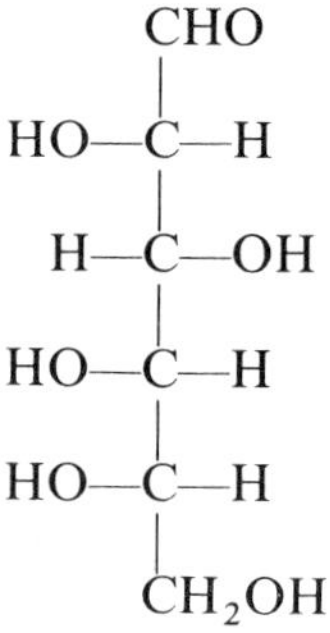

The mirror images of a sugar (L and D) are called *optical antipodes* or *enantiomorphs*. They have identical chemical and physical properties and differ only in the sign of their optical rotation.

It is noteworthy that the structures of glucose and mannose contain four asymmetric, identical groups, except for those at C_2, which are opposite in space. Such isomers are called epimers.

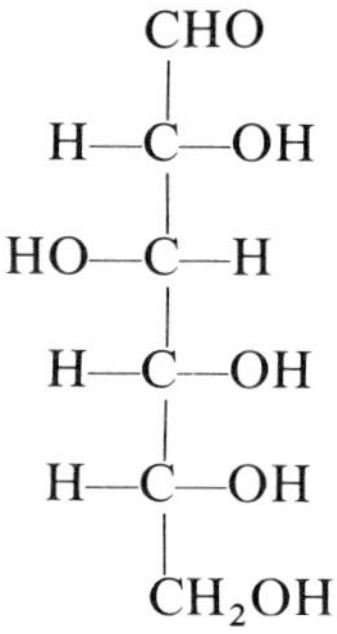

```
    CHO                CHO
H—C—OH            HO—C—H
HO—C—H            HO—C—H
H—C—OH             H—C—OH
H—C—OH             H—C—OH
   CH₂OH              CH₂OH
   Glucose            Mannose
```

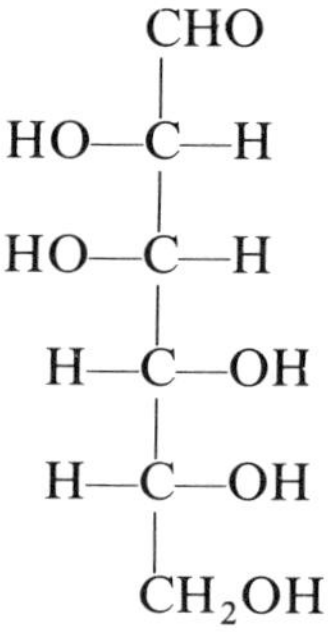

Another group of monosaccharides are the deoxy-sugars, e.g., 2-deoxy-D-ribose, found in DNA, or colitose, a 3,6-dideoxy hexose isolated from bacterial lipopolysaccharides.

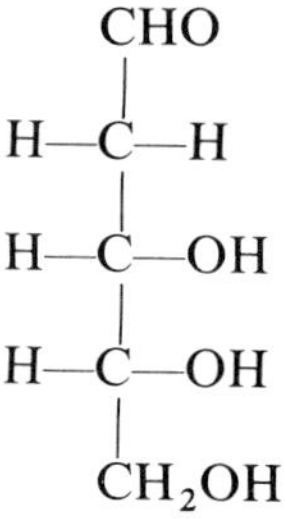

```
    CHO                CHO
 H—C—H            HO—C—H
 H—C—OH              CH₂
 H—C—OH             H—C—OH
    CH₂OH          HO—C—H
                     CH₃
2-Deoxy-D-ribose    Colitose
```

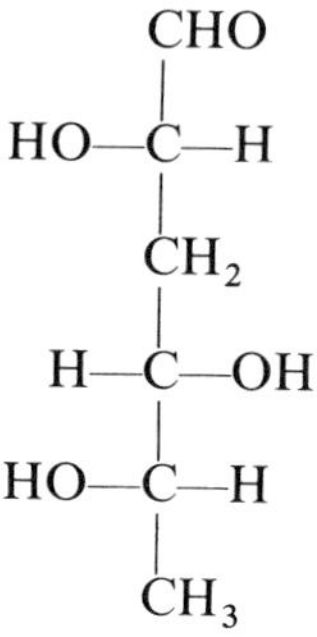

I. Mono- and Oligosaccharides 73

Streptose from streptomycin and apiose from parsley are examples of branched-chain sugars.

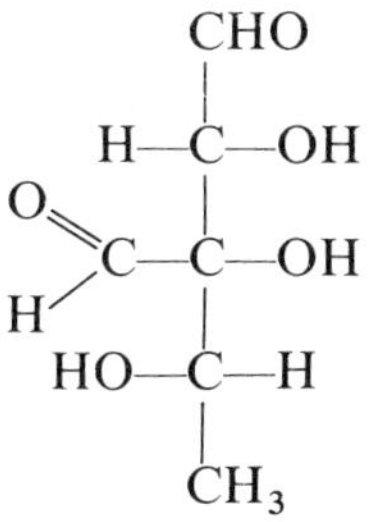

```
      CHO                          CHO
      |                            |
   H—C—OH                       H—C—OH
   O   |                            |
    \\C—C—OH                       C
   H/  |                   HO₂HC / | \ CH₂OH
    HO—C—H                        OH
      |
      CH₃
```

L-Streptose 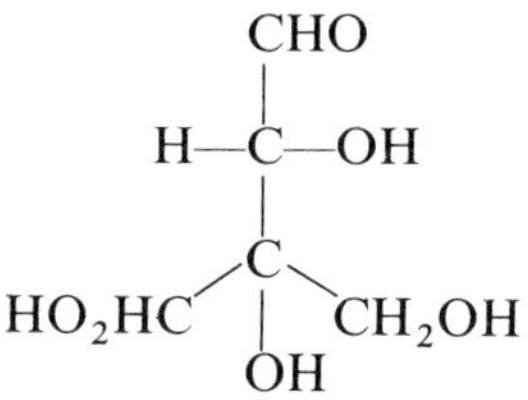   D-Apiose

As alcohols, carbohydrates form esters. The phosphoric esters of pentoses and hexoses are of great physiological importance, the former as

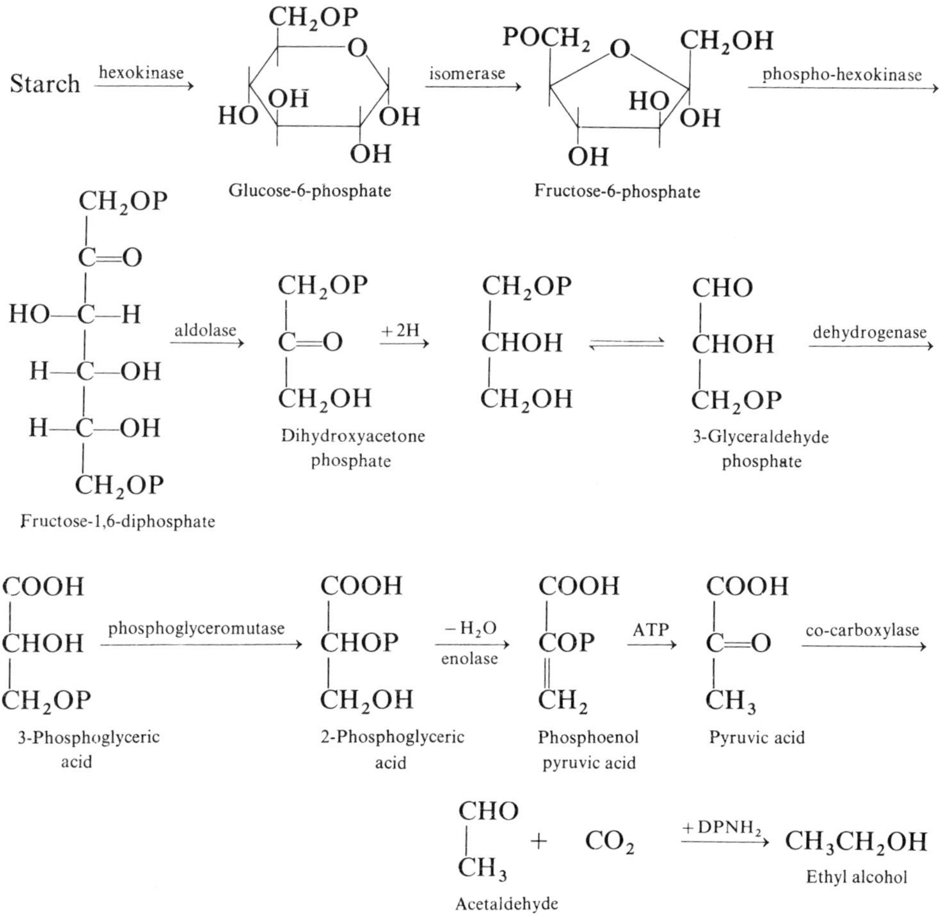

Starch —hexokinase→ Glucose-6-phosphate —isomerase→ Fructose-6-phosphate —phospho-hexokinase→

Fructose-1,6-diphosphate —aldolase→ Dihydroxyacetone phosphate —+2H→ (CH₂OP–CHOH–CH₂OH) ⇌ 3-Glyceraldehyde phosphate —dehydrogenase→

3-Phosphoglyceric acid —phosphoglyceromutase→ 2-Phosphoglyceric acid —−H₂O, enolase→ Phosphoenol pyruvic acid —ATP→ Pyruvic acid —co-carboxylase→

Acetaldehyde (CHO–CH₃) + CO₂ —+DPNH₂→ CH₃CH₂OH Ethyl alcohol

74 CHAPTER 2 Carbohydrates

constituents of nucleic acids and the latter as prerequisites for fermentation—the biological breakdown of carbohydrates.

The accepted scheme (proposed by Embden-Meyerhof) for the fermentation of starch to ethyl alcohol is shown on the previous page.

If any interference occurs at the final stage (e.g., by addition of sulfite), then reduced diphosphopyridine nucleotide ($DPNH_2$) will hydrogenate the dihydroxyacetone phosphate, forming glycerol.

A number of other fermentations are known, such as the acetone–butanol fermentation caused by *Clostridium acetobutylicum* Weizmann or that of lactose into lactic acid by *Streptococcus lactis*.

Treatment of aldo- and keto-sugars with methanol containing hydrochloric acid yields two isomeric methyl glycosides, which differ only in the configuration at C_1. These epimers are called anomers.

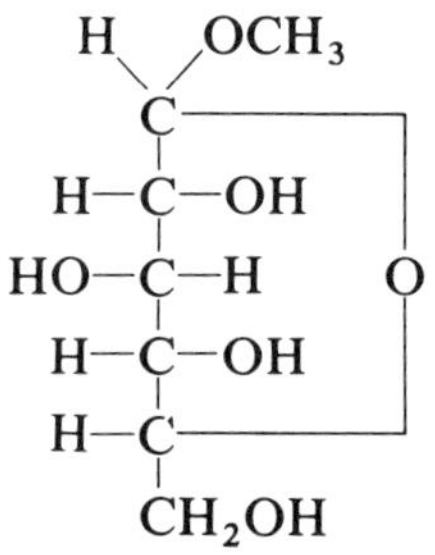

Methyl α-D-glucoside 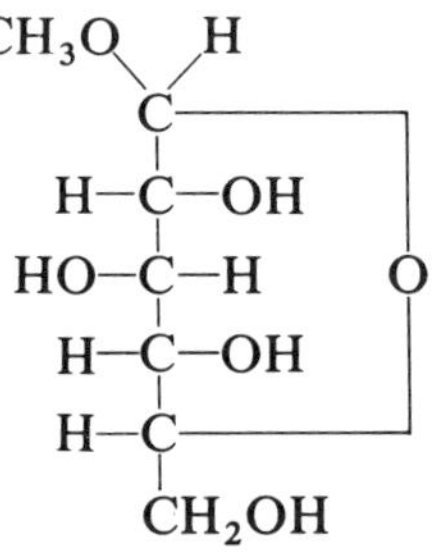   Methyl β-D-glucoside

It has been demonstrated that the free monosaccharides also exist in two forms that are structurally related to methyl glucosides. The ring may be five-membered—a derivative of tetrahydrofuran (furanose), or six-membered—a derivative of pyran (pyranose).

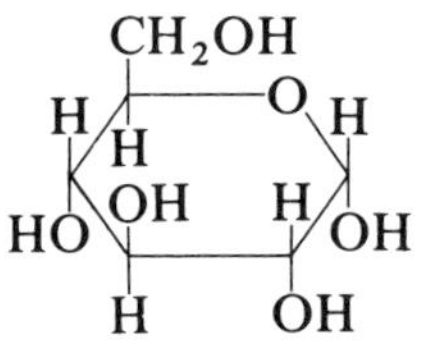

α-D-Glucopyranose 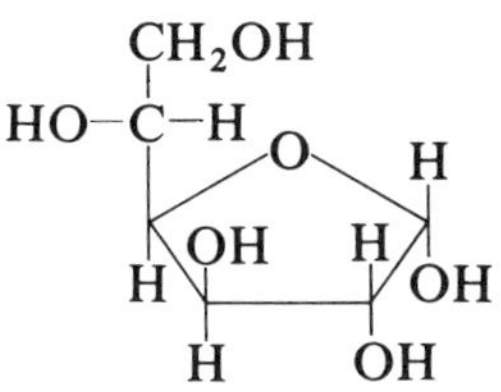   α-D-Glucofuranose

As in the case of the cyclohexane series, in which the equatorial substituent of the chair form is more stable than the axial one, pyranose rings also show a preference for the chair form and the substituent (CH_2OH in this case) occupies the equatorial position.

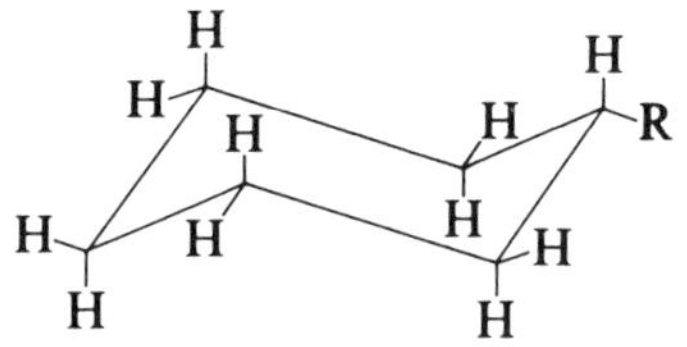

I. Mono- and Oligosaccharides 75

R equatorial

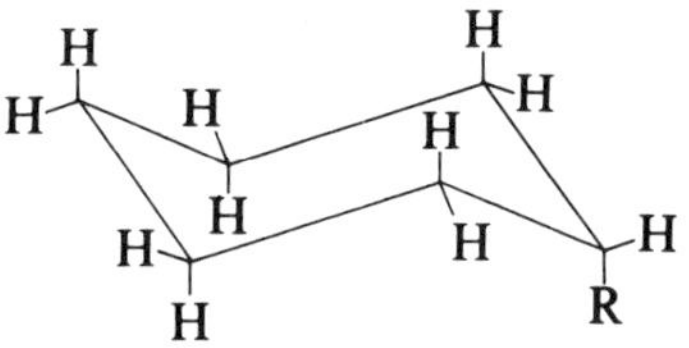

R axial

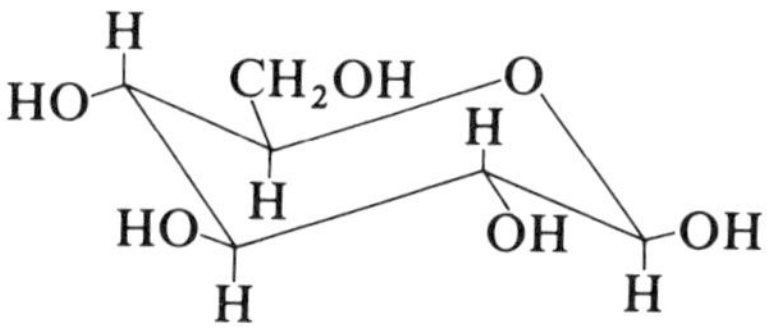

CH₂OH equatorial

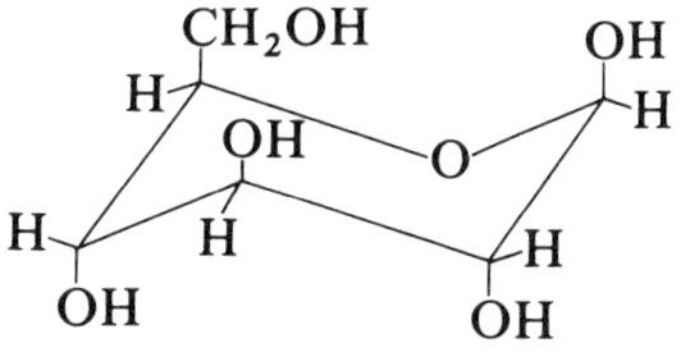

CH₂OH axial

In aqueous solution, sugars undergo changes in optical rotation—*mutarotation*; one isomeric form is probably transformed into the other via an aldehydic form.

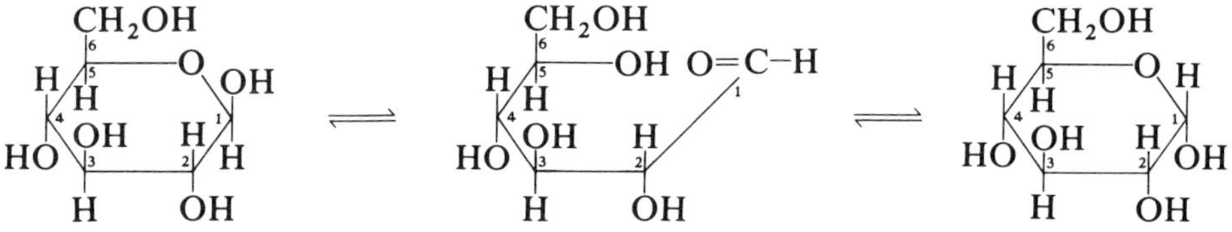

β-D-Glucopyranose ⇌ Aldehydic form ⇌ α-D-Glucopyranose

On degradation of pentoses with strong acids, 2-furfuraldehyde is formed, whereas hexoses form 5-hydroxymethyl-2-furfuraldehyde.

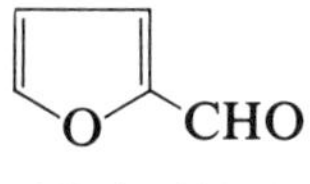

2-Furfuraldehyde 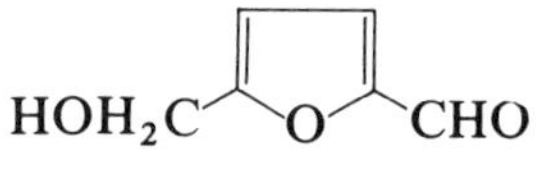   5-Hydroxymethyl-2-furfuraldehyde

In alkaline solution, the monosaccharides are unstable and undergo a variety of transformations, e.g., the Lobry de Bruyn reaction, whereby glucose is converted into fructose and mannose.

Phenylhydrazine and other substituted hydrazines are important in the identification of the various monosaccharides. When one mole of phenylhydrazine reacts with one mole of an aldose or ketose, the product is a normal hydrazone. On treating it with an excess of phenylhydrazine, an osazone is formed. Hydrolysis of the osazone with hydrochloric acid yields the corresponding osones, which are polyhydroxy-2-keto-1-aldehydes.

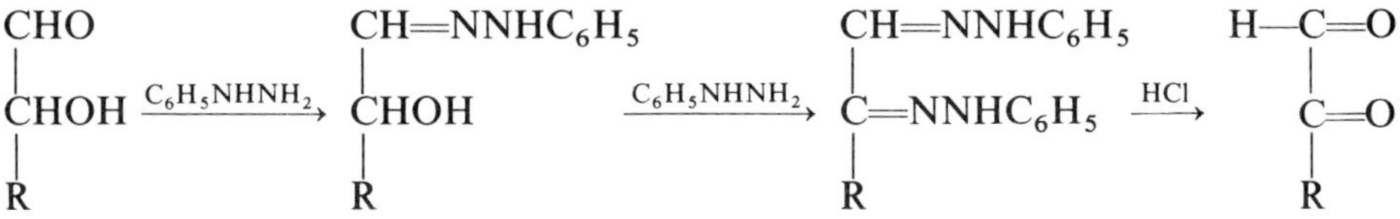

$$\underset{R}{\overset{CHO}{\underset{|}{\overset{|}{CHOH}}}} \xrightarrow{C_6H_5NHNH_2} \underset{R}{\overset{CH=NNHC_6H_5}{\underset{|}{\overset{|}{CHOH}}}} \xrightarrow{C_6H_5NHNH_2} \underset{R}{\overset{CH=NNHC_6H_5}{\underset{|}{\overset{|}{C=NNHC_6H_5}}}} \xrightarrow{HCl} \underset{R}{\overset{H-C=O}{\underset{|}{\overset{|}{C=O}}}}$$

Hydrazone   Osazone   Osone

It should be noted that the osazones formed from glucose, mannose, and fructose are identical, hence the lower carbon atoms (C_3—C_6) of these three sugars have identical configurations.

Several methods have been devised for shortening the chain of an aldose by one carbon atom: Wohl's, Ruff's, and Weerman's. Procedures used for lengthening the chain of an aldose are those of Killiani and Sowden.

Both aldoses and ketoses may be reduced to the corresponding polyhydroxy alcohols. Oxidation of aldoses with a variety of reagents gives dibasic acids, e.g., saccharic acid.

2. Oligosaccharides

The oligosaccharides are classified as di-, tri-, tetra-, etc., saccharides depending upon the number of monosaccharides produced upon acid hydrolysis. Substances composed of up to ten monosaccharide molecules are included in the group of oligosaccharides. The monosaccharide units may be identical, as

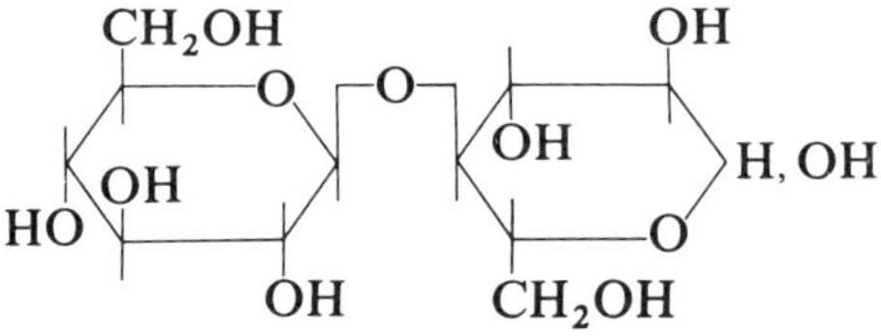

Cellobiose

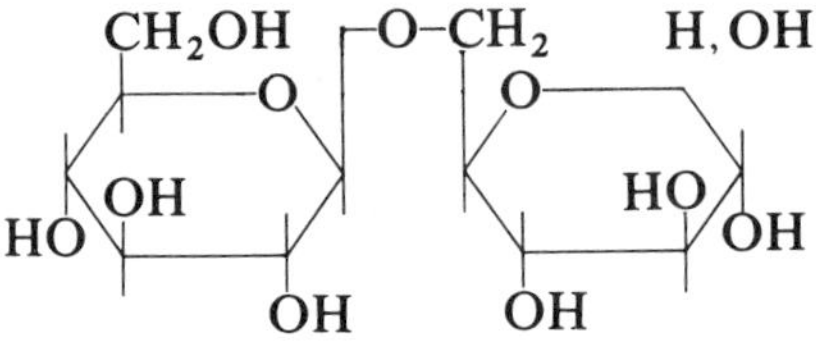

Gentiobiose

in maltose, or different, as in sucrose. If the C_1 hydroxyl groups of components (of a disaccharide) are linked to each other, the disaccharide is nonreducing (sucrose and trehalose); if the union occurs in such a way that only one of the hydroxyl groups participates, then reducing disaccharides are formed, e.g., cellobiose or gentiobiose.

Raffinose is an example of a naturally occurring trisaccharide.

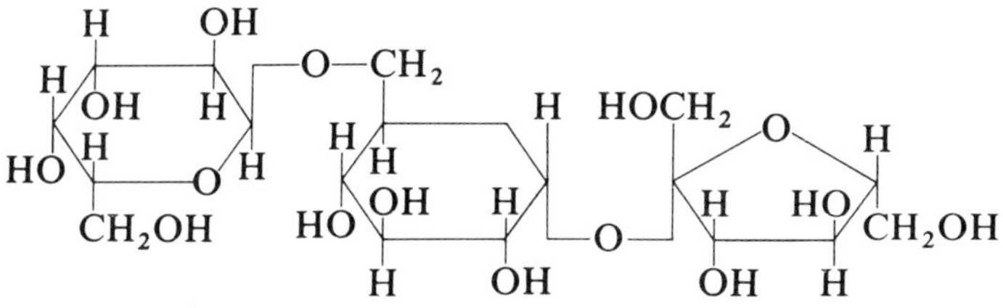

Raffinose

Amino sugars such as glucosamine and galactosamine are abundant in nature. They are obtained by hydrolysis of animal polysaccharides such as chitin and heparin, and can, of course, also be synthesized.

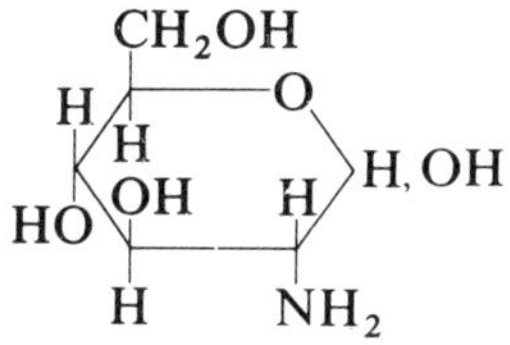

D-Glucosamine 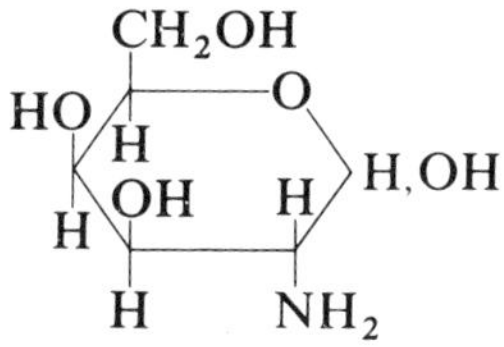   D-Galactosamine

3. Glycosides

Plant glycosides are found in leaves, seeds, and bark of trees. They are crystalline, sometimes bitter solids. The glycosides can be hydrolyzed by enzymes or by acids; the noncarbohydrate moiety is called aglycone. The phenolic glycoside, arbutin, is found in pears, salicin in willow bark, phloridzin in the bark of plum and apple trees, and coniferin in pine trees.

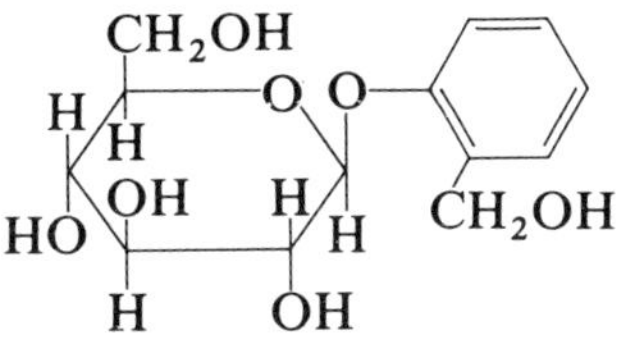

Salicin

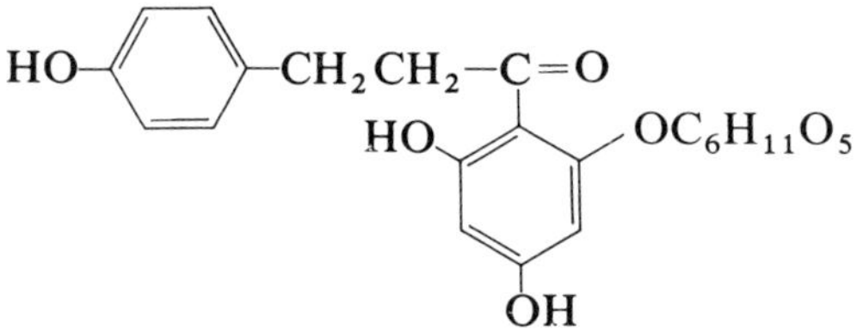

Phloridzin

Many pigments, such as those of the flavone and anthocyanin series, are of glycosidic nature. Dyestuffs such as alizarin and indigo are obtained by hydrolysis of ruberythric acid and indican, e.g., indican is hydrolyzed to indoxyl, which is then oxidized to indigo.

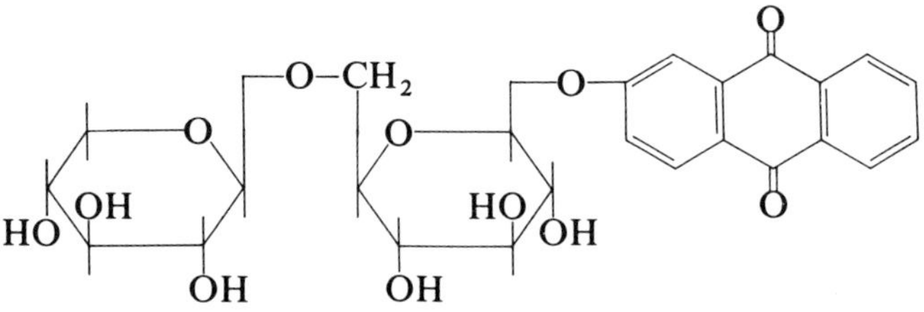

Ruberythric acid

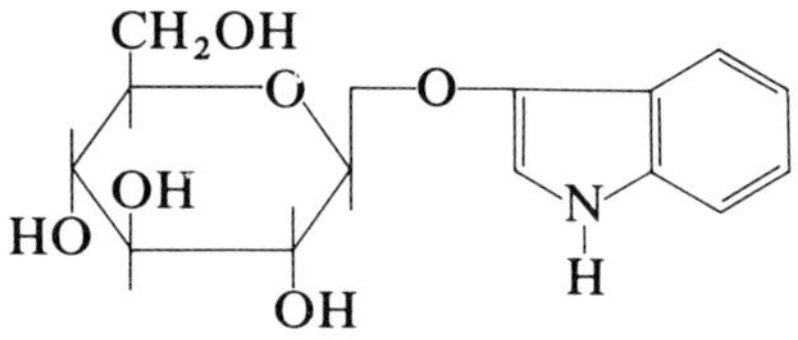

Indican

Mustard oil glycosides, such as sinigrin, liberate the pungent mustard oils, e.g., allyl isothiocyanate, when the seeds are crushed in the presence of a plant enzyme.

$$CH_2=CHCH_2N=C\begin{matrix}OSO_3K\\SC_6H_{11}O_5\end{matrix} \longrightarrow CH_2=CHCH_2N=C=S + KHSO_4$$

Sinigrin → Allyl isothiocyanate

The cardiac glycosides, such as oleandrin, are used in medicine for stimulating the heart muscle. Their sugars are mostly deoxysugars, e.g., cymarose and diginose.

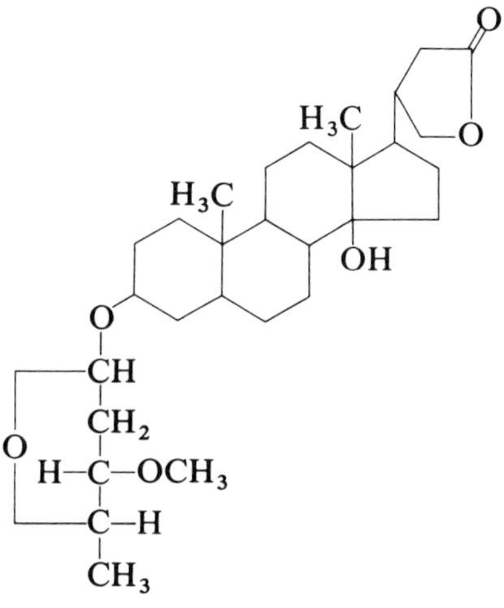

I. Mono- and Oligosaccharides

[Structures: Oleandrin; D-Cymarose; D-Diginose]

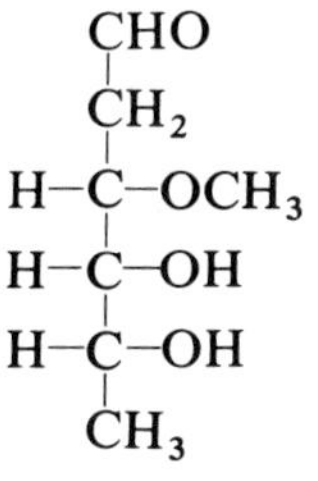

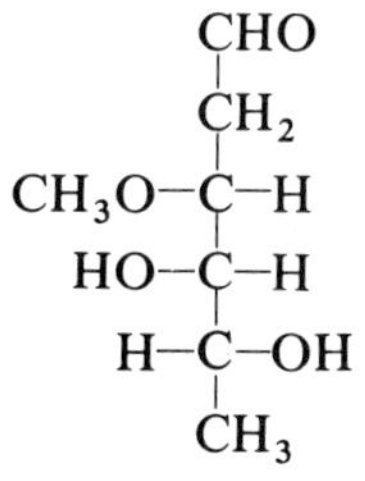

Oleandrin

D-Cymarose:
CHO
CH$_2$
H—C—OCH$_3$
H—C—OH
H—C—OH
CH$_3$

D-Diginose:
CHO
CH$_2$
CH$_3$O—C—H
HO—C—H
H—C—OH
CH$_3$

Saponin glycosides such as tigonin form a soapy foam on mixing with water, and are toxic. On hydrolysis they yield sapogenin and a sugar, composed of five sugar units (2 glucose, 2 galactose, and 1 xylose).

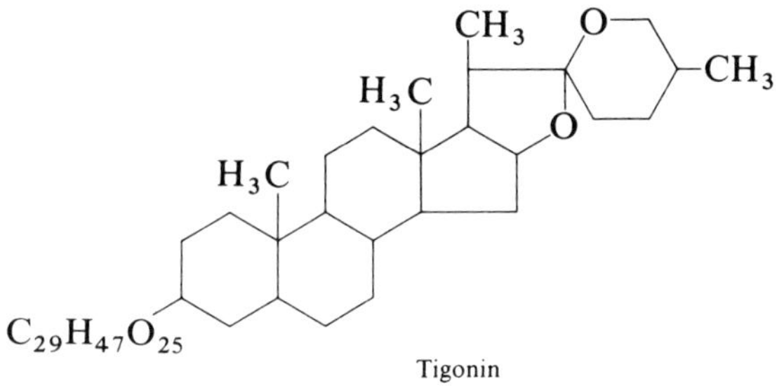

[Structure: Tigonin, $C_{29}H_{47}O_{25}$]

Tigonin

B. Isolation of D-Mannoheptulose from Avocado

1. Introduction

The first seven-carbon sugar to be discovered in nature was D-mannoheptulose, whose isolation from the avocado was announced by La Forge in 1917 (1).

The sugar is accompanied by the closely related heptitol, perseitol, and also by very small quantities of the related eight-carbon compounds, D-glycero-D-mannooctulose and D-erythro-D-galacto-octitol (2). D-Mannoheptulose has also been isolated from fresh avocado leaves (3). Avocado varieties differ greatly in their D-mannoheptulose content. In all cases, the fruit is picked green and allowed to ripen in storage; the D-mannoheptulose content does not vary under these conditions.

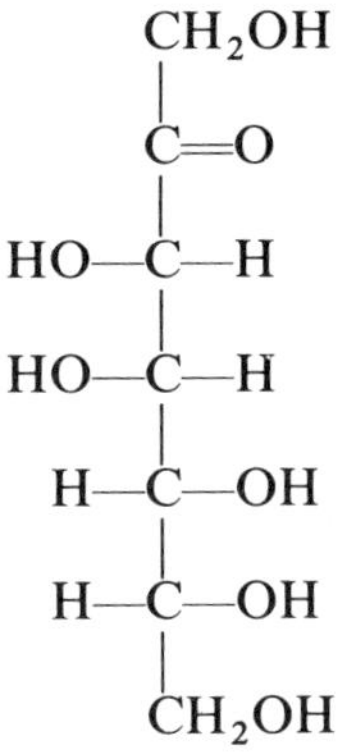

$$\begin{array}{c} CH_2OH \\ | \\ C{=}O \\ | \\ HO-C-H \\ | \\ HO-C-H \\ | \\ H-C-OH \\ | \\ H-C-OH \\ | \\ CH_2OH \end{array}$$

D-Mannoheptulose

2. Principle

According to the following procedure (4), the pulp of avocado is extracted with 20% aqueous ethanol and freed from most oil by adding Celite and filtering. The aqueous solution is then deionized by passing it through cationic and anionic ion-exchange resins. Concentration of the effluent leaves a thin syrup consisting of a mixture of perseitol and mannoheptulose, which are fractionally separated by seeding the methanolic solution first with perseitol and then with mannoheptulose. D-mannoheptulose is further purified by forming a hexacetate (5).

3. Apparatus

Columns for chromatography
Polarimeter
Vacuum distillation assembly

4. Materials

Acetic anhydride
Amberlite IR-120 [H$^+$]
Avocado
Celite
Chloroform
Duolite A-4 [OH$^-$]

Ethanol
Ether
Methanol
Perseitol
Pyridine
Sodium hydrogen carbonate

5. Time

Isolation: 5 hours
Crystallization: 2 days

6. Procedure

Avocados are quartered and peeled, the pulp is crushed and mixed thoroughly with about half its weight of 20% aqueous ethanol to make a thin soup, and the mixture is heated for 4 hr at 60°C to coagulate the pulp. The mixture is filtered through a large Buchner funnel, and the pulp is reextracted with 20% ethanol in the same manner. The combined filtrates are freed from any oil that has separated by mixing with Celite and filtering through a Buchner funnel. The solution is concentrated under reduced pressure to a thin syrup, which is poured, with stirring, into 8 volumes of methanol. The precipitated gum separates as small flakes and is filtered off and washed with methanol–water, 8:1. The filtrate is concentrated under reduced pressure to remove the methanol, and the residual aqueous solution is deionized by passing it through columns of suitable ion-exchange resins, e.g., Amberlite IR-120 [H^+] and Duolite A-4 [OH^-]. The effluent is concentrated under reduced pressure to a thin syrup, which is dissolved in 4 volumes of methanol and seeded with perseitol. The product is allowed to crystallize for several days at room temperature. Perseitol separates as white needles and is filtered off and washed with 80% methanol. The filtrate is concentrated under reduced pressure to a thick syrup, which is dissolved in methanol, seeded with perseitol, and cooled in the refrigerator. If more perseitol separates, the material is filtered off, and the filtrate is concentrated and redissolved in methanol. When no further crystallization of perseitol can be induced, the solution is seeded with D-mannoheptulose, which separates as relatively large, transparent prisms. Purified by recrystallization from methanol, D-mannoheptulose melts at 151 to 152°C; $[\alpha]_D^{20} + 29°$ (water). By way of comparison, perseitol melts at 187 to 188°C; $[\alpha]_D^{20} - 1°$ (water). The relative yields of the two sugars differ from one variety of avocado to another.

a. D-Mannoheptulose hexaacetate

Acetylation of 1 g pure, crystalline D-mannoheptulose is accomplished by dissolving the sugar in 5 ml acetic anhydride and 5 ml dry pyridine at 0°C. The solution is kept at this temperature for 48 hr and then poured into several times its volume of ice and water. The acetate is extracted with chloroform; the extract is washed with cold dilute sodium hydrogen carbonate and then with water until free of acetic acid and pyridine. The washed chloroform solution is dried and then concentrated under reduced pressure until free of solvent. The syrup crystallizes readily; the acetate may be recrystallized from warm ether as large prisms. It has a mp of 110°C; $[\alpha]_D^{20} + 39°$ (chloroform).

References

1. La Forge, F. B. (1917). *J. Biol. Chem.* **28,** 511.
2. Charlston, A. J., and Richtmyer, N. K. (1960). *J. Am. Chem. Soc.* **82,** 3428.
3. Nordal, A., and Benson, A. A. (1954). *J. Am. Chem. Soc.* **76,** 5954.
4. Richtmyer, N. K. (1962). *Methods Carbo. Chem.* **1,** 173.
5. Montgomery, E. M. (1962). *Methods Carbo. Chem.* **1,** 175.

Recommended Reviews

1. Ahmed, E. M., and Barmore, C. R. (1980). Avocado. In *"Tropical and Subtropical Fruits."* (S. Nagy and P. E. Shaw, eds.), p. 121. Avi Publishing, Westport, Connecticut.
2. Webber, J. M. (1962). Higher carbon sugars. In *"Advances in Carbohydrate Chemistry."* (M. L. Wolfrom, and R. S. Tipson, eds.) Vol.15, p. 15. Academic Press, New York.

C. α-D-Glucosamine from Crustacean Shells

1. Introduction

Chitin is a polyglucosamine and is widely distributed in invertebrates. The occurrence of chitin in the plant kingdom is confined to fungi and green algae. Because of their ready availability and low protein content, decalcified shells are the most suitable source of chitin. Crab and lobster shells are commonly employed and are available commercially.

2-Amino-2-deoxy-D-glucose hydrochloride can be isolated by the hydrolysis of crude chitin with hot concentrated hydrochloric acid. The crystalline hydrochloride was first prepared by Ledderhose from a lobster shell hydrolyzate (1). Treatment of the hydrochloride with strong bases, such as diethylamine or triethylamine, gives free D-glucosamine (2, 3).

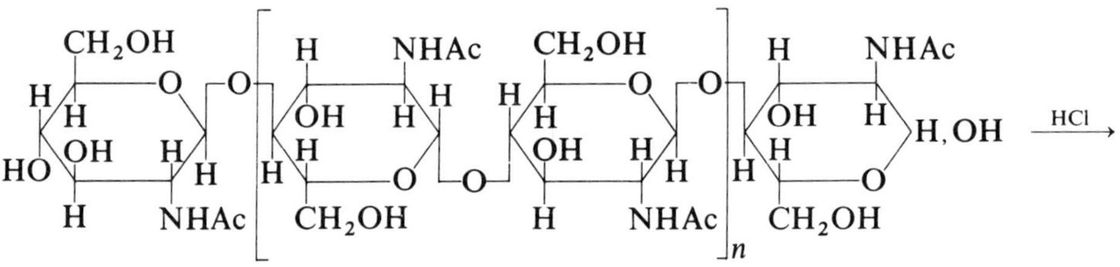

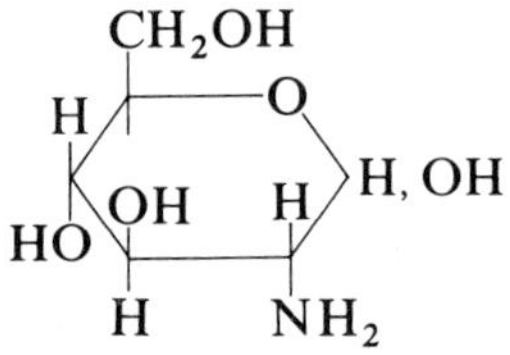

2. Principle

In the following procedure (4) crab shells are cleaned and decalcified by digestion with dilute hydrochloric acid, thus yielding crude chitin.

Degradation of chitin is accomplished by heating with concentrated hydrochloric acid. Crude glucosamine hydrochloride is purified by filtration through Celite and activated carbon. Final purification is effected by dissolving the product in hot water and adding ethanol, whereupon the α-anomer crystallizes while the β-isomer remains in the solution.

3. Apparatus

Evaporator

4. Materials

Celite
Crab shells
Ethanol, 95%
Ethanol, absolute

Ether
Hydrochloric acid
Triethylamine

5. Time

Decalcification of shells: 12 hours
Preparation of glucosamine hydrochloride: 4 hours

6. Procedure

a. Isolation of chitin

Crab shells are cleaned by washing and scraping under running water, and dried in an oven at 100°C. Dry shells (100 g) are broken into fairly large fragments, and portions containing the eyes are discarded. The fragments are then decalcified by digestion overnight with 1.5 liter $2N$ hydrochloric acid at room temperature. To speed up decalcification, the shells can be ground to a fine powder, but care must be taken to add the acid slowly to avoid loss of material due to excessive frothing. The rubbery residues are thoroughly washed with water and dried at 100°C. The yield of crude chitin is 35–40 g.

b. D-Glucosamine hydrochloride

Concentrated hydrochloric acid (400 ml) is added to the dried residues, and the mixture is heated on a boiling water bath with continuous mechanical agitation. After 2.5 hr, by which time dissolution should be essentially complete, an equal volume of water is added, and the solution is separated from the black sludge by filtration through a bed of a filter aid such as Celite. The brown filtrate is stirred for 30 min at 60°C with activated carbon, and the solution is again filtered through Celite. The filtrate should now have a pale yellow color. The solution is then concentrated at 50 to 60°C under reduced pressure to a volume of 75 ml. During this process some crystals separate. Ethanol (4 volumes) is added to the concentrate, and the mixture is stored for 24 hr at room temperature. The product is removed by filtration, washed successively with ethanol and ether, and air-dried; yield, about 20 g. Purification is effected by dissolving the product in a minimal volume of hot water and adding 4 volumes of ethanol. D-Glucosamine hydrochloride crystallizes as the α-anomer. Appreciable amounts of the β-isomer remain in the solution and may be recovered by precipitation with ether.

References

1. Ledderhose, G. (1876). *Ber.* **9,** 1200.
2. Breuer, R. (1898). *Ber.* **31,** 2193.
3. Westphal, O., and Holzmann, H. (1942). *Ber.* **75,** 1274.
4. Stacey, M., and Webber, J. M. (1962). *Methods Carbo. Chem.*, **1,** 228.

Recommended Reviews

1. Foster, A. B., and Webber, J. M. (1960). *Chitin. In* "*Advances in Carbohydrate Chemistry,*" (M. L. Wolfrom, and R. S. Tipson, eds.), Vol. 15, p. 371. Academic Press, New York.
2. Tracey, M. V. (1955). *Chitin, In* "*Modern Methods of Plant Analysis,*" (K. Paech, and M. V. Tracey, eds.), Vol. 2, p. 264. Springer-Verlag, New York.

D. Chromatographic Separation of Sugars on Charcoal

1. Introduction

For many years charcoal has been used industrially for the purification of sugars. In 1932, Hyashi (1) reported that charcoal stirred in an aqueous acetone–acetic acid solution of glucose and sucrose completely adsorbed the sucrose but left the glucose in solution. Tiselius (2) proposed that charcoal might be employed in chromatography for the separation of glucose from lactose, and later he separated mixtures of glucose and sucrose, and sucrose and maltose (3). The method of Tiselius has been used by Claesson (4) and by Montgomery, Weakley, and Hilbert (5). Recently, the versatility of carbon as an adsorbent for the segregation of sugars into molecular-weight groups was demonstrated by Whistler and Durso (6). Later, charcoal was used by Whistler and Tu (7) for the separation of sugars from partially hydrolyzed xylan, and by Whistler and Hickson (8) for the isolation of various sugars from hydrolytic products of starch. Tu and Ward (9) have improved separation of disaccharides on heated charcoal columns. Other adsorbents used for carbohydrate fractionation include cellulose powder, alumina, silicates, and ion-exchange resins.

2. Principle

The method described below is according to Whistler and Durso (6).

3. Apparatus

Columns for chromatography
Polarimeter

4. Materials

Celite	Maltose	Sodium chloride
Charcoal	Melibiose	Sucrose
Ethanol	Sodium bicarbonate	Xylose
Glucose		

5. Time

4–5 hours

6. Procedure

The adsorbent is a mixture of equal parts by weight of Darco-G-60 and Celite, which has been washed with water and dried. This material is packed into a glass chromatographic tube measuring 230×34 mm, forming a column 170 mm in length. Before the addition of the sugar solution, the column is moistened with 150 ml water. To effect the displacement of the adsorbed sugar, water and more powerful developers are used in succession. The effluent is collected in 100-ml fractions. The course of the desorption process is followed polarimetrically; a 2-dm tube is used to indicate the complete removal of each component prior to the addition of the succeeding developer.

a. Separation of mixtures and isolation of components

A mixture of 1.0 g each of glucose, maltose, and raffinose in the form of a 10% solution is adsorbed on a charcoal column, and is resolved by the successive introduction of 800 ml water, 1.5 l, 5% ethanol, and 700 ml 15% ethanol, for displacement of glucose, maltose, and raffinose, respectively. Each of these sugars is obtained in crystalline form. One recrystallization serves to purify the material sufficiently to give the accepted values of melting point and rotation. The following mixtures are separated in a similar manner: maltose and raffinose; sucrose, melibiose, and raffinose; glucose and raffinose. These mixtures are in the form of 10% solutions containing 1 g of each sugar.

b. Separation of a sugar mixture in the presence of salts

A mixture of 7 g xylose, 0.6 g maltose, 0.4 g raffinose, 4 g sodium chloride, and 0.4 g sodium bicarbonate is dissolved in water and made up to 1 liter. This mixture can be successfully separated in the above-described manner.

The salts are obtained in the water effluent. Polarimetric data indicate that the separation is complete.

References

1. Hyashi, J. (1932). *J. Biochem. (Japan)* **16**, 1.
2. Tiselius, A. (1941). *Arkiv. Kemi, Mineral Geol.* **14B** (32), 8.
3. Tiselius, A. (1943). *Kolloid Z* **105**, 101.
4. Claesson, S. (1947). *Arkiv. Kemi, Mineral Geol.* **24A** (16), 9.
5. Montgomery, E. M., Weakley, F. B., and Hilbert, G. E. (1949). *J. Am. Chem. Soc.* **71**, 1682.
6. Whistler, R. L., and Durso, D. F. (1950). *J. Am. Chem. Soc.* **72,**, 677.
7. Whistler, R. L., and Tu, C.-C. (1952). *J. Am. Chem. Soc.* **74**, 3609.
8. Whistler, R. L., and Hickson, J. L. (1954). *J. Am. Chem. Soc.,* **76**, 1671.
9. Tu, C.-C, and Ward, K. (1955). *J. Am. Chem. Soc.* **77**, 4938.

Recommended Reviews

1. Binkley, W. W. (1955). *Column chromatography of sugars and their derivatives*. In "Advances in Carbohydrate Chemistry," (M. L. Wolfrom and R. S. Tipson, eds.), Vol. 10, p. 55. Academic Press, New York.

E. Gas—Liquid Chromatography of Carbohydrates

1. Introduction

The application of gas chromatography to the separation of carbohydrates and related polyhydroxy compounds has been restricted owing to the lack of volatility of these compounds and to the difficulties involved in the preparation of volatile derivatives of carbohydrates.

The first report of the application of gas-liquid partition chromatography to carbohydrates (1) described the separation of tri-O-methyl derivatives of methyl pentapyranosides from tetra-O-methyl derivatives of methyl hexapyranosides. The separation of polyacetyl derivatives of carbohydrates was reported by Bishop and Copper (2). Kircher (3) described the separation of fully methylated anomeric methyl glycopyranosides.

More recently, acetyl derivatives were used for the resolution of amino sugars and for the separation of acetal and ketal derivatives (4). Hedgley and Overend (5) and Ferrier (6) have described the preparation and analysis of carbohydrates in the form of the remarkably volatile trimethylsilyl ethers.

2. Principle

The method described below (7) involves the solution or suspension of carbohydrates in pyridine, followed by reaction with hexamethyldisilazane $(CH_3)_3SiNHSi(CH_3)_3$ and trimethylchlorosilane $(CH_3)_3SiCl$. The resulting trimethylsilyl ethers $ROSi(CH_3)_3$ of aldoses, oligosaccharides, and glycosides

(α- and β-isomers) are subjected to gas chromatography on Gas Chrom P impregnated with 3% SE-52.

These ethers can be recovered quantitively and unchanged from the effluent gas stream, thus showing that no anomerization, hydrolysis, or degradation occurs during the separation.

3. Apparatus

> Gas chromatograph
> Stainless steel columns
> Microsyringe

4. Materials

> Gas Chrom P
> Glycosides (see procedure)
> Hexamethyldisilazane
> Oligosaccharides (see procedure)
>
> Pyridine, dry
> Sugars (see procedure)
> Trimethylchlorosilane

5. Time

> 2–3 hours

6. Procedure

a. Trimethylsilylation of carbohydrates

Ten mg carbohydrate is treated with 1 ml anhydrous pyridine, 0.2 ml hexamethyldisilazane, and 0.1 ml trimethylchlorosilane. The reaction is carried out in a 1-dram, plastic-stoppered vial or similar container. The mixture is shaken vigorously for about 30 sec and allowed to stand for 5 min at room temperature before chromatography. The solutions become cloudy on addition of trimethylchlorosilane, but this ammonium chloride precipitate does not interfere with the subsequent gas chromatography. If the carbohydrate appears to remain persistently insoluble in the mixture, the vial is warmed for 2 to 3 minutes at 75 to 85°C.

In some cases where no rearrangements are likely to occur, the sugar is first dissolved by warming with pyridine, before the addition of hexamethyldisilazane and trimethylchlorosilane. From 0.1 to 0.5 μl of the resulting reaction mixture is used for injection into the gas chromatograph.

An instrument with hydrogen flame ionization detector is used. The columns are of stainless steel, 6 ft in length and 0.25 inches in diameter, packed with Gas Chrom P impregnated with 3% SE-52.

b. Operating conditions

In general, flow rates are adjusted for optimal column efficiencies and range from 75 to 150 ml per min, with inlet pressures of 15 to 20 psi.

The retention times are listed in the following tables.

Retention Times of Aldoses[a]

Aldose	Relative Retention Time at 140°C
β-Arabinose	0.28
Ribose	0.32
α-Xylose	0.43
β-Allose	0.81
α-Galactose	0.88
α-Glucose	1.00
β-Glucose	1.57
α-Gulose	0.75
β-Mannose	1.08
α-Talose	0.86

[a] The retention times are relative to that of α-glucose, which is 20 to 22 minutes at 140°C.

Retention Times of Oligosaccharides[a]

Oligosaccharide	Relative Retention Time at 210°C
Sucrose	10.4
Lactose	10.5
β-Maltose	13.1
β-Cellobiose	16.6
Gentiobiose	22.6
Raffinose	99.0

[a] The retention times are relative to that of α-glucose, which is 1.1 to 1.3 minutes at 210°C.

Retention Times of Glycosides[a]

Glycoside	Relative Retention Time at 140°C
Methyl β-arabinoside	0.20
Methyl α-arabinoside	0.21
Methyl α-xyloside	0.31
Methyl β-xyloside	0.34
Methyl α-mannoside	0.59
Methyl β-mannoside	0.67
Methyl α-glucoside	0.92
Methyl β-glucoside	1.07

[a] The retention times are relative to that of α-glucose (see aldoses).

References

1. McInnes, A. G., Ball, D. H., Cooper, F. P., and Bishop, C. T. (1958). *J. Chromatogr.* **1,** 556.
2. Bishop, C. T., and Cooper, F. P., (1960). *Can. J. Chem.* **38,** 388.
3. Kircher, H. W. (1960). *Anal. Chem.* **32,** 1103.
4. Jones, H. G., Jones, J. K. N., and Perry, M. B. (1962). *Can. J. Chem.* **40,** 1559.
5. Hedgley, E. J., and Overend, W. G. (1960). *Chem. Ind. (London)* 378.
6. Ferrier, R. J. (1962). *Tetrahedron* **18,** 1149.
7. Sweeley, C. C., Bentley, R., Makita, M., and Wells, W. W. (1963). *J. Am. Chem. Soc.* **85,** 2497.

Recommended Reviews

1. Bishop, C. T. (1962). *Separation of carbohydrate derivatives by gas-liquid partition chromatography.* In "Methods of Biochemical Analysis," (D. Glick, ed.), Vol. 10, p. 1. Interscience, New York.
2. Berry, J. W. (1966). Gas chromatography of carbohydrates. *Adv. Chromatogr.* **2,** 171.

F. Thin-Layer Chromatography of Carbohydrates

1. Introduction

Several adsorbents have been used for thin-layer chromatography of carbohydrates. Stahl and Kaltenback (1) employed kieselgur impregnated with sodium acetate. The ability of borate ions to form complexes with polyhydroxy compounds was utilized by Pastuska (2) and Prey (3) for the separation of various sugars on plates impregnated with boric acid.

Schweiger (4) and Vomhof and Tucker (5) analyzed sugars on cellulose layers. Prey and coworkers (6) used a mixture of silica gel and kieselgur, 1:4. Wolfrom and his coworkers (7) have recently used microcrystalline cellulose for sugar analysis.

2. Apparatus

| Applicator for TLC | Glass plates, 20 × 20 cm |
| Drying oven | Spray gun |

3. Materials

Acetic acid	Glucose	Sucrose
Anisaldehyde	Naphthoresorcinol	Silica gel G
n-Butanol	Phosphoric acid	Sulfuric acid
Ethanol	Rhamnose	Xylose
Ether		

4. Time

Preparation of thin-layer plates: 1–2 hours
Development and detection: 3–4 hours

5. Procedure

a. Preparation of silica gel G plates

The suspension for five plates (20 × 20 cm) is prepared by shaking 30 g silica gel G and 60 ml water for 30 sec, and is then spread on the plates with an applicator to a thickness of 0.25 mm. The plates are then activated at 120°C for 1 hr.

b. Development

The samples of carbohydrates are dissolved in water and spotted by means of micropipettes along a line 2 cm above the rim of the plate. Development is accomplished in n-butanol–acetic acid–ethyl ether–water, 9:6:3:1.

c. Detection

Naphthoresorcinol reagent
Spraying with a solution of 0.2% naphthoresorcinol in butanol and 10% phosphoric acid, followed by heating for 10 min at 100°C, gives a pink color with ketoses, a green color with aldopentoses, and a blue color with aldohexoses.

Anisaldehyde reagent
This reagent is prepared from 0.5 ml anisaldehyde in 9 ml ethanol (95%), 0.5 ml concentrated sulfuric acid, and 0.1 ml acetic acid.

Detection of Sugars

Sugar	R_f	Color with Anisaldehyde
Sucrose	0.09	Violet
Glucose	0.22	Blue
Fructose	0.27	Violet
Xylose	0.40	Gray
Ribose	0.47	Blue
Rhamnose	0.55	Green

References

1. Stahl, E., and Kaltenbach, U. (1961). *J. Chromatogr.* **5**, 351.
2. Pastuska, G. (1961). *Z. Anal. Chem.* **179**, 427.
3. Prey, V., Verbalk, H., and Kausz, M. (1962). *Mikrochim. Acta* 449.
4. Schweiger, A. (1962). *J. Chromatogr.* **9**, 374.
5. Vomhof, D. W., and Tucker, T. C. (1965). *J. Chromatogr.* **17**, 300.
6. Prey, V., Scherz, V., and Bancher, E. (1963). *Mikrochim. Acta* **3**, 567.
7. Wolfrom, M. L., Patin, D. L., and de Lederkremer, R. M. (1965). *J. Chromatogr.* **17**, 488.

Recommended Reviews

1. Fried, B., and Sherma, J. (1982). Thin-layer chromatography of carbohydrates. *In* "Thin-Layer Chromatography," p. 246. Marcel Dekker, New York.
2. Stahl, E., and Kaltenbach, U. (1965). *Sugars and derivatives. In* Stahl, E. *"Thin-layer Chromatography,"* p. 461. Academic Press, New York.

G. Separation of Carbohydrates by High-Performance Thin-Layer Chromatography

1. Introduction

In view of the increased concern about nutrition and food labeling for a wide variety of foods and food ingredients, particularly in regard to their sugar content, a fast analytical method was needed. The separation of mono-, di- and trisaccharides on thin-layer chromatoplates was very popular, even though it is time consuming. A rapid and efficient separation of carbohydrates at ambient temperatures has been achieved on HPTLC ready-to-use Si 50.000 plates.

2. Apparatus

Developing chamber.

3. Materials

Acetone   Diphenylamine
Aniline   Si 50.000 HPTLC plates.

4. Time

1 hour

5. Procedure

The HPTLC ready-to-use plate Si 50.000 is first precleaned by developing it with a mixture of chloroform and methanol 1:1 (v/v) and allowing the solvent to run to the top of the plate. The plate is then reactivated by drying for 30 min at 110°C. The mixture of mono-, di- and trisaccharides (raffinose, lactose, sucrose, glucose, and fructose) (100 mg/ml) is now applied to the plate with a glass capillary.

Ascending one-dimensional double chromatography at room temperature is accomplished with the developing solvent acetonitrile–water, 17:3 (v/v).

After drying in a hood at room temperature the plate is sprayed with a chromogenic solution prepared by dissolving aniline (2 ml) and diphenylamine (2 g) in acetone (80 ml), then cautiously adding phosphoric acid (85%, 15 ml) and adding acetone to give 100 ml solution.

Heating the plate at 105–110°C for 15 min reveals fructose as a greyish red spot, glucose and lactose as a greyish blue spot, and raffinose and sucrose as a grey spot. By exposing the plate to ammonia fumes for 15 min, all carbohydrates appear as ocher yellow spots on a colorless background.

Recommended Reviews

1. Bragg, R. W., Chow, Y., Dennis, L., Ferguson, L. N., Howell, S., Morga, G., Ogino, G., Pugh H., and Winters, M. (1978). Sweet organic chemistry. *J. Chem. Ed.* **55**, 281–285.
2. Creammer, B., and Ikan, R. (1977). Properties and syntheses of sweetening agents. *Chem. Soc. Revs.* **6**, 431–465.
3. Davidson E. A. (1967). *"Carbohydrate Chemistry."* Holt, Rinehart & Winston, New York.
4. Guthrie, R. D., and Honeyman, J. (1964). *"An Introduction to the Chemistry of Carbohydrates."* Clarendon Press, Oxford, England.
5. Pigman, W. (1957). *"The Carbohydrates."* Academic Press, New York.
6. Van Dyke, S. F. (1960). *"The Carbohydrates."* Interscience, New York.

H. Questions

1. Distinguish between each of the following pairs of terms: (a) aldose and ketose; (b) D and L in sugar classification; (c) epimer and enantiomorph.
2. Outline the Kiliani synthesis of sugars and the Ruff method for degrading sugars.
3. Show how D-glucose reacts with reducing agents; with oxidizing agents; with cupric sulfate; with phenylhydrazine.
4. Why are aldohexoses less reactive than aliphatic aldehydes?
5. Draw the two chair conformations for the α- and β-anomers of the pyranose form of each of the following D-aldohexoses: mannose, glucose, galactose, and talose.
6. Name four disaccharides made up of glucose units only and explain how they differ from each other.
7. What are the chief carbohydrates in honey, fruit, liver, blood, milk, and cereals?
8. Discuss the occurrence of pentose-yielding substances in nature. What commercial value do they have?
9. Trehalose, a nonreducing disaccharide, is found in insects. Methylation and hydrolysis of this compound yield two moles of 2,3,4,6-tetra-O-methyl-D-glucopyranose from each mole of trehalose. If both anomeric carbon atoms have an α-configuration, what is the structure of trehalose?
10. Describe in detail the photosynthesis of glucose.

I. Recommended Books

1. Balazs, E. A., and Jeanloz, R. W. (eds.) (1965). *"The Amino Sugars."* Academic Press, New York.
2. Bollenback, G. N. (1958). *"Methyl Glycosides."* Academic Press, New York.
3. Calvin, M., and Bassham, J. A. (1962). *"The Photosynthesis of Carbon Compounds."* Benjamin, New York.

II. POLYSACCHARIDES

A. Introduction

Polysaccharides are polymers of monosaccharides. They are found in the higher plants, in ferns and mosses, in seaweed, in fungi, in bacteria, and in animals, where they serve as a structural support and as food reserves. Many of the polysaccharides are hydrolyzed by specific glycosidases. Thus, cellulose is hydrolyzed by cellulase, and the starches and glycogens by the amylases and diastases. On complete acid hydrolysis, the following monosaccharides are formed: D-glucose, D-mannose, D-fructose, D-galactose, D-xylose, L- and D-arabinose, D-glucosamine, D-galactosamine, and D-glucuronic acid.

1. Cellulose

Cellulose is the chief constituent of the fibrous parts of plants, and is used for the preparation of paper, rayon, explosives, and plastics. X-ray analysis of cellulose fibers indicates that they are arranged in a bundle called a micelle. The molecular weight of cellulose from different sources may range from 100,000 to 2,000,000. Cellulose consists of D-glucopyranose units linked through C_1 and C_4. The chain is linear and is composed of 100–200 units.

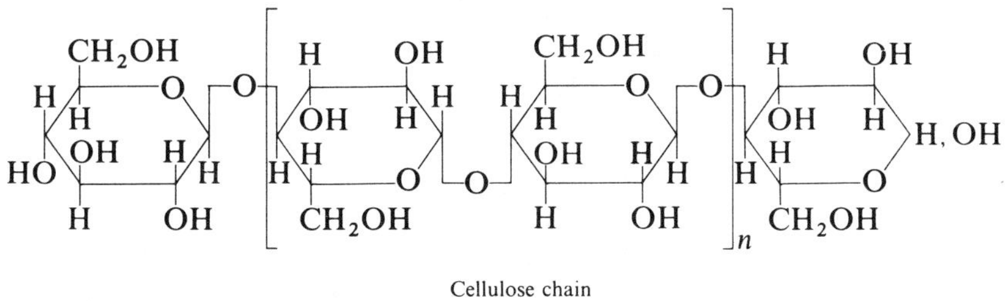

Cellulose chain

Polysaccharides extractable from the cell walls by dilute aqueous alkali are called hemicelluloses. The most abundant hemicellulose polysaccharides are the xylans, which are composed of D-xylose units. Some xylans are linear, while others have a branched structure. Xylans occur mainly in corncobs, corn stalks, grain hulls, and stems.

A xylan from corncobs consists of a linear chain of D-xylopyranose units connected by β-D-1 $\rightarrow$ 4 bonds:

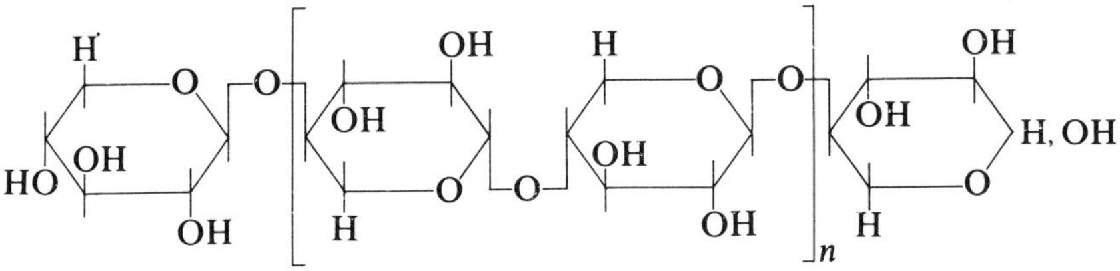

Xylan

Its molecular weight is about 30,000.

The red seaweed, *Rhodymenia palmata*, contains a xylan composed of D-xylopyranose units and has 1→3 and 1→4 linkages in the ratio of about 1:3.

Mannans are also hemicelluloses. They probably consist of a chain of D-mannose units linked 1→4, with side chains of D-galactopyranose linked 1→6.

2. Starch

Starches are mixtures of two polysaccharides, amylose and amylopectin. Amylose constitutes about 20% and is made up of approximately 300 D-glucose units linked by 1→4 α-glucosidic bonds, which form a helix with about six glucose residues per turn. In the presence of diastase, it is hydrolyzed completely to maltose; acidic hydrolysis yields D-glucose quantitatively.

Amylose

Amylopectin is the major constituent of starch. Its hydrolysis with distase gives a 55% yield of maltose, the remainder being a dextrin. Complete acid hydrolysis yields D-glucose only. In amylopectin the majority of units are linked by α-D-1→4 and α-D-1→6 glucosidic bonds at the branch points.

Many reagents, such as chloral hydrate, thymol, nitropropane, and *n*-butanol, have been used for the separation of amylose from amylopectin.

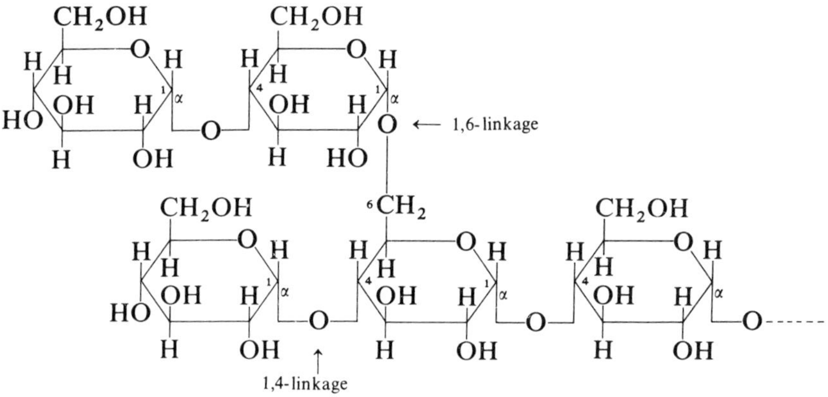

II. Polysaccharides

[Structure of Amylopectin showing 1,6-linkage and 1,4-linkage between glucose units]

Amylopectin

3. Glycogens

Glycogens are reserve carbohydrates of animals and are stored in the liver and muscles. They have highly branched chains; on enzymatic degradation they form maltose and a dextrin. Glycogens yield D-glucose on complete acidic hydrolysis.

4. Fructans

Fructans are D-fructose polymers that occur in the roots, stems, and seeds of various plants. They are subdivided into the following two types: those with a 1,2′-linkage, and those with a 2,6′-linkage.

5. Inulin

Inulin occurs in the tubers of chicory, in the Jerusalem artichoke, and in the bulbs of onion and garlic. Complete hydrolysis gives D-fructose and some

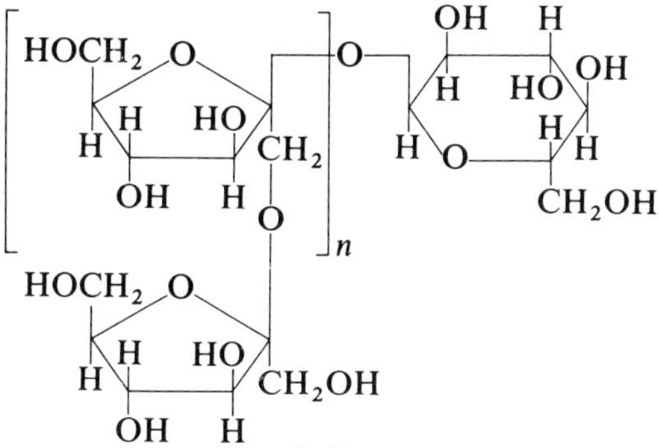

[Structure of Inulin]

Inulin

D-glucose. The chain of inulin consists of about 35 D-fructofuranose residues linked β–1 → 2. Inulin serves as a source of commercial fructose.

6. Levan

Levan is a gummy product of the attack of certain organisms on solutions of sucrose and raffinose. Levan gives a very high yield of crystalline fructose on hydrolysis.

7. Dextrans

Dextrans are produced in processes such as the manufacture of sugar from beets and wine-making, and from sucrose by the agency of bacteria such as *Leuconostoc dextranicum*.

8. Plant Gums

When a plant suffers injury by insect, bacterial, or fungal attack, it often exudes a viscous, sticky fluid that tends to seal the wounds. These exudates are mostly polysaccharides and are usually highly branched.

Gum arabic is one of the most important gums and is used in confectionery and medicaments. On complete hydrolysis, it yields L-arabinose, L-rhamnose, D-galactose, and D-glucuronic acid.

9. Pectins

Pectins are widespread in nature and are found in the pulp of citrus fruits, carrots, apples, beets, etc. They are widely used for the gelation of fruit juices. Pectin is the methyl ester of pectic acid and is composed of long chains of galacturonic units.

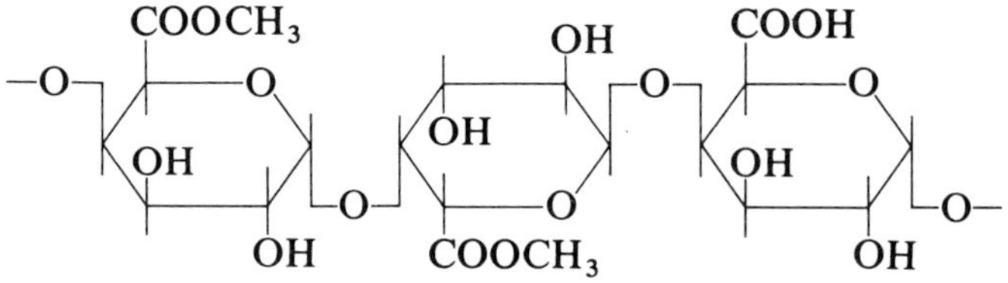

Pectin

10. Chitin

Chitin is a polyglucosamine that occurs mainly in invertebrates. It forms the exoskeletons of crustaceans and insects. Complete acid hydrolysis of chitin gives very high yields of D-glucosamine and acetic acid, whereas controlled acid or enzyme hydrolysis produces N-acetyl-D-glucosamine.

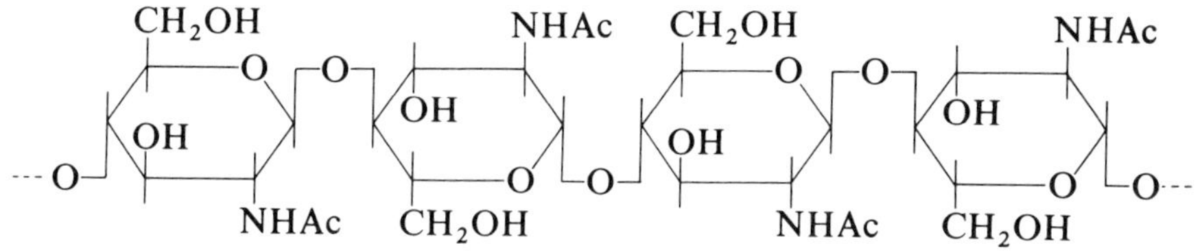

Chitin

11. Heparin

Heparin is a blood anticoagulant that is found in the liver, lungs, thymus, spleen, and blood. It is a polymer of D-glucuronic acid and D-glucosamine. The amino group and some of the hydroxyls are sulfated. The molecular weight of heparin appears to be about 17,000 to 20,000.

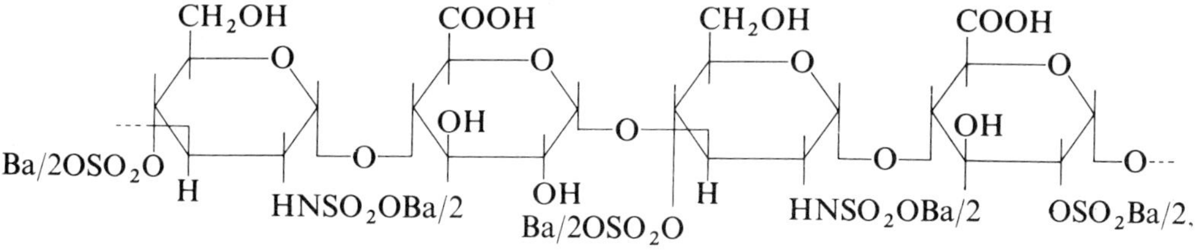

Heparin

B. Isolation of Amylopectin and Amylose from Potato Starch

1. Introduction

Many different methods have been used to fractionate starch into its branched and linear components. The method most suitable for one kind of starch is not necessarily the best for another, and much depends upon the source and pretreatment of the starch. In 1941, Schoch (1) demonstrated that slow cooling of a hot, aqueous starch solution saturated with butanol gives a microcrystalline precipitate of the linear starch component, amylose. Fractionation of starch by leaching techniques was applied by Baum and Gilbert (2), who utilized the insolubility in cold, dilute alkali of the amylopectin fraction from undamaged starch granules. Using the same freshly prepared starch and working under anaerobic conditions, these authors found that heating of starch suspensions in distilled water for 5 min at 100°C yields the amylose fraction. Another method calls for the use of chloral hydrate, which forms a complex with amylose, amylopectin remaining in the mother liquor. Fractionation by fractional precipitation is achieved by addition of salts (e.g., $MgSO_4$) or alcohols (e.g., ethanol and methanol).

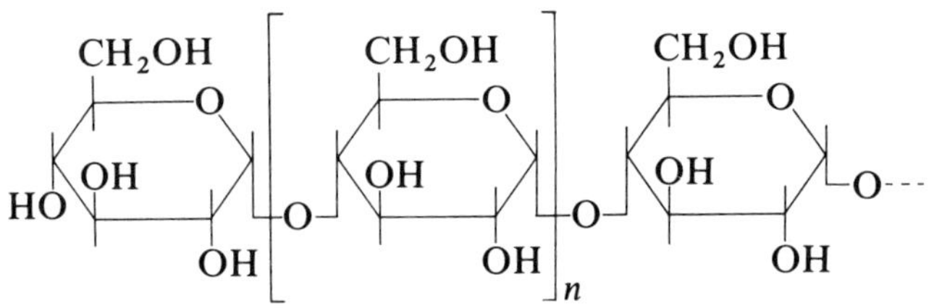

Amylose

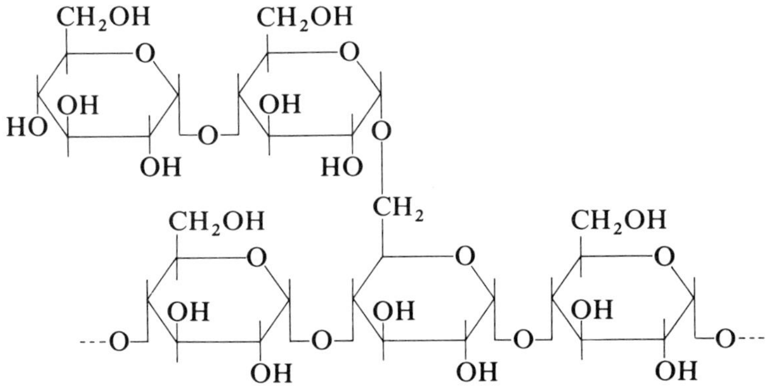

Amylopectin

2. Principle

In the following method (3) starch is separated into its principal components—amylose and amylopectin—by addition of dilute alkali followed by neutralization. Amylose remains in solution, while amylopectin forms a gel. The former is precipitated selectively from the saline solution by addition of n-butanol; removal of the butanol from the amylose–butanol complex yields pure amylose. Amylopectin is isolated by centrifugation followed by freeze-drying.

3. Apparatus

Blender
Centrifuge
Stirrer

4. Materials

Butanol
Hydrochloric acid
Potatoes

Sodium chloride
Sodium hydroxide

5. Time

Settling of a gel: overnight
Isolation: 5–6 hours.

6. Procedure

a. Preparation of potato starch

One kg undamaged potatoes is peeled, sliced, and then extracted in 250-g lots with 750 ml 1% sodium chloride solution in a mechanical food blender. Each lot is blended for about half a min, and the slurry is then filtered through fine muslin with gentle agitation. The filter cakes are bulked and reextracted for one min with 150 ml salt solution. The slurry is filtered through the muslin as before, and the filtrates are bulked. The starch granules are allowed to settle, and the supernatant, together with fragments of cellulose that have passed through the muslin, is decanted and discarded. The starch is washed three times with 1% sodium chloride and then once with $0.01N$ sodium hydroxide. The solution is finally washed three times with distilled water and stored underwater at 4°C. Precautions must be taken against freezing, as this would alter the starch; yield, about 40 g wet starch. Before samples of the starch are weighed, the supernatant is poured off, and the solid is drained for a moment; the dry weight is approximately 50% the weight of the drained solid.

b. Dispersion of the starch

To 3.3 liters $0.157N$ sodium hydroxide in a glass-walled tank is added 40 g drained, freshly prepared starch, weighed wet and suspended by gentle shaking in 260 ml water. The temperature of the dispersion must be at least 22°C, but should preferably be near 25°C. The mixture should be *stirred gently* until it clears. The method of stirring is of vital importance; motor-driven stirrers or the bubbling of gas through the mixture must not be used, since the gel structure of the amylopectin is damaged thereby, and this makes subsequent separation difficult. Gentle stirring with a plate-glass strip is sufficient. The alkaline solution should stand for a total of 5 min; 915 ml 5% sodium chloride solution is added, and the dispersion is neutralized with $1N$ hydrochloric acid to pH 6.5–7.5 with the aid of an external indicator or long-lead glass electrode assembly.

After standing overnight at room temperature, the gel settles until it occupies about one third of the total volume. There should be a sharp division between the gel and the amylose solution; the latter is then removed by siphoning or by suction, and should be filtered through a 15-cm fritted-glass filter (No. 3). This solution is then filtered through a No. 4 filter.

c. Precipitation of amylose

The amylose retrogrades if stored as a saline solution for more than a few days. It should thus be separated from this solution as soon as possible. Butanol is used to precipitate the amylose selectively from the saline solution according to the following procedure. The filtrate is saturated with redistilled butanol and stirred gently with a motor-driven stirrer for about 1 hr at room temperature. The precipitated amylose–butanol complex is allowed to settle for 2 to 3 hrs, and the clear supernatant is siphoned off. The partially sedimented solid is then centrifuged at 3000 rpm for 15 min in polypropylene buckets. The complex is then stirred in butanol-saturated water (125 ml) and reconcentrated by centrifuging as before. With some samples of starch, a further precipitation may be necessary. The complex is stable if stored in a stoppered container. It is readily soluble in water at room temperature. To remove most of the butanol from this solution, oxygen-free nitrogen is bubbled through it for 10 min in a vessel heated in a boiling water bath; 40 g wet potato starch yields 1.5–1.8 g amylose.

d. Purification of amylopectin

The amylopectin gel is centrifuged at 8000 rpm for 20 min at 20°C. The supernatant is discarded, and 1% sodium chloride solution is added to the gel (100 ml per 4 g original wet starch) with stirring. The mixture is allowed to stand for 20 hr at 18°C, and the gel is then collected by centrifuging at 8000 rpm as before. A second washing with sodium chloride is necessary. After centrifuging, the gel can be either suspended in water for immediate use (it is insoluble in cold water) or freeze-dried and stored; 40 g wet starch yields approximately 6 g amylopectin.

e. Iodine test

A few granules of the polysaccharide in 1 ml water are treated with several drops of a dilute solution of iodine in potassium iodide (prepared by dissolving 1.8 g potassium iodide in 15 ml water and then adding 1.2 g iodine). Starch gives an intense blue color; amylose exhibits a blue color; and Amylopectin develops a reddish color.

References

1. Schoch, T. J. (1941). *Cereal Chem.* **18,** 121.
2. Baum, H., and Gilbert, G. (1956). *J. Colloid Sci.* **11,** 428.
3. Gilbert, L. M., Gilbert, G. A., and Spragg, S. P. (1964). *Methods of Carbo. Chem.* **4,** 25.

Recommended Reviews

1. Muetgeert, J. (1961). *The fractionation of starch*. In "*Advances in Carbohydrate Chemistry*," Wolfrom, M. L. and Tipson, R. S. (eds.), Vol. 16, p. 299. Academic Press, New York.
2. Schoch, T. J. (1945). *The fractionation of starch*. In "*Advances in Carbohydrate Chemistry*," W. W. Pigman, and M. L. Wolfrom, (eds.), Vol. 1, p. 247. Academic Press, New York.

C. Isolation of Wood Cellulose

1. Introduction

Cellulose, as a naturally occurring fiber, is widespread in nature, mainly in woody tissues. However, it has also been found in some lower animals, in bacteria, and in certain mammalian muscle tissues.

In 1868, Fremy aand Terreil (1) treated wood with chlorine and then with aqueous potassium hydroxide to obtain a crude cellulose. This work was followed by that of Cross and Bevan (2), who also used chlorination as the first step, but then extracted the chlorinated lignin with hot aqueous sodium sulfite solution. Many other methods of delignifying wood and plants were subsequently developed; two of these are the Van Beckum and Ritter method (3) and the Jame-Wise method (4, 5). These methods are aimed at retaining the maximal amout of carbohydrate in the resulting "holocellulose," whereas the Cross and Bevan method is not.

Cellulose invariably undergoes degradation during delignification, and it thus appears to be impossible to isolate a wood cellulose showing the same degree of polymerization as that in the original wood or plant. The two major components to be removed are xylan and mannan; other polysaccharides, such as galactan, araban, and glycuronans, are easily removed. Softwoods contain large amounts of mannan and some xylan, and the resulting wood cellulose may retain appreciable amounts of mannan.

Cellulose chain

2. Principle

Delignification of wood is performed by treating wood meal with a solution of sodium chlorite at about 70°C [6]. These pH and temperature conditions are the most suitable for the removal of xylan and mannan, since the lower temperatures the rate of delignification is slower. The resulting holocellulose is then extracted with alkali to remove certain soluble carbohydrates, e.g., hemicelluloses or α- and β-celluloses, and *α-cellulose* or *alkali-resistant cellulose* is left as a residue. More thorough extraction will remove most of the nonglucan polysaccharides and leave an essentially pure wood cellulose.

3. Apparatus

   Screw cap bottle

4. Materials

Acetic acid	Sodium chlorite
Benzene	Sodium hydroxide
Ether	Wood meal

5. Time

   Chloriting treatment: 4 hours
   Alkaline extraction: 24 hours

6. Procedure

a. Preliminary treatment

Samples of wood meal 25 g are used as the starting material. The wood meal should preferably be of 40 to 60 mesh; a finer particle size is to be avoided. It can be prepared from coarse sawdust, wood chips, or thin spokeshave shavings.

In the case of conifer wood, the meal is extracted with 2:1 benzene–ethanol and then with 95% ethanol, or else with 95% ethanol and then with ether. This extraction is not necessary when dealing with most nonresinous woods.

b. Chloriting treatment

The air-dried sample is suspended in 800 ml hot water in a 2-liter Erlenmeyer flask also containing 3 ml glacial acetic acid. Technical-grade sodium chlorite, 7.5 g, is then added. This addition of sodium chlorite and subsequent operations, including washing, should be carried out in a well-ventilated hood. The flask is stoppered with an inverted 50-ml Erlenmeyer flask and heated on a steam bath. The period of heating is generally 30–60 min; the temperature of the solution will be about 70°C.

Depending on the type of wood used, this operation may have to be repeated several times. For samples of temperate hardwood, a total of 3 steps, or a total reaction time of 3 hr is generally sufficient. Tropical hardwoods will require more steps. In the case of softwood, heating periods of up to 6 hr may be necessary. The final residue should be almost white and retain the woody structure of the original sample. In some cases, the residue will have a slight yellow color.

c. Washing

When a satisfactory degree of whiteness has been attained, the mixture is filtered on a Buchner funnel, and the holocellulose is thoroughly washed with 2 liters cold water. The product is air-dried for subsequent alkaline extraction.

d. Alkaline extraction of holocelluloses of softwood

The sample (20 g air-dry holocellulose) is added to a 1-liter, wide-mouthed, screwcap bottle, 1 liter of 12% sodium hydroxide solution is added, and a stream of high-purity nitrogen is bubbled through for 1 min. The bottle is then securely closed with a polyethylene liner and screwcap, and stirred gently by rotating end-over-end at 6000 rpm for 24 hr at room temperature. The mixture is then filtered by suction on a fritted-glass Buchner funnel (C porosity) with the aid of a rubber dental dam. No attempt is made to exclude air during this filteration, but a minimal amount of air is sucked through the filter. The volume of filtrate is noted in order to determine the volume of alkali retained in the holocellulose on the filter.

The first extract will generally be brown in color because of the lignin present. Later extracts will be lighter in color, and finally colorless.

The residue is returned to the bottle, and 1 liter of 7.1% sodium hydroxide is added in three portions. In this manner, three extractions are made with the sodium hydroxide over a period of 24 hr. After the final extraction, the mixture is filtered, and the residue is washed with 5% sodium hydroxide and then with cold water. It is then stirred briefly in the funnel with 10% acetic acid and allowed to stand for 10 min. The mixture is filtered, and the wood cellulose is washed thoroughly with water. The wood cellulose is air-dried and analyzed for sugar constituents by a suitable method.

References

1. Fremy, E. and Terreil, A. (1868). *Bull. Soc. Chim. France.* **9**, 437.
2. Cross, C. F., Bevan, E. J., and Beadle, C. (1895). *"Cellulose and Outline of the Chemistry of the Structural Elements of Plants,"* p. 95. Longmans, Green and Co., London, England.
3. Van Beckum, W. G., and Ritter, G. J. (1937). *Paper Trade J.* **105** (18), 127.
4. Jame, G. (1942). *Cellulose chemie* **20**, 43.
5. Wise, L. E., Murphy, M., and D'Addieco, A. A. (1946). *Paper Trade J.* **122** (2), 35.
6. Green, J. W. (1963). *Methods Carbo. Chem.* **3**, 9.

Recommended Reviews

1. Hatch, R. S. (1954). *Bleaching and purification of wood cellulose. In "Cellulose and Cellulose Derivatives,"* Ott, E, Spurlin, H. M., and Grafflin, M. W. (eds.), Vol. 2, p. 589. Interscience, New York.
2. Polglase, W. J. (1955). *Polysaccharides associated with wood cellulose. In "Advances in Carbohydrate Chemistry,"* Wolfrom, M. L. and R. S. Tipson, (eds.), Vol. 10, p. 283. Academic Press, New York.

D. Questions

1. Compare starch and cellulose from each of the following aspects: source, intermediate and final hydrolysis products, structure of the polymer chain, their function in nature.
2. Define and give the structural formulas of the following compounds: lignin, pentosans, glycogen, collodion, gum cotton, dextrin.
3. What is pectin? Give the formulas of its hydrolytic products. Of what commercial value is pectin?
4. Write the structural formula of chitin. Where it is found in nature?
5. Compare the physical properties of mono- and polysaccharides.
6. Where are starches most commonly found? What is a soluble starch?
7. How does a plant synthesize starch? What is the difference between amylose and amylopectin?

E. Recommended Books

1. Brautlecht, C. A. (1953). *"Starch, Its Sources, Production, and Uses."* Reinhold, New York.
2. Honeyman, J. (1959). *"Recent Advances in the Chemistry of Cellulose and Starch."* Heywood and Co.
3. Sjöstrom, E. (1981). *"Wood Chemistry, Fundamentals and Application."* Academic Press, New York.
4. Thiem, J., (ed.) 1990. *"Carbohydrate Chemistry. Topics in Current Chemistry."* Vol. 154. Springer-Verlag, New York.
5. Van Dyke, S. F. (1960). *"The Carbohydrates."* Interscience, New York.
6. Whistler, R. L., and Smart, C. L. (1953). *"Polysaccharide Chemistry."* Academic Press, New York.

ISOPRENOIDS — 3

I. CAROTENOIDS

A. Introduction

The carotenoids are yellow, orange, or red pigments, which are widely distributed in the plant and animal kingdoms. They are called lipochromic pigments because they are soluble in fats. In the animal organism the carotenoids are either dissolved in fats or combined with protein in the aqueous phase. In the higher plants the carotenoids are found in the leaves together with chlorophyll; they also constitute the principal pigments of certain yellow, orange, and red flowers and of many microorganisms.

The carotenoids comprise two groups: hydrocarbons, soluble in petroleum ether; and xanthophylls, oxygenated derivatives of the carotenes. These compounds are alcohols, aldehydes, ketones, epoxides, and acids soluble in ethanol.

1. Hydrocarbons

The three isomeric orange-red carotenes isolated from carrots are α-, β- and γ-carotene; the red carotenoid hydrocarbon of tomatoes is lycopene.

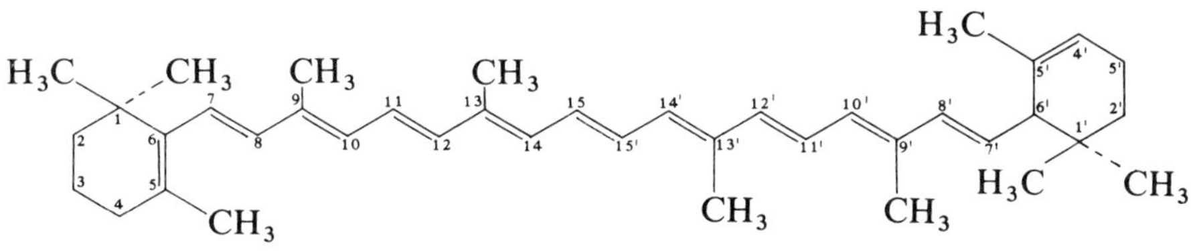

α-Carotene

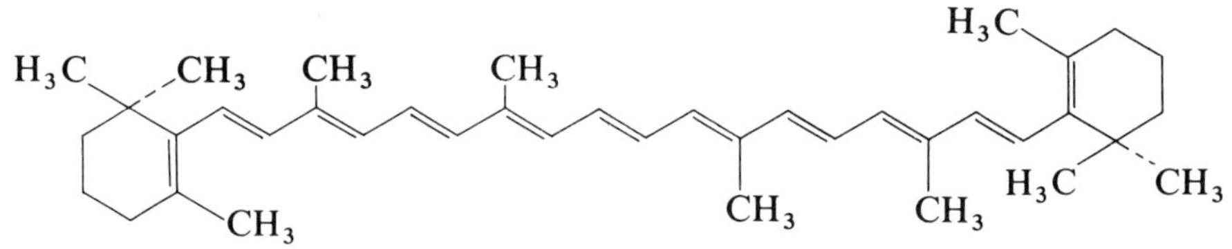

β-Carotene

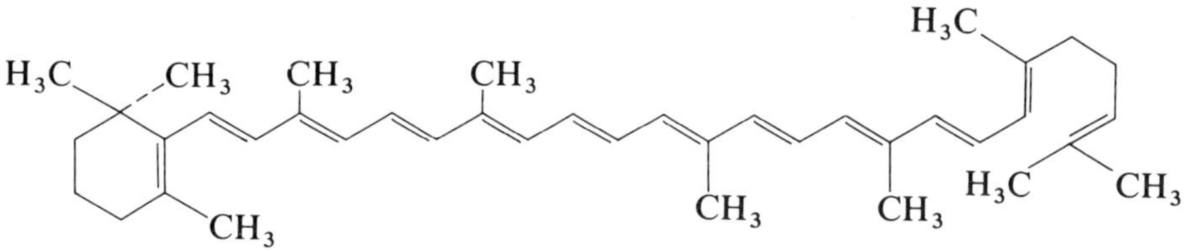

γ-Carotene

A series of partly saturated carotene hydrocarbons (e.g., phytofluene, phytoene) has recently been discovered in various plant tissues. These carotenes appear to be lycopene derivatives.

2. Xanthophylls

Xanthophylls may be classified as follows:
Hydroxylated carotenoids such as cryptoxanthin, lutein, and zeaxanthin. The hydroxyl group can exist either free or esterified.

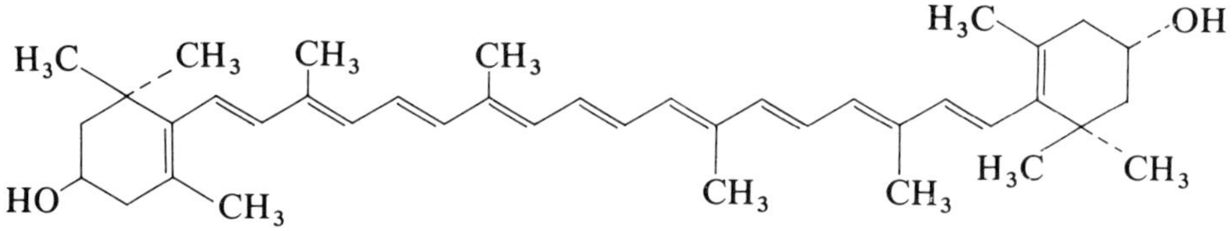

Zeaxanthin

Methoxylated carotenoids such as rhodovibrin. They are lycopene derivatives substituted at position 1.

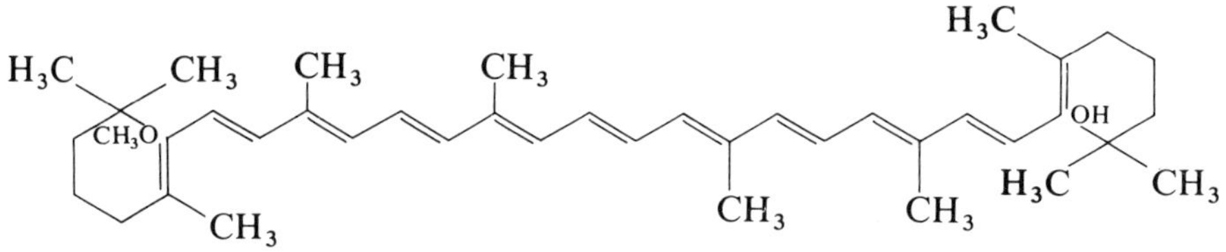

Rhodovibrin

Oxocarotenoids such as capsanthin and rhodoxanthin.

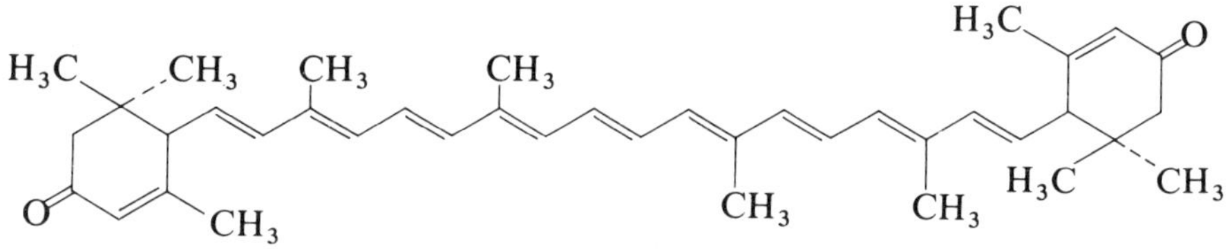

Rhodoxanthin

Epoxycarotenoids exist in nature as 5,6- and 5,8-epoxides, e.g., violaxanthin and flavoxanthin. Luteochrome contains both the 5,6- and 5,8-epoxide systems.

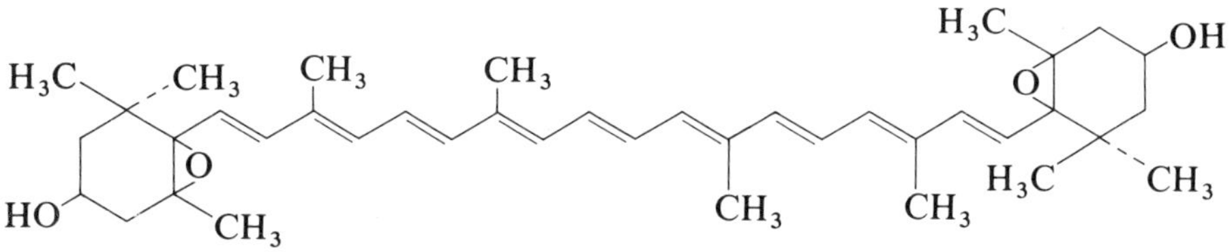

Violaxanthin

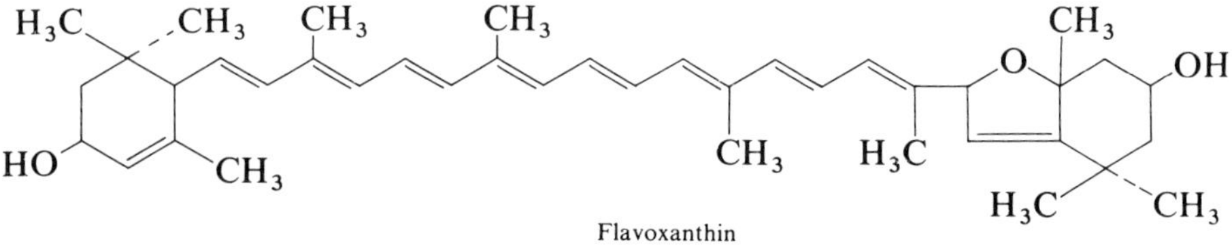

Flavoxanthin

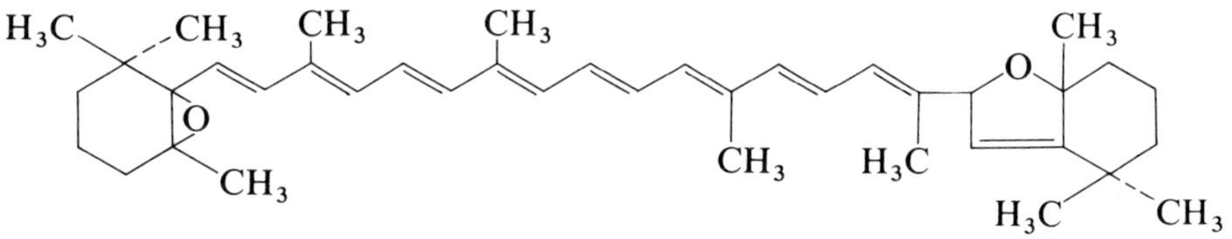

Luteochrome

Carboxycarotenoids such as bixin and crocetin.

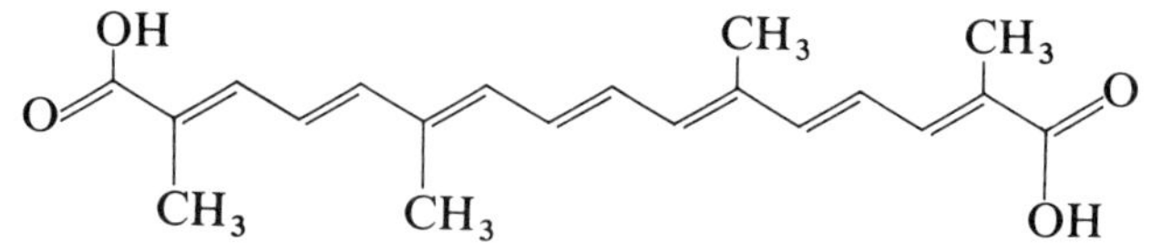

$$\text{Crocetin structure}$$

Crocetin

Almost all naturally occurring carotenoids are tetraterpenoids consisting of eight isoprenoid molecules. On the basis of X-ray and spectral analysis, it was shown that, with a very few exceptions, the double bonds of the natural carotenoids have a trans configuration. In most naturally occurring cyclic carotenoids, the linkage between the terminal ring and the central chain is via a single bond. However, by dehydrogenation it is possible to transform this linkage into a double bond. The compounds are then designated by the prefix retro.

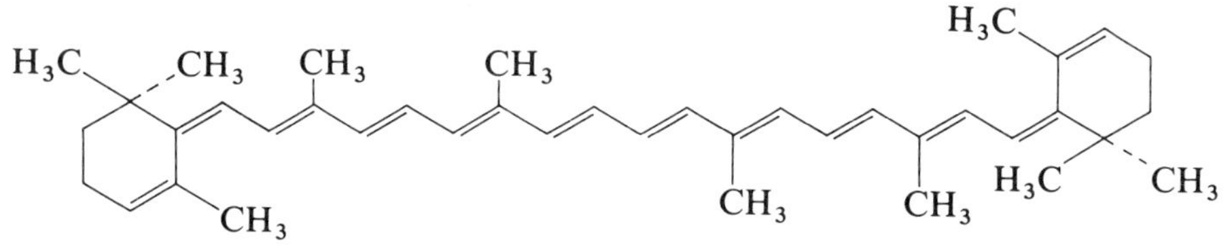

Retrodehydro-β-carotene

The methods used for elucidating the constitution of carotenoids are catalytic hydrogenation or addition of halogen, iodine chloride, or oxygen for determination of double bonds; oxidation with chromic acid of the side-chain methyls into carboxyl groups; oxidative degradation with potassium permanganate and ozone.

Chromatography (column, paper, and thin-layer) has proved to be a powerful and essential tool for the isolation and structural determination of carotenoids. The synthesis of the polyene pigments involves the union of a bifunctional unit forming the central part of the molecule with two molecules that form its terminals.

Combination of units $C_{19} + C_2 + C_{19}$.

β-Carotene, zeaxanthin, and physalien have been prepared in this way.

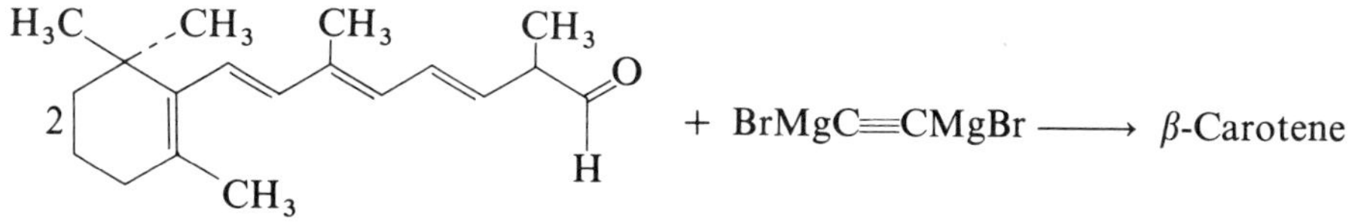

$$2 \text{ [C}_{19}\text{ aldehyde]} + \text{BrMgC}\equiv\text{CMgBr} \longrightarrow \beta\text{-Carotene}$$

I. Carotenoids

Combination of units $C_{16} + C_8 + C_{16}$.

In this process, the center of the molecule is an unsaturated diketone that reacts with two molecules of a C_{16} acetylene.

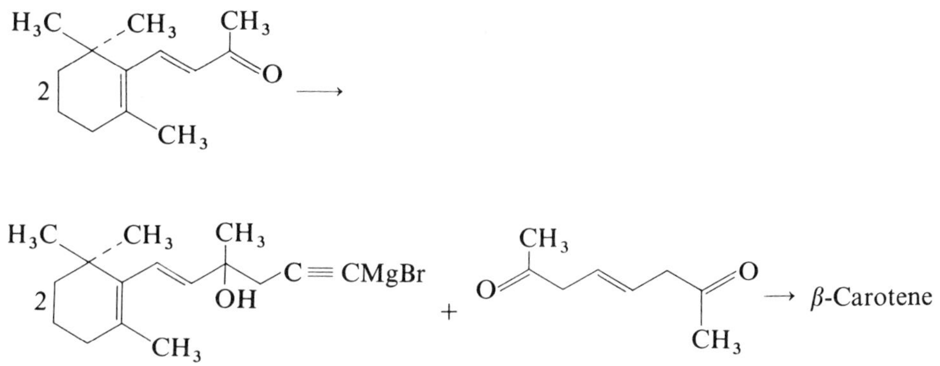

[Structure: cyclohexene ring with methyl substituents and side chain ending in C=O (CH₃ group)] →

[Structure: cyclohexene ring with side chain -CH=CH-C(CH₃)(OH)-C≡C-MgBr] + [Structure: O=C(CH₃)-CH₂-CH=CH-CH₂-C(CH₃)=O] → β-Carotene

Combination of units $C_{14} + C_{12} + C_{14}$.

β-Carotene may be synthesized by a combination of a *bis*-enol ether and the acetal of a C_{14} aldehyde.

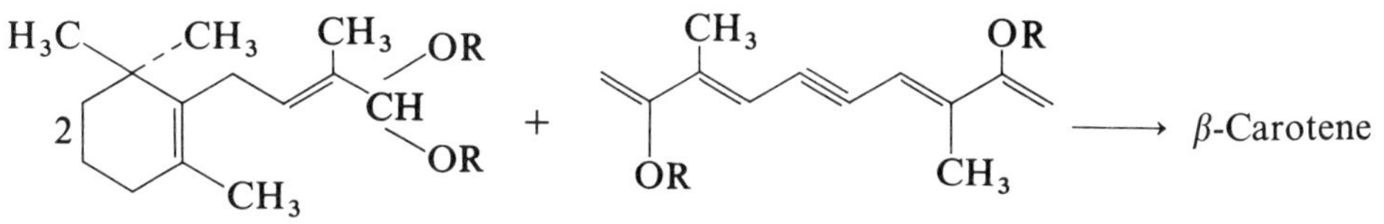

[Structure: cyclohexene ring with side chain -CH=C(CH₃)-CH(OR)(OR)] + [Structure: bis-enol ether with OR groups and CH₃ substituents] → β-Carotene

Numerous investigators in recent years have tried to establish the significance of carotenoids in plants. Some investigators claim that carotene and xanthophyll fulfil the role of light filters for chlorophyll, while others suggest that they function as protectors of enzymes in the cell.

It is noteworthy that while plants and microorganisms synthesize their carotenoids, those present in the tissues of higher animals are derived from dietary sources, e.g., β-carotene, which is converted into vitamin A in the animal organism. Furthermore, certain carotenoids play a part in the visual process.

The biosynthesis of carotenoids has been studied much more extensively than that of any of the other terpenoids. The following is a generalized scheme.

Isopentenyl pyrophosphate ⟶ Geranyl pyrophosphate $\xrightarrow{\text{IPP}}$

Farnesyl pyrophosphate $\xrightarrow{\text{IPP}}$ Geranylgeranyl pyrophosphate $\xrightarrow{\text{Dimerization}}$

Phytoene ⟶ Carotenoids.

CHAPTER 3 Isoprenoids

The main building unit from which isopentenyl pyrophosphate (IPP) is synthesized is mevalonate. It has not yet been established whether the various carotenoids are interconvertible or whether they arise via parallel paths from a common C_{40} precursor.

B. Isolation of Capsanthin from Paprika

1. Introduction

The first investigation of paprika pigments was carried out by Braconnot in 1817 (1). In 1927, Zechmeister and von Cholnoky (2) obtained the pigment of *Capsicum annuum* (paprika) in a crystalline form and proposed the name capsanthin.

Capsanthin is a rare carotenoid. It was found in the esterified form in the ripe pods of *Capsicum annuum* and *Capsicum frutescens japonicum* (3, 4). Karrer and Oswald (5) observed that the anthers of *Lilium tigrinum* contain capsanthin.

Capsanthin is well adsorbed on calcium carbonate or zinc carbonate from solutions of carbon disulfide or from a mixture of benzene and ether.

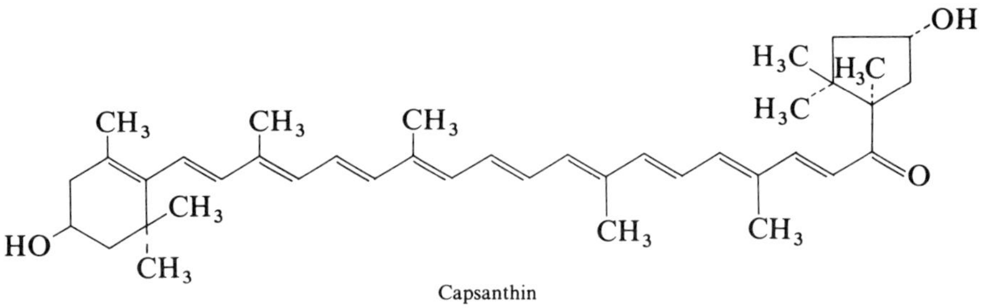

Capsanthin

2. Principle

Capsanthin, the main carotenoid of paprika, is first extracted from paprika pods with petroleum ether at room temperature. Since capsanthin, like most hydroxylated carotenoids, occurs in nature in the ester form, it is saponified by means of 30% methanolic potassium hydroxide at room temperature (6). For further purification, the crude capsanthin is subjected to column chromatography on suitable adsorbents.

3. Materials

Benzene	Ether	Petroleum ether
Calcium carbonate	Methanol	Potassium hydroxide
Carbon disulfide	Paprika pods	Sodium sulfate, anhydrous

4. Time

   14–15 hours

5. Procedure

The paprika pods are freed from their shells and seeds and dried at 35 to 40°C. Finally ground pod material (100 g) is stirred with 200 ml petroleum ether (bp 40–60°C) for 4 hr at room temperature and then filtered through a Buchner funnel; the solid is washed with 25 ml petroleum ether. The red-colored solution of petroleum ether is diluted with a threefold volume of ether, 100 ml 30% methanolic potassium hydroxide is added, the mixture is stirred for 8 hr, and the free phytoxanthins are dissolved in ether. The ethereal layer is then washed with water, dried over sodium sulfate, and concentrated to 20 ml under reduced pressure. The ethereal residue is diluted with 60 ml petroleum ether and allowed to stand in a cool place for 24 hr. The red capsanthin is filtered off and crystallized from a small amount of carbon disulfide.

The separation of capsanthin from the accompanying carotenoids, such as zeaxanthin and capsorubin, is effected by chromatography on calcium carbonate or zinc carbonate. Carbon disulfide or a mixture of benzene and ether (1:1) is employed as a solvent for developing the chromatogram. Crystallization from carbon disulfide yields carmine red spheres, mp 176°C. The pigment crystallizes from petroleum ether as needles, and from methanol as prisms.

Care should be taken when working with carbon disulfide, since it is highly inflammable.

a. Color reactions
1. On treating a solution of capsanthin in chloroform with concentrated sulfuric acid, the latter assumes a deep blue coloration.
2. Capsanthin gives a deep blue coloration with antimony trichloride in chloroform.

b. Capsanthin

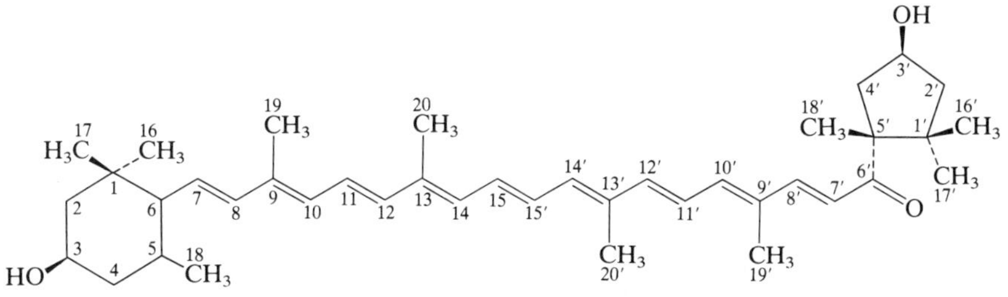

c. ¹NMR (nuclear magnetic resonance) (400 MHz)
δ_H 0.840[s, 3H, CH$_3$ (16')],
1.075[s, 6H, Me(16), Me(17)],
1.207[s, 3H, Me(17')].
1.367[s, 3H, Me(18')],
1.736[s, 3H, Me(18)],
1.957[s, 3H, Me(19'),
1.974[s, 6H, Me(19), Me(20)],
1.989[s, 3H, Me(20')],
2.39[ddd, J = 17.6, ca 1.5, 1H, Hα—C(4)],
2.96[dd, J = 15.5, 9, 1H, Hα – C(4')],
ca. 4.00[br, m, 1H, Hα – C(3)],
4.52[m, 1H, Hα – C(3')],
6.13[s, 2H, H—C(7), H—C(8)],
6.16[d, J = 11.6, 1H, H—C(10)],
6.26[d; J = 11; 1H, H—C(14)];
6.35[d, J = 11, ca. 1H, H—C(14')],
6.36[d, J = 15, 1H, H—C(12)],
6.45[d, J = 15, 1H, H—C(7')];
6.52[d, J = 15, 1H, H—C(12')],
6.55[d, J = 11, 1H, H—C(10')],
ca 6.6–6.8[m, 4H, H—C(11), H—C(11'), H—C(15); H—C(15')],
7.33[d, J = 15, 1H, H—C(8')]

d. ¹³C NMR (100, 6MHz)
δ_c 12.75[C(19)], 48.61[C(2)], 132.39[C(14)],
12.79[C(20)], 51.06[C(2')], 133.69[C(9')];
12.84[C(19')], 59.01[C(5')], 135.24[C(14')];
12.90[C(20')], 65.41[C(3)], 135.93[C(13')],
21.39[C(18')], 70.44[C(3')], 137.46[C(12)],
21.63[C(18)], 121.04[C(7')], 137.60[C(13)],
25.16[C(17')], 124.13[C(11')], 137.85[C(6)],
25.95[C(16')], 125.58[C(11)], 138.51[C(8)],
28.80[C(16)], 125.93[C(7)], 140.63[C(10')],
30.32[C(17)], 126.30[C(5)], 141.97[C(12')],
42.69[C(4)], 129.74[C(15')]; 146.86[C(8')],
44.01[C(1')], 131.27[C(10)], 202.82[C(6')].
45.49[C(4')], 131.68[C(15')],

e. Mass spectrum
Capsanthin C$_{40}$H$_{56}$O$_3$ Mol. wt. 584.
m/z 584(75%)   127(36%)   105(44%)
    478(62%)   109(100%) 91(65%)
    429(6%)   106(31%)   83(56%)
    145(51%)

f. UV spectrum of capsanthin

$\lambda_{max}^{benzene}$ 486, 520 mμ

Capsanthin is a polyene with 9 conjugated C=C double bonds. The appearance of a peak at 520 mμ (as compared with 504 mμ for lycopene) is due to the introduction of a carbonyl group in conjunction with the system of the C=C double bond. One such carbonyl thus has a more pronounced bathochromic effect than two C=C bonds.

References

1. Braconnot, H. (1817). *Ann. Chim.* **6**, 122, 133.
2. Zechmeister, L., and von Cholnoky, L. (1927). *Ann.* **454**, 54.
3. Zechmeister, L., and von Cholnoky, L. (1931). *Ann.* **487**, 197.
4. Zechmeister, L., and von Cholnoky, L. (1931). *Ann.* **489**, 1.
5. Karrer, P., and Oswald, A. (1935). *Helv. Chim. Acta* **18**, 1303.
6. Zechmeister, L., and von Cholnoky, L. (1934). *Ann.* **509**, 269.

Recommended Reviews

1. Goodwin, T. W. (1955). *Carotenoids. In "Modern Methods of Plant Analysis"* (K. Paech and M. V. Tracey, eds.), Vol. 3, p. 272. Springer-Verlag, Berlin.
2. Goodwin, T. W. (1965). *Distribution of Carotenoids. In "Chemistry and Biochemistry of Plant Pigments"* (T. W. Goodwin, ed.), p. 127. Academic Press, New York.

C. Isolation of Lycopene from Tomatoes

1. Introduction

Lycopene was first isolated from *Tamus communis* by Hartsen in 1873 (1). Recent investigations employing highly defined chromatographic methods have shown that the tomato pigment is widely distributed in nature. It has been prepared from a variety of fruits and berries (2, 3). The first modern preparation is that of Willstätter and Escher (4), who processed 75 kg tomato concentrate and obtained 11 g once-recrystallized lycopene.

Porter and Zscheile (5) have described the chromatographic separation of carotenes from various species and strains of tomatoes on a magnesium oxide–Super Sel column.

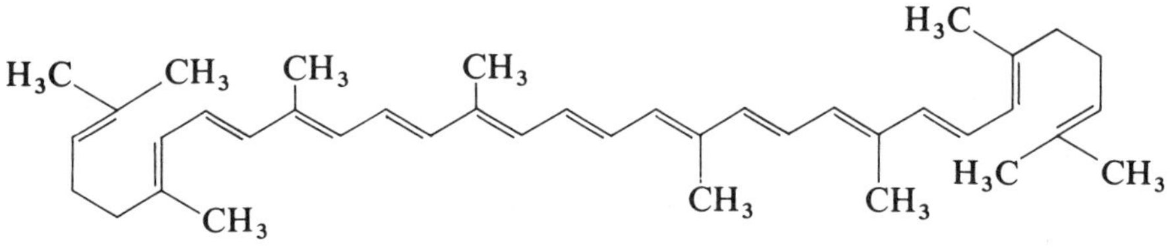

Lycopene

Recently, (6) four dominant tomato paste carotenoids, phytofluene, β-carotene, phytoene, and lycopene were identified using high performance liquid chromatography (HPLC) with UV-VIS photodiode array detection.

Carotenoid Identified (nm)	Structure
Phytoene (275 286 297)	
Phytofluene (331 348 368)	
Beta-carotene (426 448 476)	
Zeta-carotene (378 398 424)	
Gamma-carotene (435 461 490)	
Lycopene (444 471 501)	

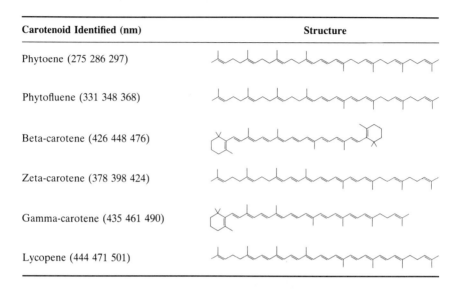

2. Principle

Tomato paste is dehydrated with methanol, and lycopene is extracted from the residue with methanol–carbon tetrachloride. The crude product is crystallized twice from benzene by the addition of methanol, giving lycopene of 98 to 99% purity (7).

Further purification is achieved by a chromatographic procedure, using calcium hydroxide as the adsorbent.

Lycopene in solution undergoes isomerization even at 20°C. Crystalline lycopene is not isomerized but has a tendency to autoxidation, especially in light. It should be kept in the dark in evacuated glass tubes.

3. Apparatus

Centrifuge

4. Materials

Benzene
Carbon tetrachloride
Methanol
Sodium sulfate, anhydrous
Tomato paste

5. Time

   5–6 hours

6. Procedure

Canned tomato paste, 50 g, in a 3-liter-wide-mouthed bottle is dehydrated by adding 65 ml methanol. The mixture is immediately shaken vigorously to prevent the formation of hard lumps. A small sample of the suspension is tested by hand; if it has a glutinous consistency, more methanol is added to the main portion to avoid the possible clogging of filters. The mixture is allowed to stand for 1 to 2 hr and is then shaken vigorously. The thick suspension is filtered on a Buchner funnel (diameter 25–20 cm). The yellow filtrate is discarded. The dark red cake is returned to the bottle and shaken with a mixture of 35 ml methanol and 35 ml carbon tetrachloride. The stopper of the bottle must fit well and should be lifted for a moment after the mixing, to release any built-up pressure. Brief shaking followed by opening of the bottle is repeated until no more excess pressure is noticed. The suspension is shaken for 10 to 15 min and separated by filtration on a large Buchner funnel. The filtrate consists of a lower, very dark red, carbon tetrachloride phase and an orange aqueous-methanolic layer. The slightly colored tomato residue is crushed by hand to form a nearly uniform powder. It is then reextracted with 35 ml of each solvent as described, and the suspension is filtered. The extraction is now almost complete. The filtrates are combined, the methanol layer is transferred to a 2-liter separatory funnel, and 1 volume of water is added. A white emulsion appears in the upper phase. If the emulsion is reddish, it is stirred with a glass rod until the droplets of carbon tetrachloride join the lower layer. The phases are separated and the carbon tetrachloride phase is washed several times with water. The carbon tetrachloride solution is drained into a 1-liter Erlenmeyer flask and dried over anhydrous sodium sulfate. The extract is then poured through a folded filter into a 1-liter round-bottomed flask equipped with a standard taper. The solvent is evaporated with the aid of a water pump to about 5 ml in a water bath at 60°C. The solution is transferred to a similar flask of 10 ml capacity using a few ml carbon tetrachloride to rinse the larger flask. The solvent is then removed completely *in vacuo*, leaving a dark oily residue, which is diluted with a few ml benzene and evaporated again to remove the carbon tetrachloride completely. The partly crystalline, dark residue is transferred quantitatively with 1 ml benzene to a 25 ml Erlenmeyer flask. A clear solution is obtained by immersing the flask in a hot water bath. Boiling methanol is added in portions, using a dropper, to the benzene solution, with stirring after each addition, until 1 ml methanol has been introduced. Crystals of crude lycopene begin to appear immediately. The crystallization is completed by keeping the liquid first at room temperature and then in ice water. After standing for 1 to 2 hr, the crystals are collected on a small Buchner funnel and washed with 2 ml boiling methanol.

The lycopene crystals are transferred to a tared 10-ml centrifuge tube, the last portion being removed from the funnel with small quantities of boiling benzene. Benzene is added to the centrifuge tube to make up the volume to 1 ml. The crystals are dissolved by dipping the tube into hot water and stirring the contents. When a clear solution is obtained, boiling methanol is introduced in small portions with a dropper, and the solution is stirred with a glass rod until crystals begin to appear. The centrifuge tube is kept at room temperature for a short time and then an ice bath. More methanol is added in small portions, with stirring, to the cold solution. The total volume of methanol present should not exceed 1 ml.

The mixture is allowed to stand for 2 hr in the ice bath, and the crystals are separated by a brief but strong centrifuging. The mother liquor is decanted and discarded. The crystals are treated in the centrifuge tube with 1 ml boiling methanol. The mixture is stirred and the methanol is removed by centrifuging before it cools. The methanol is decanted, and the washing is repeated at least twice more. If the crystallization and purification were satisfactory, long, red lycopene prisms are observed under the microscope. No colorless substance should be present. The centrifuge tube and its contents are dried *in vacuo* at room temperature for a few hours and then weighed. The yield, which is dependent on the quality of the tomato paste, is about 1.5 mg.

a. Color reactions
1. Lycopene dissolves in concentrated sulfuric acid, imparting to the solution an indigo blue color.
2. On adding a solution of antimony trichloride in chloroform to a solution of lycopene in chloroform, an intense unstable blue color is produced.

b. Lycopene—(All trans.)

c. ^{13}C NMR

C-1 131.64	C-8 135.54	C-15 130.17
C-2 124.12	C-9 136.15	C-16 25.66
C-3 26.83	C-10 131.64	C-17 17.70
C-4 40.30	C-11 125.21	C-18 16.97
C-5 139.30	C-12 137.46	C-19 12.90
C-6 125.94	C-13 136.54	C-20 12.81
C-7 124.87	C-14 132.71	

d. ^{1}H NMR [220 MHz–CDCl$_3$]

H-2	H-3/H-4	H-6	
δ5.11	2.11	5.95	
H-7	H-8	H-10	
6.49	6.25	6.19	
H-11	H-12	H-14	H-15
6.64	6.35	6.23	6.63

e. Mass spectrum

Lycopene C$_{40}$H$_{56}$, Mol. wt 536.

m/z 536(21.70%)　　105(47%)　　81(36%)
　　145(38%)　　　　93(36%)　　 69(100%)
　　119(34%)　　　　91(47%)　　 41(57%)

f. UV spectrum of lycopene

λ_{max}^{EtOH} 443, 472, 504 mµ.

Lycopene is a polyene with 11 conjugated C=C double bonds (22π-electrons). With an increase in the number of π-electrons in such a system, there is a bathochromic shift of the longest band, which in the case of carotenoids reaches the visible part of the spectrum.

References

1. Hartsen (1873). *Compt. rend.* **76**, 385.
2. Karrer, P., Rubel, F., and Strong, F. M. (1936). *Helv. Chim. Acta* **19**, 28.
3. Kuhn, R., Bielig, H., and Dann, O. (1940). *Ber.* **73**, 1080.
4. Willstätter,, R. and Escher, H. H. (1910). *Z. Physiol. Chem.* **64**, 47.
5. Porter, J. W. and Zscheile, F. B. (1946). *Arch. Biochem.* **10**, 537.
6. Tan, B. J. (1988). *Food Sci.* **53**, 954.
7. Sandoval, A., and Zechmeister, L. (1949). *Biochem. Prepns.* **1**, 57.

Recommended Review

1. Goodwin, T. W. (1962). *Carotenoids, structure, distribution, and function.* In "*Comparative Biochemistry*," (M. Florkin, and H. S. Mason, eds.), Vol. 4, p. 643. Academic Press, New York.

D. Thin-Layer Chromatography of Carotenoid Pigments in Oranges

1. Introduction

The complete analysis of carotenoid pigments involves extraction, saponification, separation by chromatographic methods and, finally, identification of each pigment. Using TLC, the preliminary fractionation into hydrocarbons, monools, diols and polyols is effected on silica-gel plates developed with the solvent system acetone–petroleum ether (3:7). Each group can be further separated into individual carotenoids by rechromatography on MgO–Kieselgur (1:1) plates using the above solvent system and increasing the amount of acetone according to group polarity.

2. Principle

In the following procedure (1) orange peels are blended with acetone in the presence of BHT (an antioxidant) and calcium carbonate (a neutralizing agent).

The extract of cartenoids is then saponified and the free carotenoids separated on thin layers of MgO-Kieselgur plates. Their identity is determined by chromatographic (R_f values) and spectroscopic data.

3. Apparatus

Ultra Turrax Homogenizer
UV-Vis Spectrophotometer
TLC applicator

4. Materials

Acetone
Butylated hydroxytoluene (BHT)
Calcium carbonate
Chloroform
Diethyl ether
Ethanol, absolute
Kieselgur (for TLC)

Magnesium oxide (for TLC)
Methanol
Orange peels
Petroleum ether
Potassium hydroxide
Silica gel–coated TL plates
Sodium chloride

5. Time

6 hours

6. Procedure

a. Extraction

Orange peels (20 g) or other material are homogenized—cooling with ice, in an Ultra Turrax homogenizer—with acetone in the presence of BHT as antioxidant and calcium carbonate as neutralizing agent. After filtering under reduced pressure, the residue is re-extracted with the same solvent until colorless. The acetonic extract is mixed with an equal volume of diethylether (peroxide-free), and sufficient saturated saline solution is added to form two layers. The upper layer (epiphase) contains all the pigments. It is washed several times with distilled water.

b. Saponification and removal of sterols

The diethylether pigment extract is evaporated to dryness in vacuum (at 40°C), absolute ethanol being added to remove the water. The residue is extracted with diethylether (2–3 ml), and an equal volume of 10% KOH in ethanol is added. The mixture is kept for 2–3 hr in a nitrogen atmosphere at room temperature. After diluting with ether, the ethereal solution is washed free of alkali and evaporated in vacuum. The amount of total carotenoids is determined by spectroscopy. The nonsaponifiable matter is dissolved in a minimal volume of methanol and kept overnight in a freezer at $-20°C$, during which period the sterols precipitate. They are removed by centrifugation in a refrigerated centrifuge.

c. TLC

After evaporation of the methanolic supernatant solution under reduced pressure, the residue is dissolved in a few drops of chloroform and petroleum ether. The extract is applied as a line on a silica-gel plate (0.4 mm thick), which is developed with 30% acetone in petroleum ether. When the upper band of carotenes reaches a distance of 2 cm from the top of the plate, it is rapidly scraped off into a beaker containing acetone. The plate is further developed with 30% acetone in petroleum ether until a good separation of all zones is obtained.

Each band is further rechromatographed on thin layers of MgO–Kieselgur (1:1). The less-polar band is developed with 2 to 5% acetone in petroleum ether. Thus, phytofluene and α-, β- and γ-carotene are readily separated and identified by their visible color: phytofluene shows blue-green fluorescence in UV light, α-carotene shows yellow; β-carotene shows orange; and γ-carotene shows lemon yellow.

The monol group is developed with 10% acetone in petroleum ether, and the more polar groups of diols and triols with 30% acetone in petroleum ether.

Identification of the pigments and colorless polyenes is made on the basis of their chromatographic and spectroscopic properties. The absorption spectrum of each pigment is recorded in a DB Spectrophotometer in the range of 220 to 550 nm (see table below).

Visible Absorption Maxima of Carotenoids

Carotenoid	Ethanol (nm)	Carotenoid	Ethanol (nm)
Phytofluene	327, 348, 367	Lutein	420, 443, 472
α-Carotene	420, 443, 372	Zeaxanthin	425, 450, 478
β-Carotene	425, 450, 475	Anteraxanthin	418, 444, 470
γ-Carotene	378, 400, 424	Luteoxanthin	400, 424, 448
OH-α-Carotene	418, 440, 468	Violaxanthin	416, 438, 468
Cryptoxanthin	425, 450, 475	Citraurin	454

For some functional groups, chemical tests, like the epoxide test and the carbonyl-reduction test, may be used.

d. Epoxide test

Isomerization of the 5,6 epoxides into 5,8 epoxides occurs rapidly even in the presence of a trace amount of a mineral acid. The reaction is very fast, and the rearrangement produces a hypsochromic shift in the visible absorption spectrum as the chromophore is shortened.

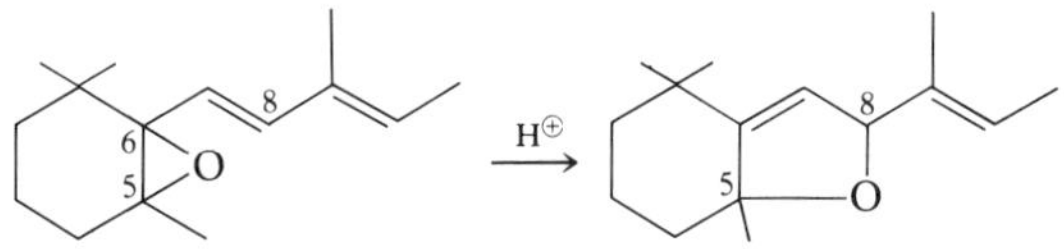

e. Carbonyl reduction test

The reduction of the carbonyl function can be carried out with $LiAlH_4$ in ether or $NaBH_4$ in ethanol. A hypsochromic shift of about 6 to 7 nm is indicative of an in-ring carbonyl. A shift of 25 to 35 nm indicates an in-chain carbonyl.

Reference

1. Ikan R., (1982). "Chromatography in Organic Microanalysis." Academic Press, New York.

Recommended Reviews

1. Gross, J. (1980). A rapid separation of citrus carotenoids by thin-layer chromatography. *Chromatographia* **13,** 572.
2. Stewart, I. (1980). Color as related to quality in citrus. *In*: "*Citrus Nutrition and Quality*" (S. Nagy, and J. A. Attaway, eds.), p. 129. American Chemical Society Symp. Series 143.

E. High-Performance Liquid Chromatography of Carotenoids in Orange Juice

1. Introduction

Certain carotenoids such as β-cryptoxanthin and α- and β-carotene are highly colored compounds that also exhibit provitamin A activity. β-Carotene has the highest vitamin A activity because it can be split through central enzymatic cleavage into two molecules of vitamin A. Because of structural differences in part of the molecule, only half of each molecule of α-carotene, γ-carotene and β-cryptoxanthin has the required intact β-ionone ring structure for vitamin A activity. (1) Therefore, these carotenes exhibit lower vitamin A activity. Nevertheless, they are important dietary sources of vitamin A.

Carotenoids, located in juice vesicle plastids, are also the major source of color in orange juice and as such are an important quality factor. Orange juices with the expected deep, yellow-orange color are preferred by most consumers and may even be perceived as being sweeter.

2. Principle

The following method (2) may be used to analyze the citrus juice carotenoids for taxonomy and regulatory or quality-control studies. It is based on methanol extraction of carotenoids from centrifuged juice solids. A C_{18} solid-phase extraction column is used to clean up the saponified sample before HPLC analysis. Nonaqueous reversed-phase HPLC employing acetonitrile–methylene chloride–methyl alcohol mixture and a C_{18} column is used to separate β-cryptoxanthin and α- and β-carotene from other juice components.

3. Apparatus

Centrifuge
High-pressure liquid chromatograph
Magnetic stirrer

4. Materials

Acetonitrile
Ethyl acetate
Methanol
Methylene chloride

Orange juice
Potassium hydroxide
Syringe
Zerbax ODS column

5. Time

3 hours

6. Procedure

a. Sample preparation

A 10 ml sample of orange juice is centrifuged for 5 min at 3000 g. This operation produces a pellet containing the carotenoids. The clarified juice is then decanted and discarded, and the wet pellet is stirred with 2 ml methanol. The slurry is recentrifuged, and the supernatant, discarded. The pellet is then extracted with 3 ml methanol using an Omni-Mixer. The resulting slurry is centrifuged, and the supernatant, saved. This operation is repeated two more times, and the methanol extracts are combined. All work with extracted carotenoids is done under nitrogen and subdued light.

b. Saponification

Methanolic potassium hydroxide, 4.5 ml, (of 10 g/100 ml methanol solution) is added to the combined methanol extracts of carotenoids, and the resulting mixture is allowed to stand in the dark without stirring for 1 hr at 23°C. The yellow reaction mixture is transferred to a 125-ml separatory funnel with 30 ml methylene chloride. The potassium hydroxide is removed by washing with 4×20 ml water. The yellow organic layer containing the carotenoids is evaporated to dryness under nitrogen at 35°C.

c. Sample cleanup

The above residue is dissolved in 2 ml methanol and placed on the top of a 6-ml C_{18} extraction column preconditioned with 5 ml methanol. A syringe with an adapter is used to push the liquid slowly through the column. The absorbed carotenoids are then eluted with 4.5 ml methanol–water (95:5; v/v; fraction 1), followed by 3 ml methylene chloride (fraction 2). Each fraction is separately evaporated to dryness. The resulting two residues are each redissolved in 2 ml of the chromatographic mobile phase (acetonitrile–methylene chloride–methanol; 65:25:10) and placed in an amber 4-ml vial before injection into HPLC apparatus.

d. HPLC conditions

Zorbax ODS analytical column (4.6 mm × 250 mm) is used with the following solvent system:

1. acetonitrile–methylene chloride–methanol (65:25:10);
2. acetonitrile–methylene chloride–methanol (70:20:10);
3. acetonitrile–tetrahydrofuran–methanol (70:20:10); and
4. acetonitrile–ethyl acetate–methanol (70:20:10).

Sample injection volume is 50 μl. The eluant is monitored at 450 nm. Column temperature is kept at 22 to 24°C. Flow rate of the solvent is 1.5 ml/min. After several injections, the column is washed with methylene chloride and then equilibrated with the mobile phase.

Notes
1. Since the carotenoids of interest are of similar polarity, an isocratic rather than a gradient system for the HPLC is used.
2. The first washing of the wet pellet with methanol removes most of the water and only the most polar materials. Only the second methanol extract solubilized α- and β-carotenes and cryptoxanthin.

3. The first water wash from the saponified sample is yellow. The color is probably due to the presence of chalcones formed by the reaction of potassium hydroxide and juice flavanones. Acidification of the yellow fraction results in a colorless solution, which is typical of chalcones but not of carotenoids.
4. If desired, the cryptoxanthin can be separated from the carotenes on the solid-phase extraction column by eluting with 97% methanol–3% water instead of methylene chloride.
5. The time required from the collection of a single juice to final carotenoid concentration values is about 4 hr. The majority of the time savings is owing to a 1-hr saponification step and the employment of an isocratic chromatographic system.
6. Chromatographic peaks are identified by chromatographic characteristics using standard carotenoids and spectroscopic data. A small absorbance band at 365 nm may be due to the minor presence of the cis-isomer of β-carotene.

References

1. Goodwin, T. W. (1980). The Biochemistry of the Carotenoids, England. Chapman and Hall, Vol. 1.
2. Fisher, J. F., and Rouseff, R. L. (1986). *J. Agric. Food Chem.* **34**, 985–989.

Recommended Reviews

1. Ruedi, P. (1985). HPLC—a powerful tool in carotenoid research. *Pure Appl. Chem.* **57**, 793.
2. Nelis, H.J.C.F., and DeLeenheer, A.P. (1983). Isocratic nonaqueous reverse-phase liquid chromatography of carotenoids. *Anal. Chem.* **55**, 270.
3. Schwartz, S. J., and von Elbe, J. H. (1982). HPLC of plant pigments. *J. Liq. Chrom.* **5**, 43.

F. Determination of β-Carotene and Chlorophylls in Plants

1. Introduction

All green plant tissues contain a mixture of carotenes, xanthophylls, and chlorophylls. A variety of solvents may be used to extract these pigments, acetone and methanol being those most commonly employed. If fresh tissues are used, they are frequently shredded with the solvent in a blender, or ground together with the extractant and sand in a mortar. This should be carried out in dim light. Heating should be avoided during the extraction stage, as the pigments have a tendency to undergo rapid oxidation and isomerization when exposed to heat and light (1).

Using a properly activated adsorbent, the separation of the carotenes from the xanthophylls and other noncarotene pigments becomes a relatively simple matter. The absorption spectrum of a pigment, together with its adsorptive power or relative position on a chromatographic column, provides a good indication of its identity. Various adsorbents have been used for the separation of chloroplast pigments. These include aluminum oxide, magnesia, inulin, starch, cellulose (2–4), calcium phosphate (5), polyethylene (6), hydrated limes (7), and powdered sugar (8).

2. Principle

After the pigments have been extracted, chlorophyll can be determined by various methods based on colorimetry, spectrophotometry, fluorimetry, and estimation of magnesium. The spectrophotometric determination of chlorophyll is in accordance with the Beer-Lambert law, and various equations have been derived for determining the concentration of total chlorophyll and of chlorophylls *a* and *b* separately (9, 10).

In the following procedure (11), the crude plant extract is introduced at the top of the column filled with magnesia and diatomaceous earth, 1:1, and eluted with a mixture of acetone–hexane, 1:9. The purified carotenes are then collected and determined spectrophotometrically, while chlorophylls and xanthophylls remain on the adsorbent.

3. Apparatus

Blender
Colorimeter

4. Materials

Acetone	Magnesia
Diatomaceous earth	Magnesium carbonate
Hexane	Sodium sulfate

5. Time

3–4 hours

6. Procedure

a. Extraction

A 2- to 5-g sample of the fresh tissue is placed in a high-speed blender, and 40 ml acetone, 60 ml hexane, and 0.1 g magnesium carbonate are added. The mixture is blended for 5 min, filtered on a Buchner funnel, and washed with 25 ml acetone and 25 ml hexane. The extractions are transferred to a separatory funnel and washed with five 100-ml portions of water. The upper layer is transferred to a 100-ml volumetric flask containing 9 ml acetone and made up to the mark with hexane. Alcohol (80 ml in conjunction with 60 ml hexane) may be used instead of acetone for extraction.

b. Determination of chlorophyll

According to this method, the chlorophyll is measured in the original extract in the presence of all the extracted yellow pigments. A portion of the original extract (regardless of the solvent used) is placed in the absorption cell of a photoelectric colorimeter, and the transmission of the solution is compared with that of the pure solvent.

The transmission of the solution is converted to concentration by means of a standard calibration curve obtained with pure chlorophyll. A red filter or a combination of filters transmits radiation in the region in which chlorophyll has maximal absorption, thus making possible a precise measurement of chlorophyll concentration.

c. Determination of β-carotene

The chromatographic column (2 × 18 cm) is packed to a height of 15 cm with 1:1 mixture of activated magnesia and diatomaceous earth, the adsorbent is gently compacted, and a 1-cm layer of anhydrous sodium sulfate is placed above the adsorbent. The extract is poured into the column and developed with 50 ml acetone–hexane (1:9), or more if necessary, to wash the visible carotenes through the adsorbent. The top of the column should be covered with a layer of solvent throughout the operation. The carotenes pass rapidly through the column, and the bands of xanthophylls, carotene oxidation products, and chlorophylls should remain on the column when the development is complete. The eluate is transferred to a 100-ml volumetric flask and made up to the mark with acetone–hexane, 1:9. The carotene content is determined photometrically. When determinations are made with a properly calibrated spectrophotometer at 436 mμ, the concentration of β-carotene is given by the following formula,

$$C = \frac{\text{absorbance} \times 454}{196 \times L \times W}$$

where, C = concentration of carotene (mg/lb) in original sample; L = cell length in cm; W = gram sample/ml final dilution.

The results are reported as mg β-carotene/lb or multiplied by 2.2 to give ppm.

References

1. Zechmeister, L. (1958). *Chem. Revs.* **34**, 267.
2. Duranthon, J., Galmiche, J. M., and Roux, E. (1958). C.R. *Acad. Sci. Paris* **246**, 992.
3. Laborie, M. E. (1963). *Ann Physiol. Veg.* **5**, 89.
4. Angapindu, A., Silberman, H., Tantivatana, P., and Kaplan, I. R. (1958). *Arch. Biochem. Biophys.* **75**, 56.
5. Moore, L. A. (1940). *Ind. Eng. Chem., Anal. Ed.* **12**, 726.
6. Anderson, A. F.H., and Calvin, M. (1962). *Nature* (*London*) **194**, 285.
7. Bickoff, E. M., Atkins, M. E., Bailey, G. F., and Stitt, F. (1949). *J. Am. Offic. Agric. Chemists* **32**, 766.
8. Perkins, H. J., and Roberts, D. W. A. (1962). *Biochim. Biophys. Acta* **58**, 486.
9. Maclachlan, S., and Zalik, S. (1963). *Can. J. Bot.* **41**, 1053.
10. Bruinsma, J. (1961). *Biochim. Biophys. Acta* **52**, 576.
11. *Assoc. Offic. Agric. Chemists*, (1960). p. 654. Washington 4, D.C.

Recommended Reviews

1. Holden, M. (1965). *Chlorophylls*. In "Chemistry and Biochemistry of Plant Pigments," (T. W. Goodwin, ed.), pp. 462. Academic Press, New York.
2. Davies, B. H. (1965). *Analysis of carotenoid pigments*. In Chemistry and Biochemistry of Plant Pigments. (T. W. Goodwin, ed.), p. 489. Academic Press, New York.

G. Questions

1. Which are the main natural sources of carotenoids?
2. How can one prove the structure of a carotenoid?
3. Which methods may be used for trans → cis isomerization of carotenoids?
4. Why should considerable amounts of yellow vegetables and green leafy plants be included in the diet?
5. How are zeaxanthin, lutein, and capsanthin related to α- and β-carotenes?
6. What is the biogenetic relationship between carotenoids and terpenoids? Outline the biosynthesis of lycopene.

H. Recommended Books

1. Bauernfeind, J. C., (1981). *"Carotenoids as Colorants and Vitamin A. Precursors."* Academic Press, New York.
2. Goodwin, T. W. (1954). *"Carotenoids."* Chemical Publishing Co., New York.

3. Goodwin, T. W. (ed.) (1965). *"Chemistry and Biochemistry of Plant Pigments."* Academic Press, New York.
4. Karrer, P. and Jucker, E. (1950). *"Carotenoids,"* Elsevier, New York.
5. Krinsky, N. I. (1990). *"Carotenoids, Chemistry and Biology,"* Plenum, New York.
6. Liaaen-Jensen, S. (1980). Stereochemistry of natural carotenoids. *Prog. Chem. Org. Nat. Prod.* **39**, 123.
7. Rando, R. B. (1990). The chemistry of vitamin A and vision. *Angew. Chem. Int. Ed.* **19**, 461.

II. STEROIDS

A. Introduction

Many naturally occurring substances, such as sterols, bile acids, sex hormones, adrenal cortical hormones, cardiac glycosides, toad poisons and sapogenins, contain the cyclopentanoperhydrophenanthrene ring system or, in very rare cases, a modification of it.

The sterols are widely distributed in nature. They are found both in the free state and combined as esters or glycosides. Very often they occur in the form of complex mixtures, the components of which are difficult to separate. Paper, column, thin-layer and vapor-phase chromatography make possible the qualitative and quantitative analysis of sterols and steroids. Other methods are ultraviolet and infrared spectra, rotatory dispersion, nuclear magnetic resonance, X-ray crystallographic analysis, and mass spectroscopy.

Most of the steroids are hydroxylated at C_3, while some also have OH groupings in other positions of the ring system or the side chain. In cholesterol and cholestanol, the —OH group on C_3 is on the same side as the —CH_3 group on C_{10}, i.e., both project toward the front of the molecule (β-orientation). When the OH of C_3 lies on the side of the ring opposite to that of the —CH_3 at C_{10}, the configuration at C_3 is α. The stereochemistry of the side chain, the angular methyl groups, and the B/C and C/D trans ring fusions of bile acids is the same as that of cholestanol. The only difference is at C_5, since the bile acids are A/B–cis (5β) compounds, while the sterols are A/B–trans (5α) compounds.

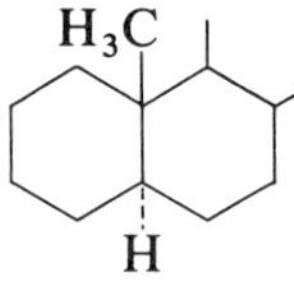

5α 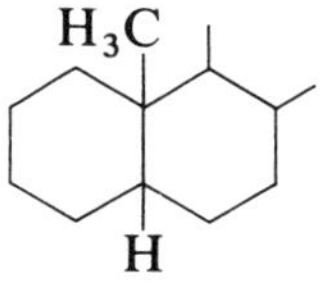   5β

Cholestanol contains three fused cyclohexane rings, A, B, and C, and a terminal cyclopentane ring. Rings B and C are locked in the chair conformation by transfusion to rings A and D. In coprostanol, rings B, C, and D are the same as in cholestanol, but ring A is cis-fused and has a chair conformation. The conformational representations of cholestanol and coprostanol are presented in the following figures:

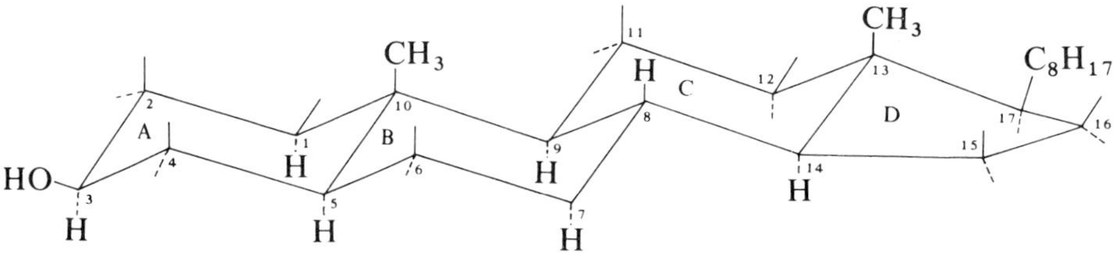

Cholestanol

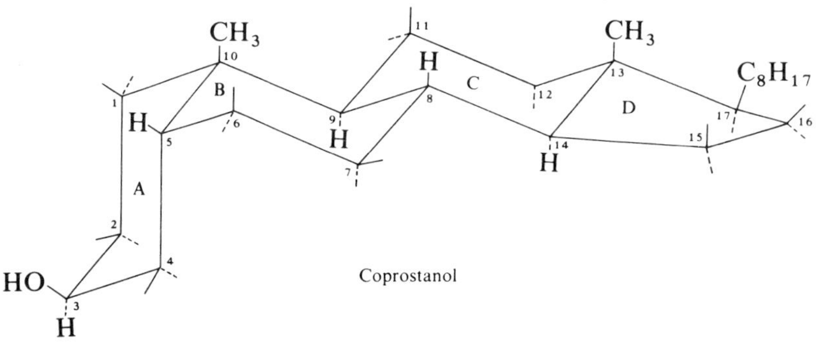

Coprostanol

A number of general rules governing the stability and reactivity of equatorial and axial positions have been elaborated, mainly by Barton.

1. Sterols

Cholesterol is one of the most abundant compounds. It is present in vertebrates and invertebrates and was recently found in plants. Cholesterol is often accompanied by closely related sterols such as cholestan-3-β-ol, coprostan-β-ol, and others.

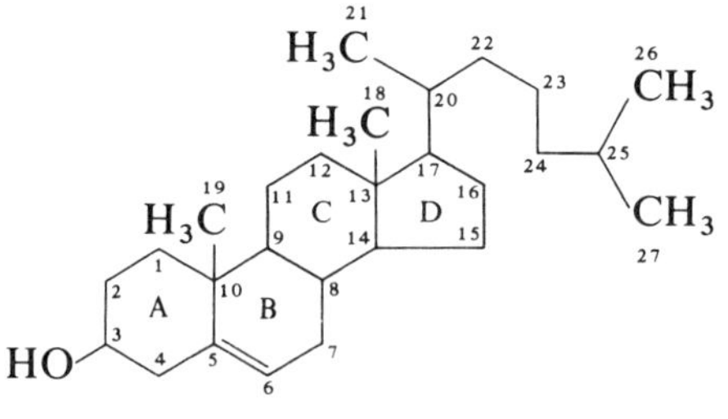

Cholesterol

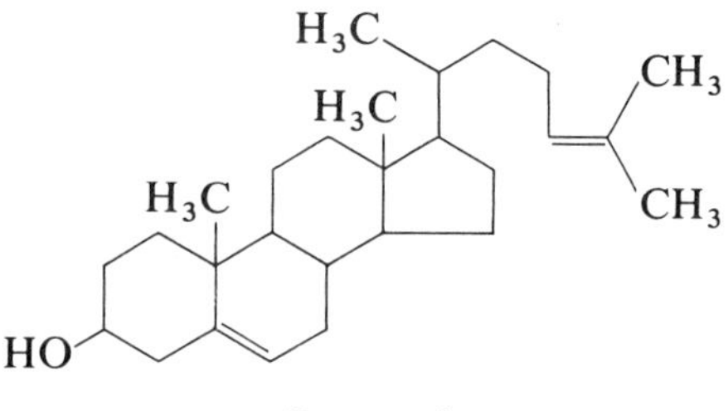

Desmosterol

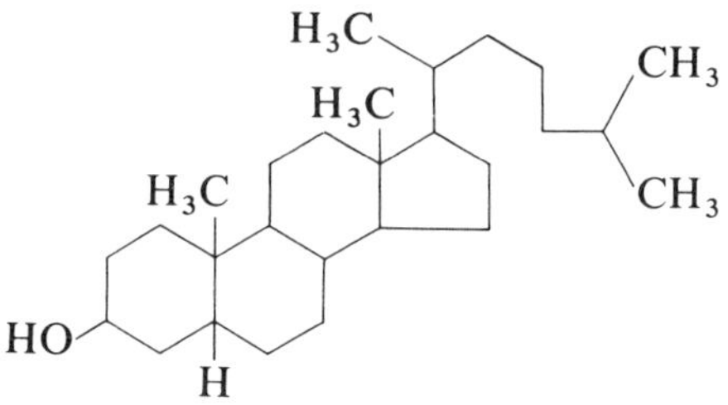

Coprostanol

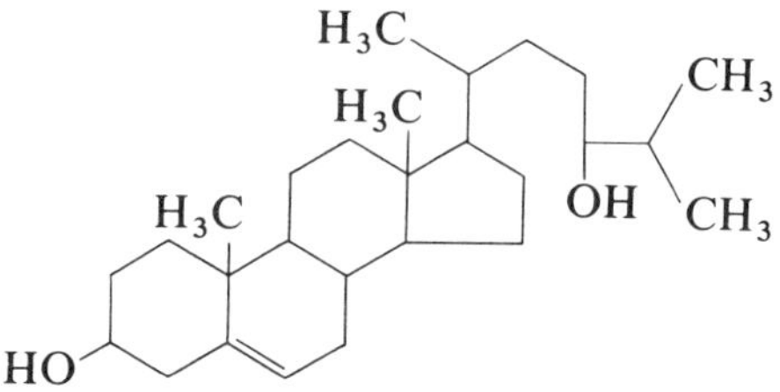

Cerebrosterol

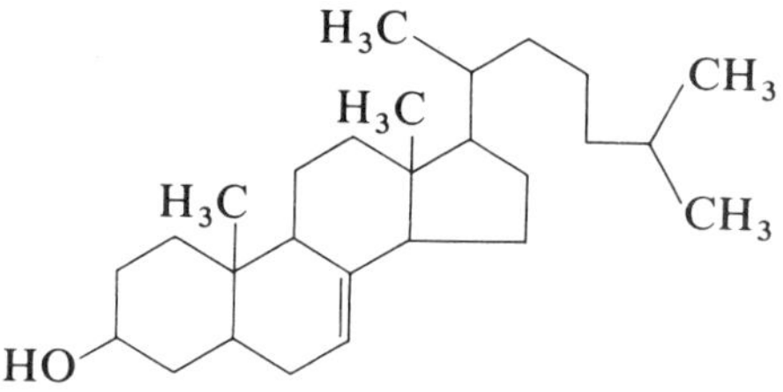

Lathosterol

The main steps in the biosynthesis of cholesterol are the conversion of acetate to squalene, the concerted cyclization of squalene to lanosterol, and the demethylation of lanosterol to cholesterol.

a. Sterols of marine invertebrates

The sterols in the lower forms of animal and plant life are of interest, as they suggest a relationship between the stage of evolution and the type of sterol formed. In many sterols of invertebrates, the eight-carbon side chain of cholesterol is replaced by a nine-carbon unit with a methyl branch at C_{24} or a 10-carbon unit with an ethyl group.

The typical marine sterols are

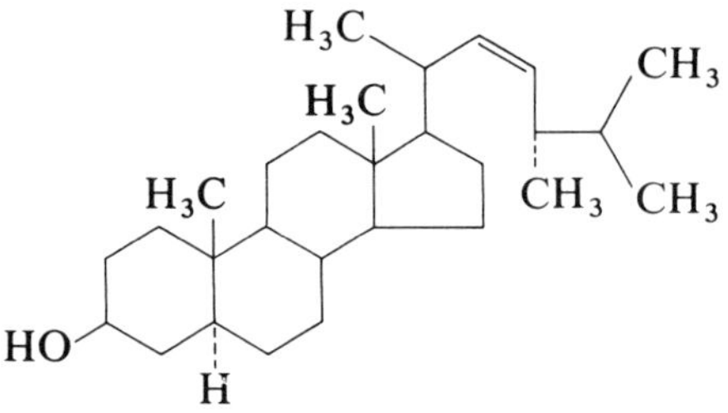

Spongesterol

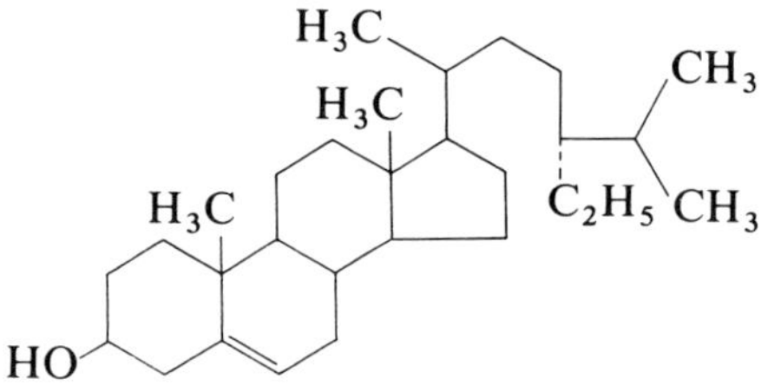

Clionasterol

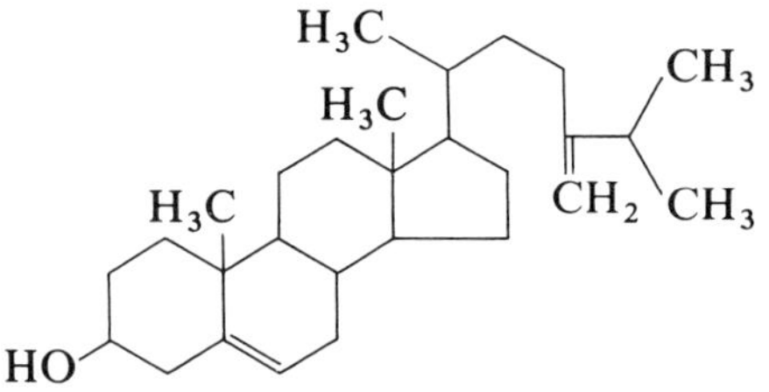

24-Methylenecholesterol

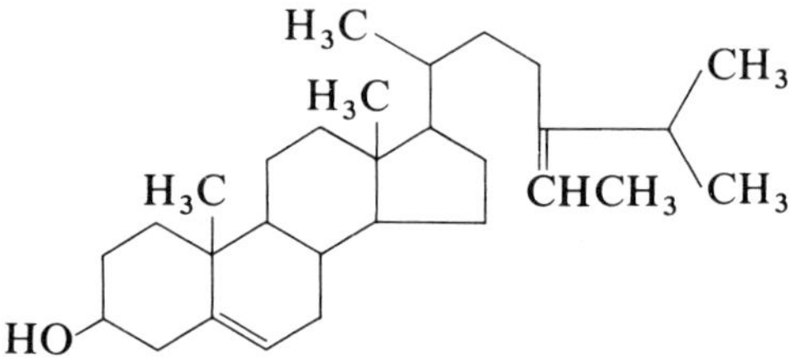

Fucosterol

b. Plant sterols

Sitosterol and stigmasterol are the most abundant plant sterols and occur in complex mixtures; they also have C_9 or C_{10} side chains.

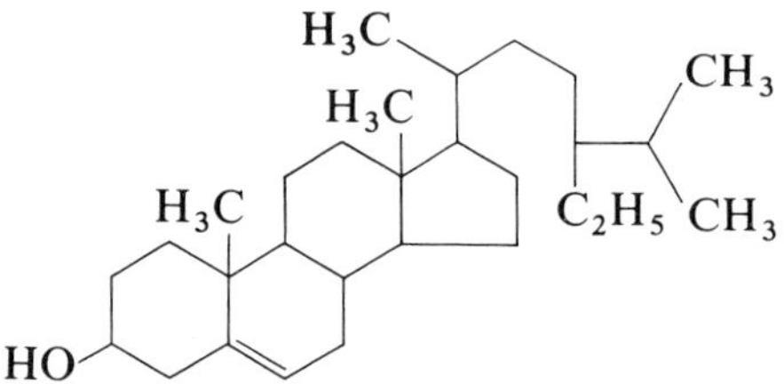

β-Sitosterol

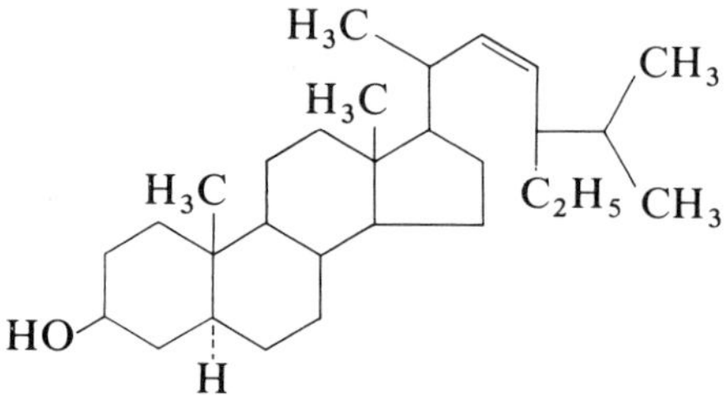

α-Spinasterol

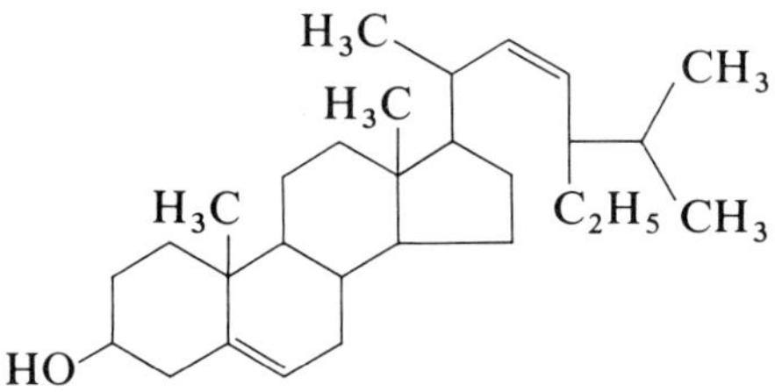

Stigmasterol

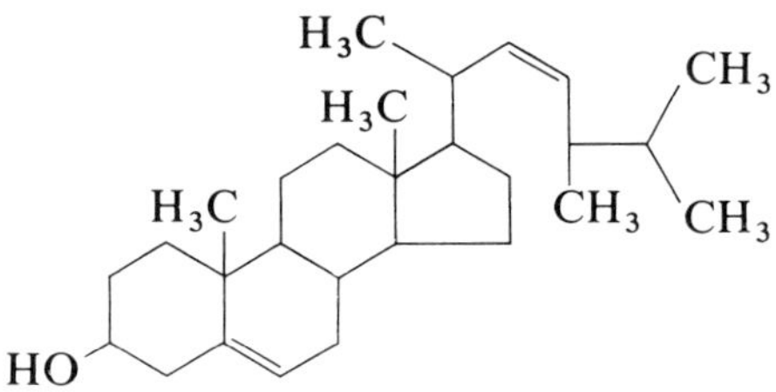

Brassicasterol

II. Steroids

c. Yeast sterols

The most abundant sterols of this type are ergosterol and zymosterol.

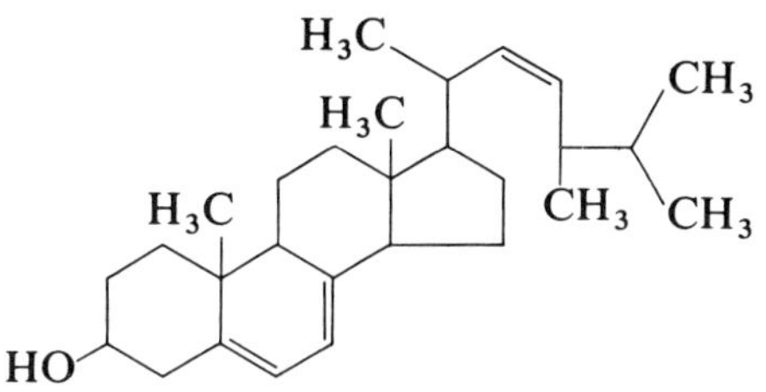

Ergosterol

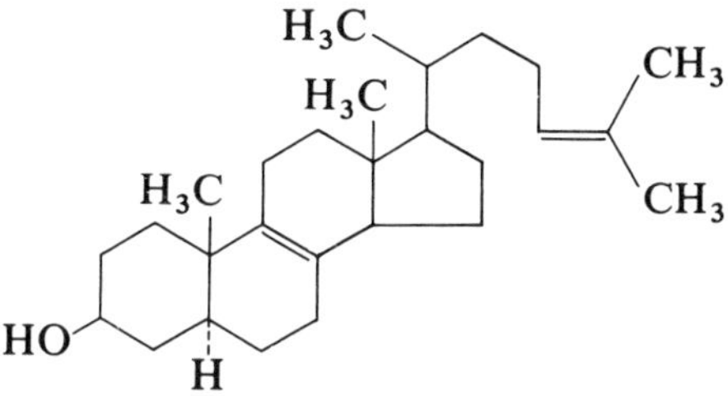

Zymosterol

Ascosterol and fecosterol rank among the minor yeast sterols.

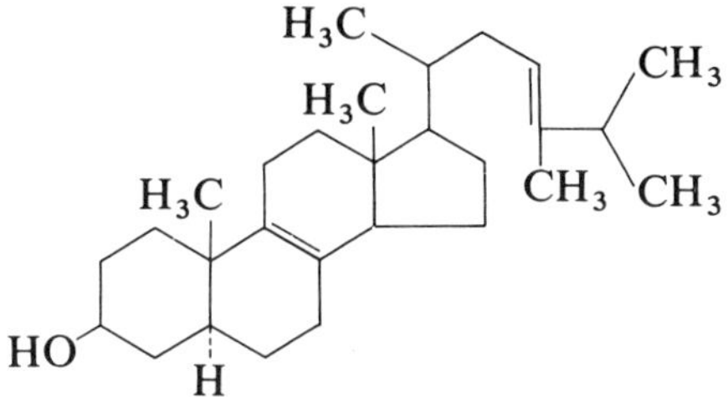

Ascosterol

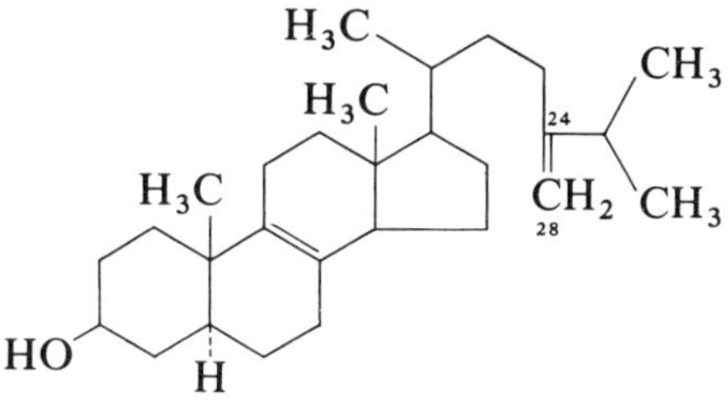

Fecosterol

2. Bile Acids

Bile acids are isolated from the bile of higher animals, where they are found as sodium salts of peptidic conjugates with taurine and glycine. In all these acids the hydroxyl groups are α-oriented, and the juncture of rings A and B is cis. The side chain is usually made up of five carbon atoms and bears the carboxyl group. The bile acids resemble cholestanol in the orientation of the side chain and of the angular methyl groups in the B/C and C/D ring fusion. There are two main classes: C_{27-28} acids and C_{24} acids. The C_{27-28} acids have

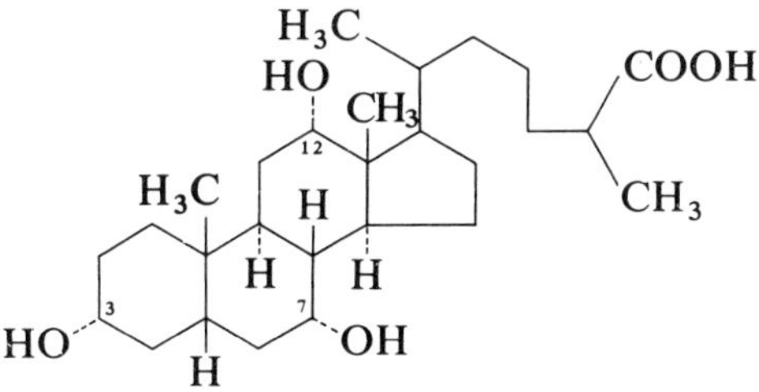

3α,7α,12α -Trihydroxycoprostanic acid

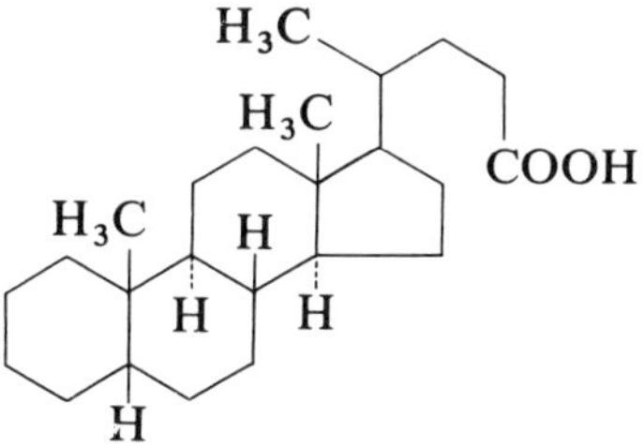

Cholanic acid

been found in the bile of reptiles and amphibia. The best known are the 25α- and β-isomers of 3α, 7α, 12α-trihydroxycoprostanic acid. Turtle and tortise bile contains tetrahydroxysterocholanic and tetrahydroxyisosterocholanic acids, probably $C_{27}H_{46}O_6$.

Numerous naturally occurring, substituted cholanic acids are known, such as:

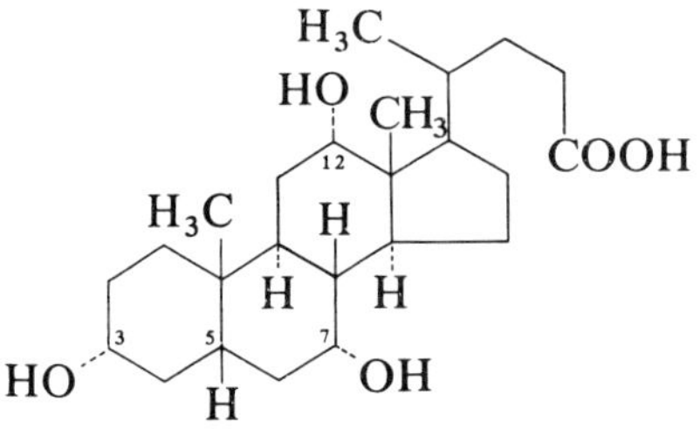

Cholic acid

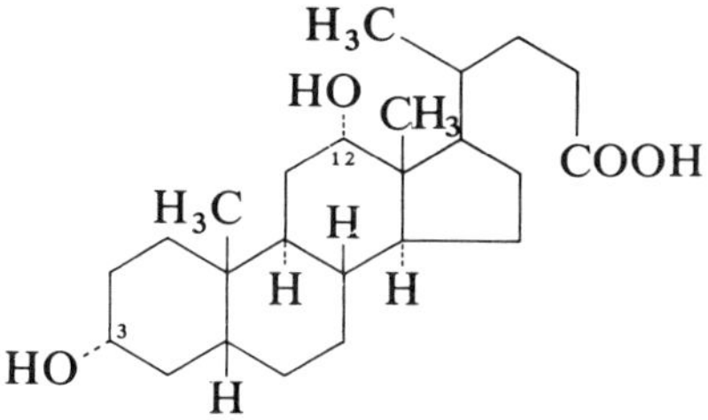

Desoxycholic acid

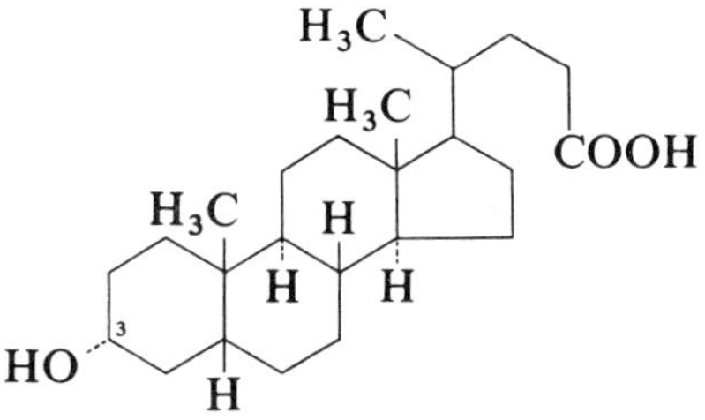

Lithocholic acid

The salts of the bile acids lower the surface tension of water and are good emulsifying agents. It has been shown that bile acids are made in the liver mainly from cholesterol.

3. Steroid Hormones

Hormones are substances secreted by specific glands, and exert control over various body processes. The steroid sex hormones are divided as follows: female hormones, such as the estrogens (estradiol, estrone, and estriol); progestational hormones, such as progestrone; and androgens—testosterone and androsterone—that have male hormone activity.

The naturally occurring estrogens differ from other steroid hormones in the presence of an aromatic ring A, the absence of a methyl group at C_{10}, and the presence of a keto- or hydroxy- group at position 17 and sometimes at position 16.

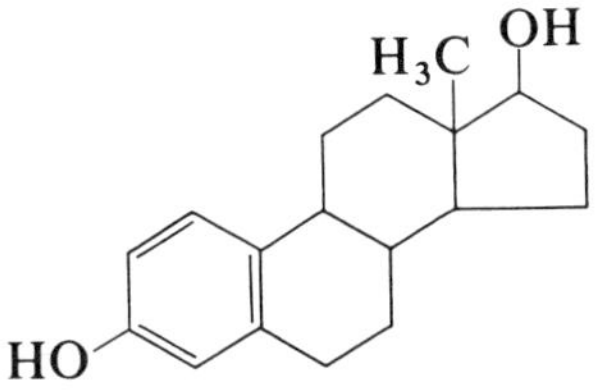

II. Steroids

Estradiol

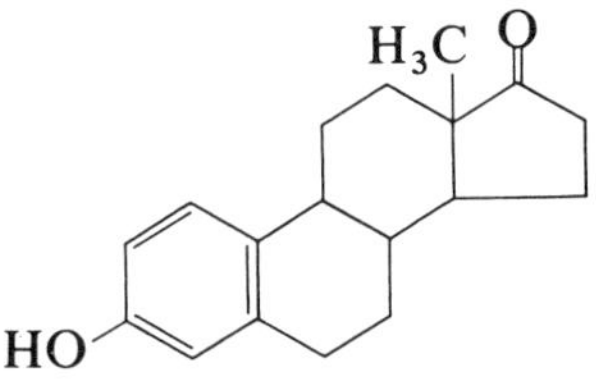

Estrone

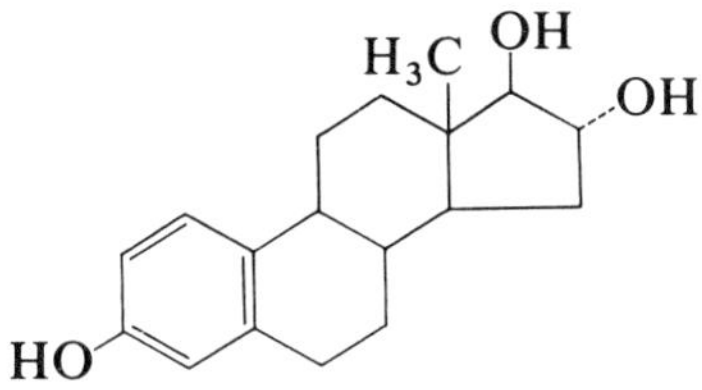

Estriol

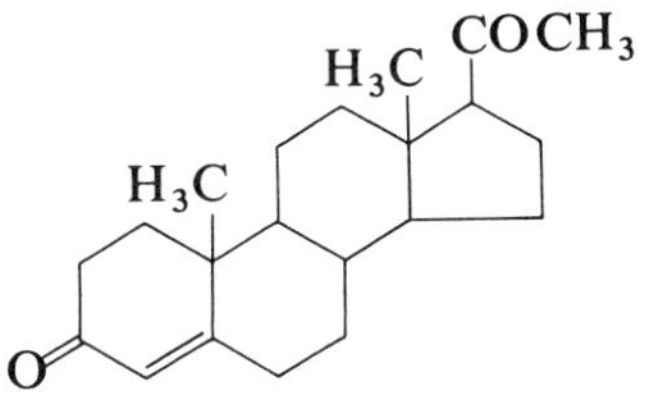

Progesterone

The adrenocortical hormones have 21 carbon atoms, ketonic groups at C_3 and C_{20}, a double bond at the 4(5)-position, and a hydroxyl group C_{21}. These compounds may be divided into the highly active glucocorticoids, e.g., cortisone and hydrocortisone, and highly active mineralocorticoids, e.g., desoxycorticosterone and aldosterone.

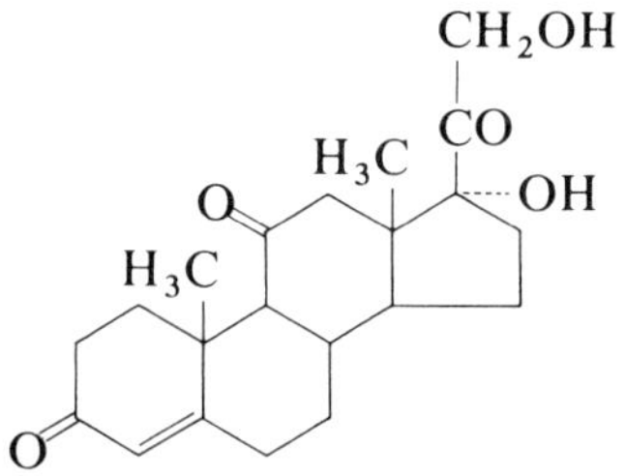

Cortisone

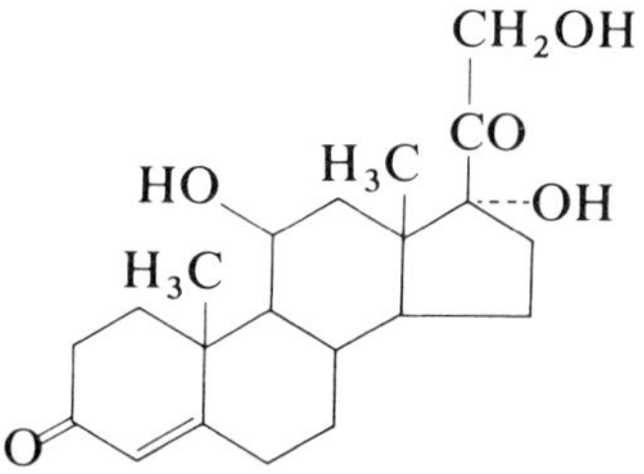

Cortisol

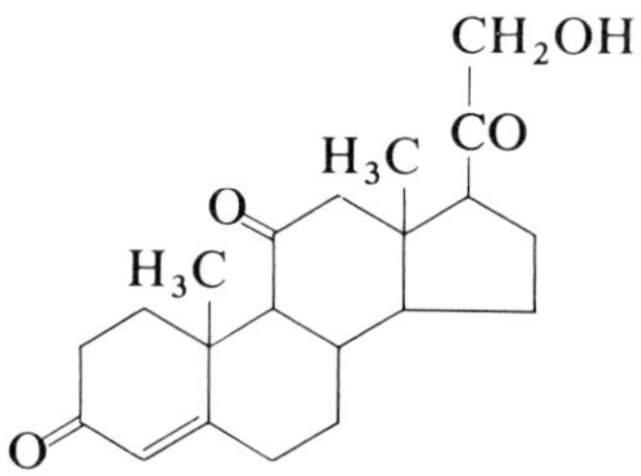

11-Dehydrocorticosterone

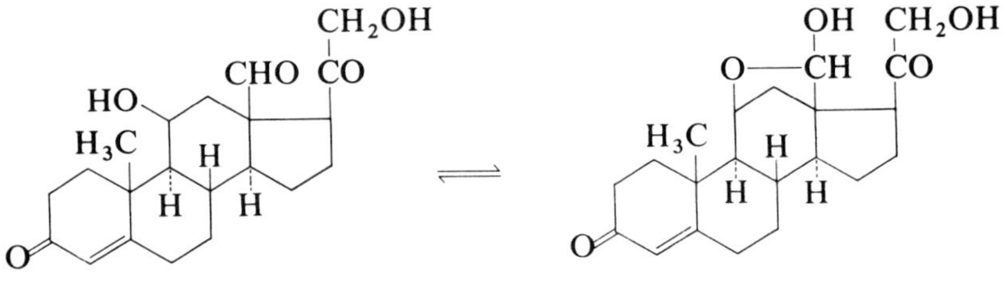

[Aldosterone structure]

Aldosterone

The naturally occurring androgens such as androsterone have 19 carbon atoms. The principal androgenic steroid is the male sex hormone testosterone.

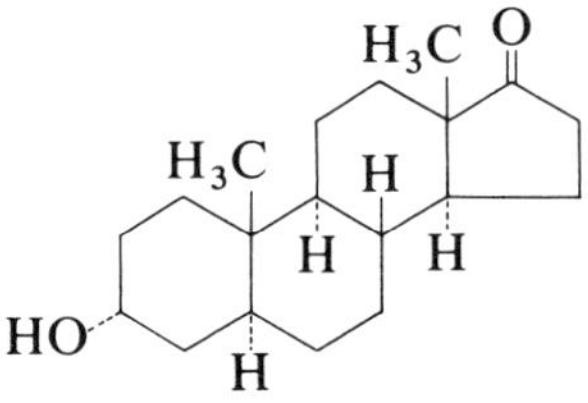

Androsterone

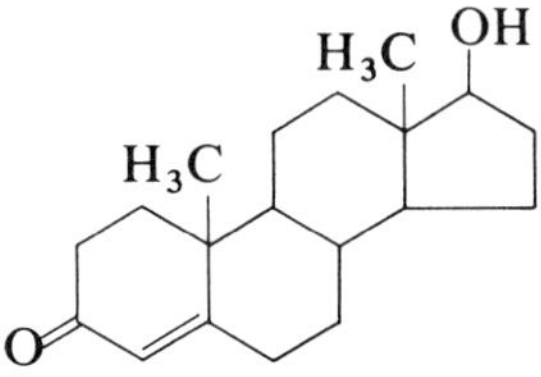

Testosterone

Recently, a series of steroidal insect-molting hormones, such as ecdysone, was discovered.

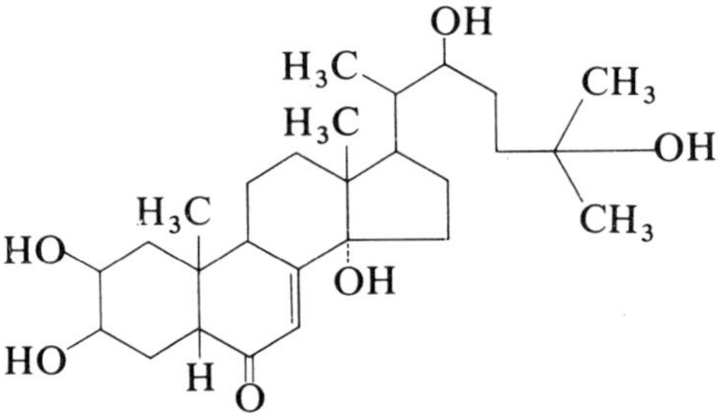

Ecdysone

4. Cardiac-Active Principles

Certain steroids exert a specific and powerful effect on the cardiac muscle and are called cardiac-active (cardiotonic) principles. Many cardiac glycosides have been isolated from plants in tropical regions and have been employed by natives of Africa and South America for the preparation of arrow poisons. Some occur in the secretions of poisonous toads. The aglycons are of two types: cardenolides and bufadienolides. The cardenolides are C_{23} steroids having as side chain an α, β-unsaturated, five-membered lactone ring and a C$_{14}$ hydroxyl group. They may be exemplified by the following:

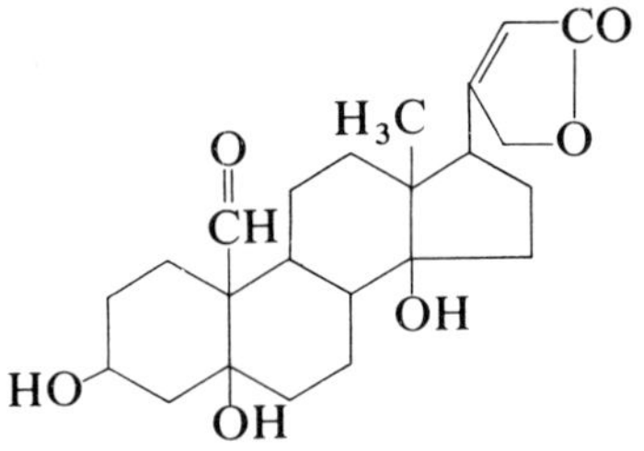

Strophanthidin

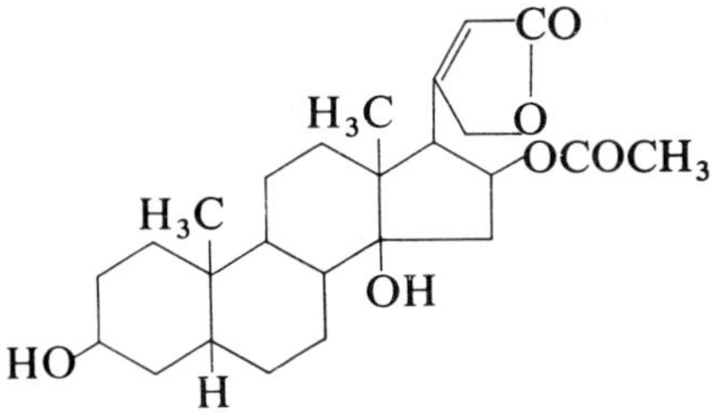

Oleandrigenin

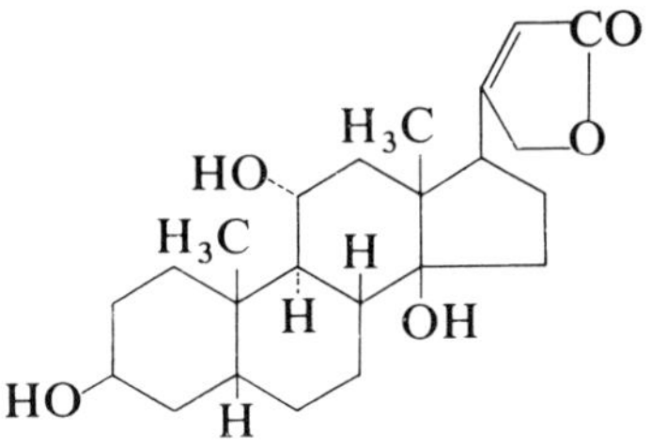

Sarmentogenin

The bufadienolides are C$_{24}$ steroids having as side chain a doubly unsaturated, six-membered lactone ring and a 14β-hydroxyl group or its modification.

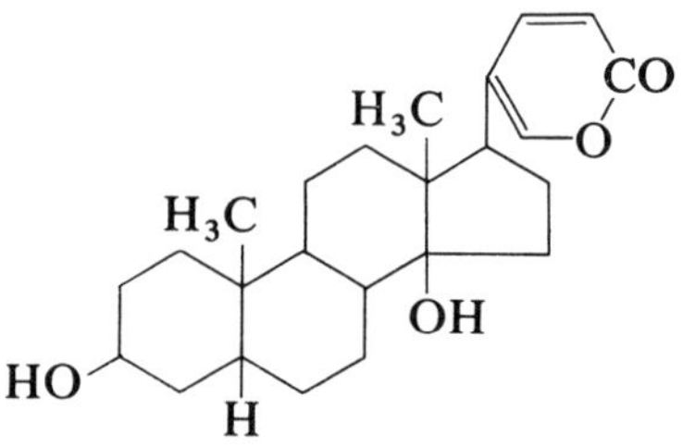

Bufalin

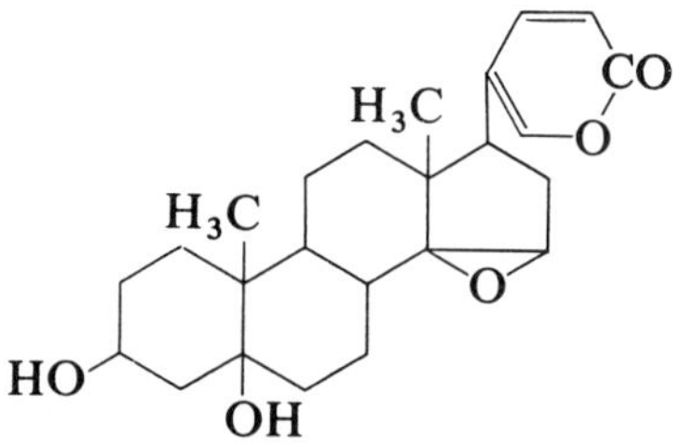

Marinobufagin

5. Sapogenins

Plant glycosides that have the property of forming a soapy lather in water are called saponins. One of them, digitonin, is used for selective precipitation of 3β-hydroxysteroids. The following are examples of sapogenins, the sugar-free moieties of the saponins:

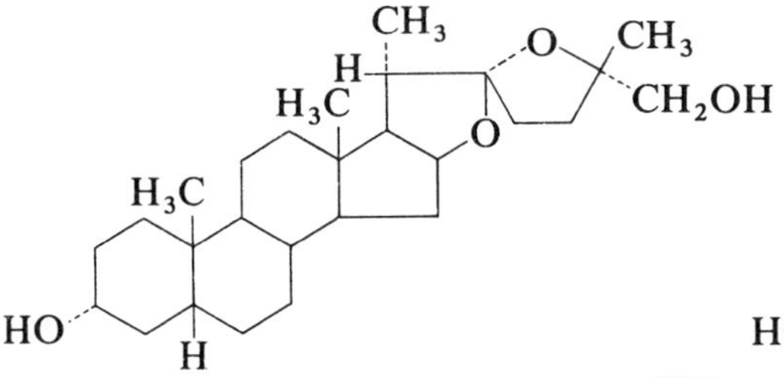

Cholegenin

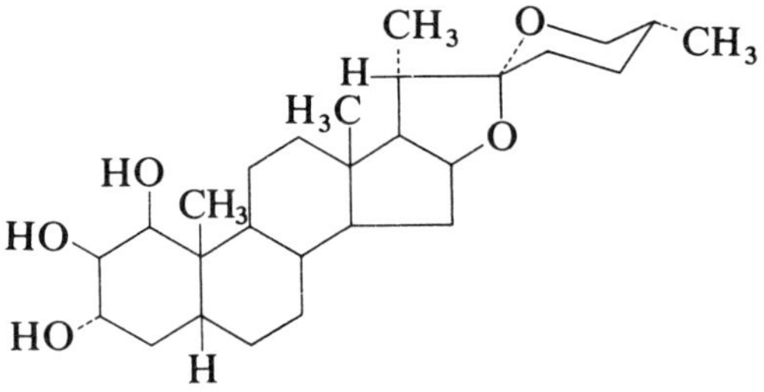

Tokorogenin

B. Isolation of Stigmasterol from Soybean Oil

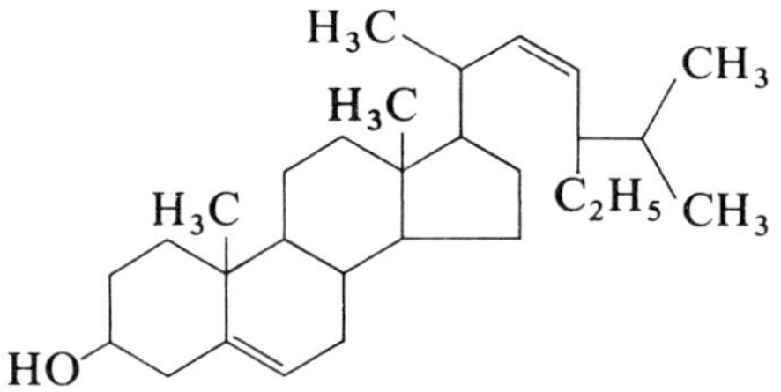

Stigmasterol

1. Introduction

Stigmasterol was first isolated by Windaus and Hauth (1) from the seeds of *Physostigma venenosum*, the Calabar bean. Other sources are rice bran fat (2), soybeans (3), coffee oil (4), ergot oil (5), tea oil (6), flue-cured tobacco leaves (7), dried potatoes (8), coconut oil (9), and many other plants.

Recent investigation of the steroid constituents of soybean oil (11) by gas-liquid chromatography has revealed, in addition to stigmasterol, the presence of campesterol and β-sitosterol.

2. Principle

The following procedure (10) consists of saponification of crude soybean oil. Ether is used to extract the unsaponifiable materials. The ether is then distilled off and replaced by petroleum ether saturated with steam, whereupon the crude sterol fraction is precipitated. It is then acetylated and brominated. The resulting stigmasteryl acetate tetrabromide is sparingly soluble in acetic acid, acetone, and ether, and can be separated from the far more soluble sitosteryl acetate dibromide. The bromine is then removed by treatment with zinc dust, and the acetate is hydrolyzed with alcoholic potassium hydroxide. Stigmasterol is obtained by crystallization from 95% ethanol solution.

3. Apparatus

   Steam Bath

4. Materials

Acetic acid	Ethanol, absolute	Sodium carbonate
Acetic anhydride	Ether	Sodium sulfite
Bromine	Methanol	Soybean oil
Chloroform	Petroleum ether	Zinc dust
Ethanol, 95%	Potassium hyroxide	

5. Time

   9–10 hours

6. Procedure

a. Isolation

To 165 g crude soybean oil are added 800 ml absolute ethanol and 150 g potassium hydroxide dissolved in a minimal volume of water. This mixture is placed in a 2-liter round-bottomed flask fitted with a condenser and refluxed for 2 hr on a steam bath. After saponification, 800 ml water is added, and the solution is extracted with ethyl ether (5 × 300 ml). The extraction of the unsaponifiable materials from the soaps is accomplished by stirring the mixture for an hour with an air stirrer, allowing the ether layer to separate, and siphoning off the ether. The extracts are concentrated to about half their original volume and washed free of alkali with distilled water. The remainder of the solvent from the ether extraction is then distilled under reduced pressure at about 40°C. The residue—the unsaponifiable material—is dissolved in 100 ml petroleum ether (bp 40–60°C), and steam is passed into the solution until the saturation point is nearly reached. The solution is allowed to stand overnight. A precipitate consisting of about 1 g colorless crystals of mp 138–144°C is obtained.

b. Acetylation

The crude material is acetylated with 20 ml acetic anhydride by refluxing for $1\frac{1}{2}$ hr. The mixture is then cooled for 1 hr at 20°C, and the crude acetates are filtered.

c. Bromination

One g of the acetates is dissolved in 10 ml ethyl ether, and to this solution is added 12 ml bromine-acetic acid solution (5 g bromine in 100 ml glacial acetic acid). After cooling, the insoluble tetrabromides are filtered off and washed with cold ethyl ether. About 300 mg of mp 190–194°C is obtained. The bromides are recrystallized from chloroform–methanol; mp of crystals 194–196°C.

d. Debromination

To a solution of 300 mg recrystallized bromides in 4 ml glacial acetic acid is added 300 mg zinc dust. The mixture is refluxed for $1\frac{1}{2}$ hr, filtered hot, diluted with about 10 ml water, and extracted with ethyl ether. The ethereal solution is washed with dilute aqueous sodium sulfite solution, then with water, and the ether is removed by evaporation. The product obtained weighs about 150 mg. It is recrystallized once from ethanol and then from methanol–chloroform, 2:1; mp 139–140°C.

e. Stigmasterol

Stigmasteryl acetate (100 mg) is hydrolyzed for 1 hr with about 10 ml 10% alcoholic potassium hydroxide. About 10 ml water is added, and the mixture is extracted with ethyl ether. The ethereal solution is washed with dilute aqueous sodium carbonate and then with water. After removal of the ether, the residue is recrystallized three times from 95% ethanol. About 30 mg of the product is obtained; mp of crystals 168–169°C.

f. Stigmasterol

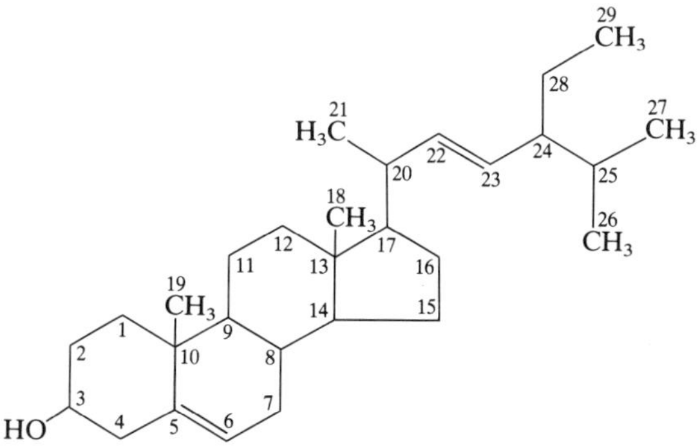

g. C^{13} NMR [CDCl$_3$]

C-1	37.31	C-4	42.35	C-7	31.94
C-2	31.69	C-5	140.80	C-8	31.94
C-3	71.81	C-6	121.69	C-9	50.20

C-10	36.56	C-17	56.06	C-24	51.29
C-11	21.11	C-18	12.07	C-25	31.94
C-12	39.74	C-19	19.42	C-26	21.26
C-13	42.35	C-20	40.54	C-27	19.02
C-14	56.91	C-21	21.11	C-28	25.44
C-15	24.39	C-22	138.37	C-29	12.27
C-16	28.96	C-23	129.32		

h. Mass spectrum

Stigmasterol $C_{29}H_{48}O$, Mol wt. 412.

m/z 412(29%)   83(65%)   55(100%)
     271(75%)   81(75%)   43(64%)
     93(50%)   69(58%)   41(58%)

i. UV spectrum of stigmasterol

λ_{max}^{EtOH} 204 mμ (logε 3.58)

Stigmasterol is a diene with two isolated double bonds. Thus, absorption in the far UV region cannot be measured with the normal spectrophotometer, but only *in vacuo*.

References

1. Windaus, A. and Hauth, A. (1906). *Ber.* **39**, 4378.
2. Nebenhauer, F. P., and Anderson, P. J. (1926). *J. Am. Chem. Soc.* **48**, 2972.
3. Thornton, M. H., Kraybill, H. R., and Mitchell, J. H. (1940). *J. Am. Chem. Soc.* **62**, 2006.
4. Bauer, K. H., and Neu, R. (1944). *Fette und Seifen* **51**, 343.
5. Ruppol, E. (1949). *J. Pharm. Belg.* **4**, 55.
6. Goguadze, V. P. (1950). *Izvest. Akad. Nauk SSSR, Otdel Khim. Nauk* 185.
7. Grossman, J. D., and Stedman, R. L. (1958). *Tobacco* **147**, (11), 22.
8. Schwartz, J. J., and Wall, M. E. (1955). *J. Am. Chem. Soc.* **77**, 5442.
9. Andersen, B, and Krawack, B. (1957). *Acta Chem. Scand.* **11**, 997.
10. Byerrum, R. U., and Ball, C. D. (1960). *Biochem. Prepns.* **7**, 86.
11. Thompson, M. J., Robbins, W. E., and Baker, G. L. (1963). *Steroids* **2**, 505.

Recommended Review

1. Daubert, B. F. (1950). *Chemical composition of soybean oil*. In "*Soybeans and Soybean Products*" (K. S. Markley, ed.), Vol. 1, p. 157. Interscience, New York.

C. Degradation of Stigmasterol

1. Introduction

Windaus, Werber, and Gschaider (1) have established the empirical formula of stigmasterol as $C_{29}H_{48}O$. Guiteras (2) isolated ethylisopropylacetaldehyde as a product of ozonization of stigmasterol, and established the presence of an ethyl group at C_{24} and a double bond at $C_{22}=C_{23}$. The position of the hydroxyl group was established by Fernholz and Chakravorty (3) by chromic acid oxidation of stigmastanyl acetate and isolation of 3β-acetoxy-bisnorallocholanic acid.

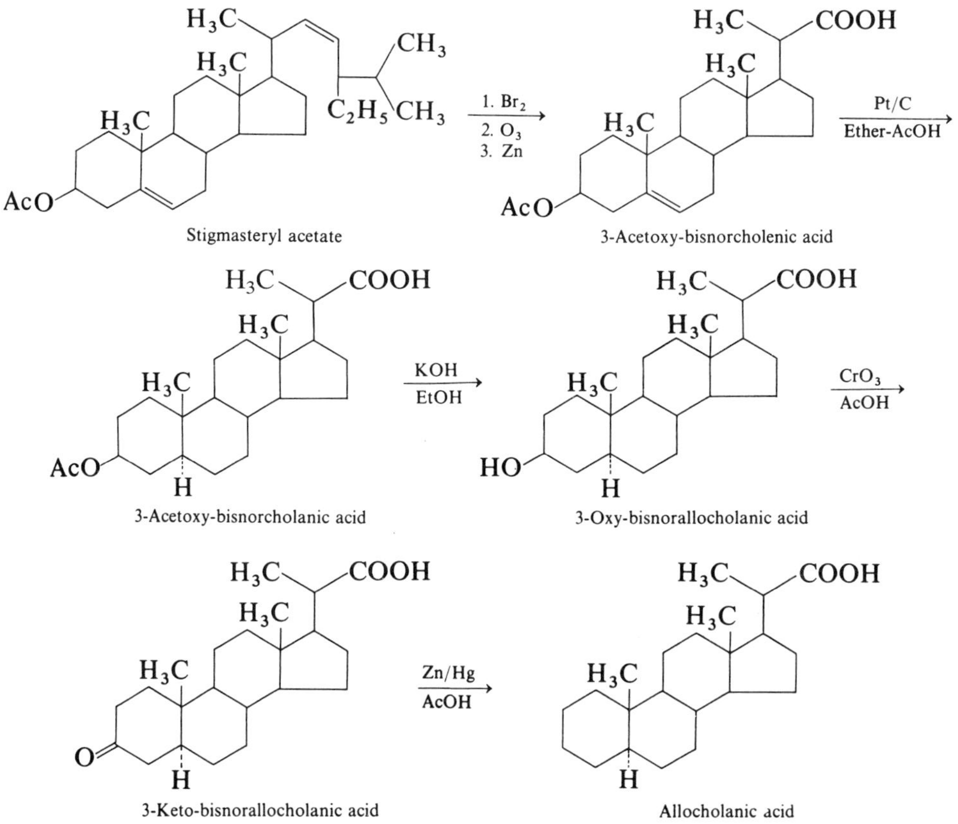

2. Principle

In the following degradation (4), stigmasteryl acetate is brominated to form the 5,6-dibromide. Subsequent ozonization and debromination with zinc in acetic acid yields 3β-acetoxy-Δ^5-bisnorcholenic acid, which is converted by catalytic hydrogenation, alkaline hydrolysis, oxidation with chromic acid, and Clemmensen reduction to bisnorallocholanic acid.

3. Apparatus

Catalytic hydrogenation apparatus
Ozonator

4. Materials

Acetic acid
Acetic anhydride
Acetone
Bromine
Chloroform
Chromium trioxide
Dioxane
Ethanol
Ether
Hydrochloric acid
Methanol
Ozone
Potassium hydroxide
Platinum on charcoal (5%)
Pyridine
Sodium hydroxide
Stigmasterol
Sulfuric acid
Xylene
Zinc, amalgamated
Zinc, dust

5. Time

15 hours

6. Procedure

a. Stigmasteryl acetate

Stigmasterol (5 g), 20 ml acetic anhydride, and a few drops of pyridine are refluxed for 1 hr. After cooling, methanol is added to hydrolyze excess acetic anhydride, and the solvents are removed under reduced pressure. Stigmasteryl acetate is filtered off and crystallized from ethanol as flakes of mp 141°C.

b. 3-Acetoxy-bisnorcholenic acid

To a cold solution of stigmasteryl acetate in 100 ml chloroform is added dropwise a solution of 2.1 g bromine in 50 ml chloroform. After the addition is complete, ozone is bubbled through the solution for 2 to 3 hr. Chloroform is removed under reduced pressure at room temperature, 6 g zinc powder and 100 ml acetic acid are added to the residue, and the mixture is heated on a steam bath until solution is complete. The solution is then left on a steam bath for a further period of 1 hr, diluted with water, and extracted with ether. The ethereal solution is extracted with $2N$ aqueous sodium hydroxide, whereupon the sodium salt of the acid separates at the interface. The salt is acidified with dilute sulfuric acid and extracted with ether. The ethereal solution is then treated with activated charcoal and filtered, and the ether is distilled off. The Liebermann-Burchard test gives a yellow color.

c. 3-Acetoxy-bisnorcholanic acid

3-Acetoxy-bisnorcholenic acid (0.8 g) is dissolved in a mixture of ether and glacial acetic acid and hydrogenated in the presence of platinum on charcoal. The catalyst is filtered off and the ether is distilled. Addition of water to the residue precipitates the acid. Recrystallization from dilute acetic acid or from acetone gives needles of mp 194°C; yield, 0.5 g.

d. 3-Oxy-bisnorallocholanic acid

3-Acetoxy-bisnorcholanic acid (0.52 g) is refluxed for 2 hr with 5% ethanolic potassium hydroxide, diluted with water, and acidified with dilute hydrochloric acid. The acid is filtered off and recrystallized from acetic acid as flakes of mp 270°C; yield, 0.45 g.

e. 3-Keto-bisnorallocholanic acid

3-Oxy-bisnorallocholanic acid (0.5 g) is dispersed in 20 ml glacial acetic acid, 0.15 g chromium trioxide is added, and the mixture is left in a refrigerator overnight. The keto acid is precipitated by addition of water, filtered through a Buchner funnel, and recrystallized from aqueous dioxane as needles, mp 244–246°C; yield, 0.37 g.

f. Allocholanic acid

3-Keto-bisnorallocholanic acid (0.35 g), 20 g amalgamated zinc, and 50 ml acetic acid are refluxed for 8 hrs. During this period, three 100-ml portions of concentrated hydrochloric acid and a few ml xylene are added. The cold solution is diluted with water and extracted with a mixture of ether and petroleum ether. This solution is shaken thoroughly with $2N$ aqueous sodium hydroxide, whereupon the sodium salt is precipitated at the interface. It is filtered off and acidified with dilute sulfuric acid. Recrystallization from acetic acid yields plates of mp 214°C.

References

1. Windaus, A., von Werber, F., and Gschaider, B. (1932). *Ber.* **65**, 1006.
2. Guiteras, A. (1933). *Z. Physiol. Chem.* **214**, 89.
3. Fernholz, E., and Chakravorty, P. N. (1934). *Ber.* **67**, 2021.
4. Fernholz, E. (1933). *Ann.* **507**, 128.

D. Isolation and Identification of Algal Sterols

1. Introduction

Since their isolation from prokaryotic organisms, i.e., bacteria (1,2) and blue-green algae (cyanobacteria) (2,3), sterols have been isolated from all major groups of living organisms. Cholesterol is the primary sterol of all higher animals, lower animals may contain cholesterol and/or a complex mixture of C_{27}-, C_{28}-, and C_{29}-carbon sterols.

In general, sterols of algae differ from those of the higher plants. Most eukaryotic organisms contain sterols that are used as architectural components of the cellular membranes (4), although their roles in reproduction (5), or other developmental or hormonal roles have been demonstrated in a number of cases (4). In the higher plants, these sterols are in most cases 24α-methyl or 24α-ethyl cholesterol and closely related Δ^5-sterols, although recently a few higher plant families have been shown to contain Δ^7-sterols (6,7). Sterols of algae have the 24β-methyl or 24β-ethyl configuration (8).

The great majority of red algae contain only cholesterol and other 27-carbon sterols as principal sterol components. These sterols have no alkyl group and thus no asymmetry at C-24. Brown algae contain fucosterol as their principal sterol. Its 24-ethylidene group eliminates the possibility of α/β isomers at C-24. Other algal taxa have their own characteristic sterol composition; the green algae, as expected, most closely resemble higher plants.

For many years *Chlorella*, a unicellular green alga, has been an important research organism for plant physiologists. In addition, it might be a source of food for man and domestic animals. The nutritive factors, such as carotenoids and sterols, of this (and other) algae have received much attention in recent years.

The first sterol isolated from *Chlorella pyrenoidosa* was ergosterol (9). Sterols with the $\Delta^{5,7}$-double bond system (such as ergosterol) are known to have provitamin D activity (10).

The species *Chlorella ellipsodea* and *Chlorella saccharophila* have been found to contain only Δ^5-sterols (11), while *Chlorella vulgaris* contains only Δ^7-sterols (12).

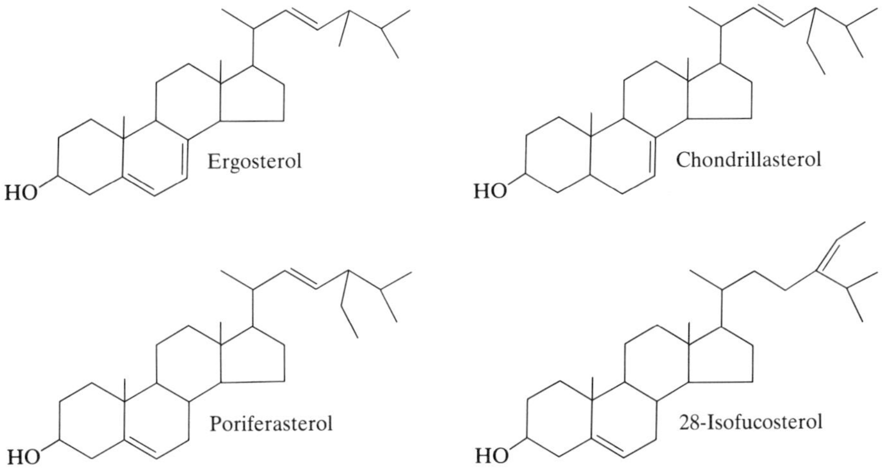

Sterols of green algae.

2. Principle

Cell material of a Chlorella species is freeze-dried and extracted with chloroform–methanol (2:1). The lipid obtained is saponified, and the nonsaponifiable fraction (NSF) is fractionated on a column of silica gel. The sterol fraction is acetylated, and the steryl acetates are identified by chromatographic (TLC, GC) and spectroscopic techniques. Gas chromatography–mass spectrometry (GC-MS) is a powerful tool for confirming the structures of sterols. In certain steryl acetates (such as the Δ^5 ones), the M^+ peak is absent in the mass spectrum, and the highest mass-peak appears at M-60.

The experimental part is based on the work of Ikan and Seckbach (13).

3. Apparatus

Gas chromatograph   Soxhlet extractor
Lyophilizer

4. Materials

Acetic anhydride   Liquid phases: XE-60; DEGS; OF-1
Algae cells   Methanol
Alumina   Silver nitrate
Chloroform   Thin-layer plates

5. Time

8 hours

6. Procedure

a. Harvesting of algae cells

The algal cells (which may be obtained from a plant physiologist) are harvested by centrifugation at 5000 rpm for 10 min, and the pellets are washed in deionized water to remove traces of soluble nutrients.

b. Extraction of lipids

The algae are extracted with chloroform–methanol (2:1) in a Soxhlet apparatus for 3 hr. The extract is filtered and concentrated under reduced pressure, yielding a green viscous residue.

c. Saponification

The green viscous residue is refluxed for 4 hr with 7% methanolic potassium hydroxide. The solvent is removed *in vacuo*, water is added, and the residue is extracted with ether. Evaporation of ether leaves the NSF.

d. Isolation of sterols

The NSF is dissolved in light petroleum and chromatographed on a silica gel column (height 15 cm; diameter 1.5 cm). The compounds are eluted with 40 ml of each of the following concentrations of light petroleum in benzene, 10, 25, 50% (v/v), followed by ether and 25% methanol in ether, and finally with 20% methanol in chloroform. (**Carry out the elution of the column in a well-ventilated hood!**). The fractions (15 ml each) are evaporated to dryness. Examination of each fraction by qualitative TLC (on silica gel GC-254, developing with light petroleum– ethyl ether– acetic acid, 20:4:1, and spraying with 50% H_2SO_4, and charring at 180°C), reveal the presence of sterols at R_f 0.43. The sterol fractions are pooled and subjected to preparative thin-layer chromatography (PTLC). The zones of sterols are scraped off the plates and extracted with chloroform–ethyl ether (1:1).

e. Steryl acetates

To 50 mg of the isolated sterols, 2 ml of dry pyridine and 0.3 ml of acetic anhyride is added, and the mixture is allowed to stand at room temperature overnight (or until the next laboratory work). It is then dissolved in 20 ml ether and washed in succession with two 20-ml portions each of water, $0.5N$ H_2SO_4, $0.5N$ NaOH, and again water. Then it is dried over anhydrous sodium sulfate, filtered, and the ether evaporated.

f. Gas chromatography of steryl acetates

The GC analysis may be carried out in a 6 ft × 0.25 inch glass or stainless steel column. It may be filled with 3% SE-52 on 100 to 120 mesh Gas Chrom P and maintained at 240°C, nitrogen flow, 30 ml/min, on 1% QF-1 on Gas Chrom P at 275°C, or on 1% DEGS on Gas Chrom P at 205°C. Cholestane is injected as internal standard.

g. Gas chromatography–mass spectrometry

A Finnigan 400 quadrupole mass spectrometer interfaced with Finnigan gas chromatograph with a DB-5 column may be used. Temperature programming at 4°C/min to 280°C, followed by 2°C/min to 310°C. Em voltage 1500V; electron energy 70 eV; source temperature 240°C. The compounds are identified by means of mass spectrometric comparison with standards and by coinjection of authentic samples.

References

1. Schubert, K., Rose, G., Wachtel, H., Hoerhold, C., and Ikekawa, N. (1968). *Eur. J. Biochem.* **5**, 246.
2. Reitz, R. C., and Hamilton, J. G. (1968). *Comp. Biochem. Physiol.* **25**, 401.
3. De Souza, N. J., and Nes, W. R., (1968). *Science* 162.

146 CHAPTER 3 Isoprenoids

4. Nes, W. R., and McKean, M. (1977). "Biochemistry of Steroids and Other Isoprenoids." University Park Press, Baltimore, Maryland.
5. Nes, W. R., Patterson, G. W., and Bean, G. A. (1980). *Plant Physiol.* **66**, 1008.
6. Salt, T. A., and Adler, J. H. (1986). *Lipids* **21**, 754.
7. Xu, S., Patterson, G. W., Lusby, W. R., Schmid, K. M., and Salt, T. A. (1990). *Lipids* **25**, 61.
8. Patterson, G. W. (1971). *Lipids* **6**, 120.
9. Klosty, M., and Bergmann, W. (1952). *J. Amer. Chem. Soc.* **74**, 1601.
10. Fieser, L. F., and Fieser, M. (1959). "Steroids." Reinhold, New York.
11. Patterson, G. W., and Krauss, R. W. (1965). *Pl. Cell. Physiol.* **6**, 211.
12. Ikan, R., and Seckbach, J. (1972). *Phytochem.* **11**, 1077.

Recommended Reviews

1. Patterson, G. W. (1971). The distribution of sterols in algae. *Lipids* **6**, 120.
2. Thompson, M. J., Patterson, G. W., Dutky, S. R., Svoboda, J. A., and Kaplanis, J. N. (1980). Techniques for the isolation and identification of steroids in insects and algae. *Lipids* **15**, 719.

E. Isolation of Hecogenin from Agaves and Its Transformation to Tigogenin and Rockogenin

1. Introduction

Hecogenin is a sapogenin possessing a keto group at C_{12}, which can be transposed to C_{11}, as is the case for the 12-keto bile acids. Hecogenin can therefore serve as starting material for cortisone preparations.

Hecogenin was first isolated in 1943 by Marker and his coworkers (1) from several *Agave* species of Texas, Arizona, California, and Mexico. Callow, Cornforth, and Spensley (2) found hecogenin in sisal. Gedeon and Kincl (3) have isolated hecogenin together with tigogenin from *Agave cantala* and *Agave vera-cruz*.

Rockogenin was isolated from *Agave gracilipes* along with large amounts of hecogenin. The presence of hecogenin together with tigogenin and rockogenin in certain plants points to their structural similarity.

2. Principle

In the following procedure (1), agave is extracted with ethanol. The extract is treated with acid and alkali to hydrolyze the glycosides and the fatty esters. Hecogenin may then be characterized by preparation of 2,4-dinitrophenylhydrazone.

Hydrazine hydrate reduction of hecogenin leads to formation of tigogenin, while catalytic reduction affords rockogenin. Further correlation between hecogenin and rockogenin can be established by their transformation to hecogenone.

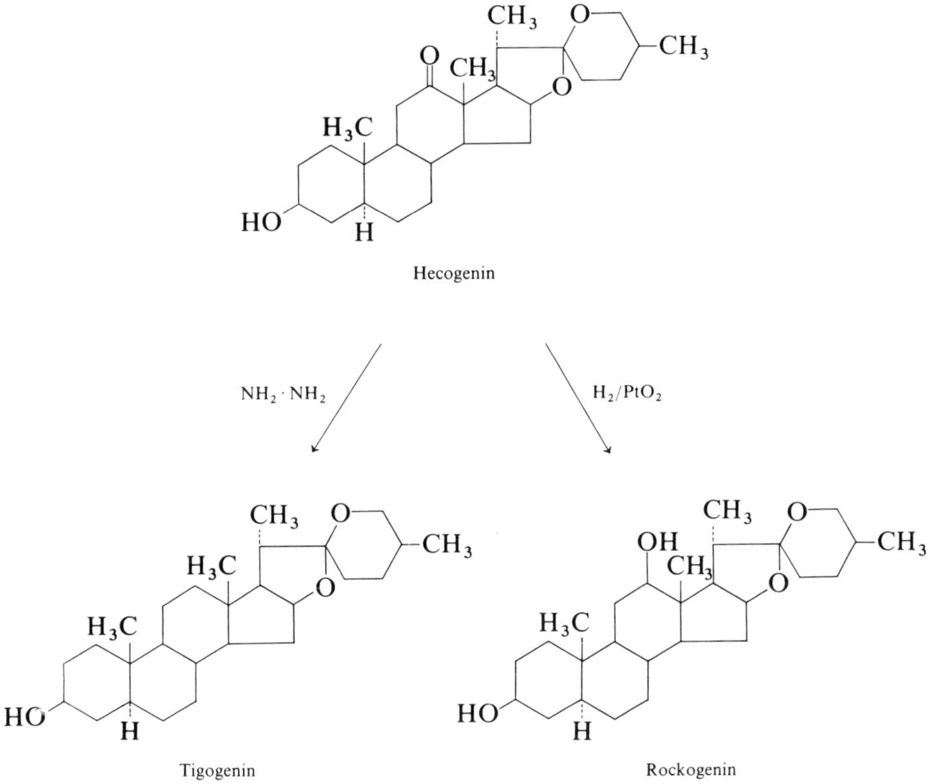

II. Steroids

Hecogenin

— NH$_2$·NH$_2$ → Tigogenin

— H$_2$/PtO$_2$ → Rockogenin

3. **Apparatus**

   Catalytic hydrogenation apparatus

4. **Materials**

   Acetic acid
   Acetic anhydride
   Acetone
   Agaves
   Chromium trioxide
   2,4-Dinitrophenylhydrazine
   Ethanol, 95%
   Ethanol, absolute

   Ether
   Hydrazine hydrate
   Hydrochloric acid
   Methanol
   Platinum oxide
   Potassium hydroxide
   Sodium, metal
   Sodium hydroxide

5. **Time**

   30 hours

6. Procedure

a. Isolation of hecogenin

A 5-kg sample of ground, undried agaves is extracted with 6 liters 95% ethanol on a steam bath for 12 hr. The slurry is filtered hot and washed with hot ethanol. The solvent is removed under reduced pressure and the residue is hydrolyzed by refluxing for 2 hr with 1.5 liter $2N$ ethanolic hydrochloric acid. The reaction mixture is cooled and filtered and diluted with 2 liters ether. The solution is then washed first with water, then with 5% aqueous sodium hydroxide, and again with water, and evaporated. The fatty esters are hydrolyzed by refluxing the residue with three volumes of 10% alcoholic potassium hydroxide for 30 min. The cooled hydrolysis mixture is extracted with ether, and the ethereal solution is washed with water and evaporated to a small volume. The crude sapogenin fraction that separates is dissolved in acetone and treated with Norit; mp of hecogenin 256–260°C.

b. 2,4-Dinitrophenylhydrazone of hecogenin

To a solution of 0.2 g hecogenin acetate in 30 ml hot ethanol is added a solution of 0.2 g 2,4-dinitrophenylhydrazine in 50 ml ethanol containing 1 ml concentrated hydrochloric acid. Within a few minutes, needles start to form. After standing at room temperature for 3 hr, the crystals are filtered off and washed with cold ethanol; mp 275–276°C (dec.). Recrystallization from ethanol gives orange needles, mp 281–282°C.

c. IR spectrum of hecogenin

1702 cm^{-1}: C=O stretching (in chloroform)
1055: C—O stretching of alcoholic OH (in carbon disulfide)
1250–900: vibrations associated with the spiroketal side chain
(12 bands), (in carbon disulfide)

d. Transformation of hecogenin to tigogenin

To a solution of 3 g sodium in 60 ml absolute ethanol in a bomb-tube are added 0.5 g hecogenin acetate and 5 ml 85% hydrazine hydrate. The tube is sealed and heated for 12 hr at 200°C. After cooling, it is opened, and the reaction mixture is poured into water. The precipitate is taken up in ether. The ethereal solution is washed with water, dilute hydrochloric acid, and water, and then concentrated until crystals start to appear. After cooling, the crystalline material is filtered. Crystallization from methanol gives white needles, mp 203–206°C; yield, 0.3 g.

e. Transformation of hecogenin to rockogenin

An ethereal solution of 0.3 g hecogenin containing several drops of acetic acid is shaken with hydrogen and 0.2 g Adams' catalyst for 2 hr at room temperature and under three atmospheres pressure. After filtering off the catalyst, the solvent is removed. The oily residue is refluxed with acetic anhydride for 1 hr. The acetic anhydride is removed *in vacuo* on a steam bath, and the 12-dihydrohecogenin diacetate is crystallized from methanol as long needles, mp 204–206°C; yield, 0.12 g. When hydrolyzed with 5% alcoholic potassium hydroxide for 20 min on a steam bath, the diacetate is converted to rockogenin, which crystallizes from methanol as thick needles, mp 208–210°C. Repeated crystallization from ether gives crystals of mp 217–220°C.

f. Conversion of hecogenin and rockogenin to hecogenone

A solution of 0.1 g hecogenin in 30 ml acetic acid is mixed with a solution of 0.1 g chromium trioxide in 5 ml 80% acetic acid. After standing 30 min at 25°C, water is added and the product is extracted with ether, washed with dilute aqueous sodium hydroxide and water, and dried. The ethereal solution is concentrated and cooled to give white needles, mp 237–240°C. Mild oxidation of rockogenin with chromium trioxide in acetic acid under the conditions outlined above also gives hecogenone.

References

1. Marker, R. E., Wagner, R. B., Ulshafer, P. R., Wittbecker, E. L., Goldsmith, D. P. J., and Rouf, C. H. (1943). *J. Am. Chem. Soc.* **65**, 1199.
2. Callow, R. K., Cornforth, J. W., and Spensley, P. C. (1951). *Chem. Ind. (London)* 699.
3. Gedeon, J., and Kincl, F. A. (1953). *Arch. Pharm.* **286**, 317.

Recommended Review

1. Stoll, A., and Jucker, E. (1955). *Phytosterine, steroid saponine und herzglykoside.* In *"Modern Methods of Plant Analysis,"* (K. Paech, and M. V. Tracey, eds.), Vol. 3, p. 141. Springer-Verlag, New York.

F. Preparation of Δ^5-Cholesten-3-one and Δ^4-Cholesten-3-one from Cholesterol

1. Introduction

Cholesterol dibromide has been prepared by unbuffered (1) and buffered (2) bromination of cholesterol, and oxidized to $5\alpha, 6\beta$-dibromocholestan-3-one with acid permanganate (1), sodium dichromate (2), and chromic acid (3).

Debromination of 5,6-dibromocholestan-3-one with zinc dust in acetic acid yielded Δ^4-cholesten-3-one (1), which is also formed by using sodium iodide in refluxing benzene-ethanol (4). Butenand and Schmidt-Thome (3) found that by using a neutral or weakly ionizing medium such as zinc in

150 CHAPTER 3 Isoprenoids

boiling ethanol, Δ^5-cholesten-3-one is formed. The nonconjugated ketone is easily isomerized to Δ^4-cholesten-3-one by a trace of mineral acid or alkali.

Δ^4-Cholesten-3-one has also been prepared by debromination of the dibromoketone with ferrous chloride (5) or chromous chloride (6), by modification of the Windaus method (2), by Oppenauer oxidation of cholesterol (7), and by catalytic dehydrogenation (8).

Δ^5-Cholesten-3-one has been prepared by oxidation of Δ^5-stenols according to the Jones procedure (9, 10).

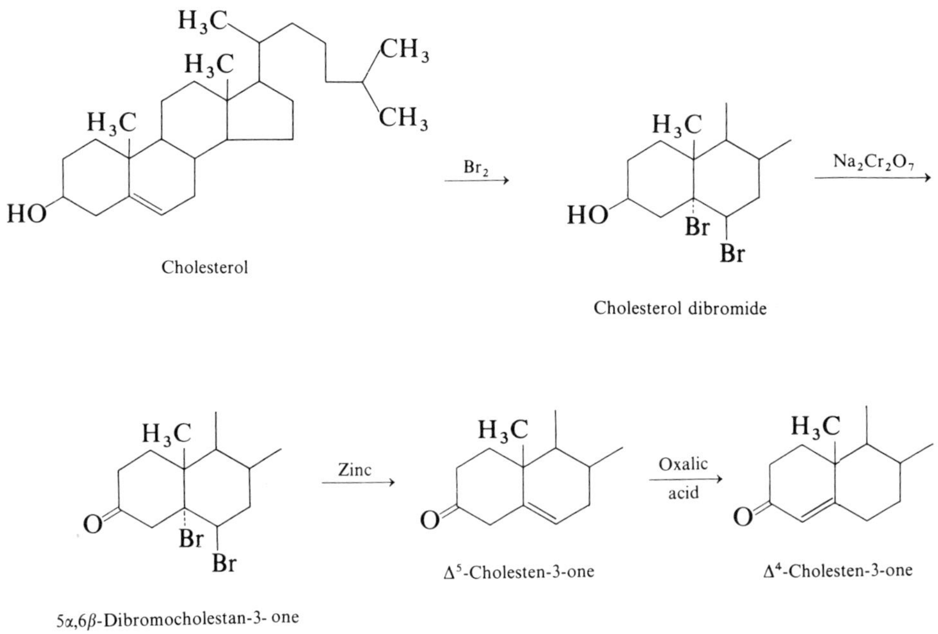

[Reaction scheme: Cholesterol → (Br$_2$) → Cholesterol dibromide → (Na$_2$Cr$_2$O$_7$) → 5α,6β-Dibromocholestan-3-one → (Zinc) → Δ^5-Cholesten-3-one → (Oxalic acid) → Δ^4-Cholesten-3-one]

2. Principle

In the following procedure (11), the double bond of cholesterol is protected by bromination, the dibromide is oxidized with sodium dichromate in acetic acid, forming the dibromoketone, and debromination is accomplished with ethanol–zinc dust. Isomerization to Δ^4-cholesten-3-one is effected by the action of oxalic acid in hot ethanol.

3. Apparatus

Stirrer

4. Materials

Acetic acid	Ether, dry	Sodium bicarbonate
Bromine	Methanol	Sodium dichromate
Cholesterol	Oxalic acid	Zinc dust
Ethanol	Sodium acetate	

5. Time

3–4 hours

6. Procedure

a. Cholesterol dibromide

Commercial cholesterol 15 g is dissolved in 100 ml absolute ether in a 500-ml beaker by warming on the steam bath and stirring with a glass rod; the solution is then cooled to 25°C. A second solution is prepared by adding 0.5 g powdered anhydrous sodium acetate to 60 ml acetic acid, stirring the mixture, and breaking up the lumps with a flat stirring rod; 6.8 g bromine is then added, and the solution is poured, while stirring, into the cholesterol solution. The solution turns yellow and promptly solidifies as a stiff paste of dibromide. The mixture is cooled in an ice bath to 20°C, and the product is then collected on a Buchner funnel. The cake is pressed and washed with acetic acid until the filtrate is completely colorless; 50 ml is usually sufficient. Dibromide moistened with acetic acid is satisfactory for most transformations; dry dibromide, even when highly purified by repeated crystallization, begins to decompose within a few weeks. When the material prepared as described is dried to constant weight at room temperature, it is obtained as the 1:1 dibromide/acetic acid complex; yield 17–18 g.

b. 5α, 6β-Dibromocholestan-3-one

The moist dibromide obtained from 15 g cholesterol is suspended in 20 ml acetic acid in a 500-ml round-bottomed flask equipped with a stirrer and mounted over a bucket of ice and water. This ice bath can be raised to room temperature (25–30°C). A solution (preheated to 90°C) of 8 g sodium dichromate dihydrate in 200 ml acetic acid is introduced through a funnel. The mixture reaches a temperature of 55 to 58°C during the oxidation, and all the solid dissolves within 3 to 5 min. After an additional 2 min, the ice bucket is raised until the flask is immersed. Stirring is then stopped, and the mixture is allowed to stand in the ice bath for 10 min to allow the dibromoketone to separate as readily filterable crystals. On resuming stirring, the temperature is first brought to 25°C and then, after addition of 40 ml water, to 15°C. The product is collected on a Buchner funnel, and the filter cake is drained until the flow of filtrate amounts to no more than 25 drops per min. The suction is released, the walls of the funnel are washed down with methanol, and 20 ml methanol is added. After allowing the funnel to stand for a few minutes, suction is again applied, and the crystals are drained thoroughly of solvent before they are washed as described above with 20 ml fresh methanol. Dried dibromoketone consists of shiny, white crystals, mp 73–75°C (dec.); yield, about 17 g.

c. Δ^5-Cholesten-3-one

The moist $5\alpha,6\beta$-dibromocholestan-3-one obtained from 15 g cholesterol is transferred to a 500-ml round-bottomed flask and covered with 200 ml ether and 2.5 ml acetic acid. The suspension is stirred mechanically, an ice bath is raised into position, and the temperature is brought to 15°C. The ice bath is then lowered, and 0.5 g fresh zinc dust is added. With the onset of the exothermic reaction of debromination, the temperature is maintained at 15 to 20°C by cooling during the addition (in about 5 min) of 3.5 g zinc dust. The ice bath is then lowered, and the ethereal solution containing suspended zinc dust is stirred for an additional 10 min. With continued stirring, 4 ml pyridine is added, bringing about the precipitation of a white zinc salt. The mixture is filtered through a Buchner funnel, and the filter cake is washed well with ether. The colorless filtrate is washed with three 60-ml portions of water and then shaken thoroughly with 60 ml 5% aqueous sodium bicarbonate solution until free from acetic acid, as indicated by testing the ethereal solution with moist, blue litmus paper. The solution is dried over magnesium sulfate and evaporated to a volume of about 100 ml. Methanol (50 ml) is then added, and the evaporation is continued until the volume is approximately 120 ml. Crystallization is allowed to proceed at room temperature, then at 0 to 4°C. The large colorless prisms are collected by suction filtration; yield, 8.7–9.4 g; mp 124–129°C.

Concentration of the mother liquor gives a second batch of crystals weighing 1.2–1.9 g, mp 117–125°C, and suitable for conversion to Δ^4-cholesten-3-one. Total yield, 10.6–10.8 g.

d. IR spectrum of Δ^5-cholesten-3-one
(region 1300–1500 cm^{-1} in CCl$_4$; other regions in CS$_2$)
1724 cm^{-1}: C=O stretching, six-membered ring ketone (nonconjugated)
1682: C=C, stretching, trisubstituted double bond
1436: unsubstituted CH$_2$ group in a six-membered ring adjacent to an unsaturated bond
1377: angular methyl groups
1368: terminal gem-dimethyl group of side chain
960: out-of-plane bending of olefinic C—H bond
830: out-of-plane bending vibrations of H atom of a trisubstituted olefin

II. Steroids

e. **Δ⁴-Cholesten-3-one**

A mixture of 10 g Δ⁵-cholesten-3-one, 1 g anhydrous oxalic acid and 80 ml 95% ethanol is heated on a steam bath until all the solid has dissolved (15 min) and heating is then continued for an additional 10 min. The solution is then allowed to stand at room temperature. If crystallization has not started after a period of several hours, the solution is seeded or the container is scratched. After crystallization has proceeded at room temperature and then at 0 to 4°C, the large, colorless, prismatic needles are collected by suction filtration; yield, 9 g; mp 81–82°C.

f. **Δ⁴-Cholesten-3-one**

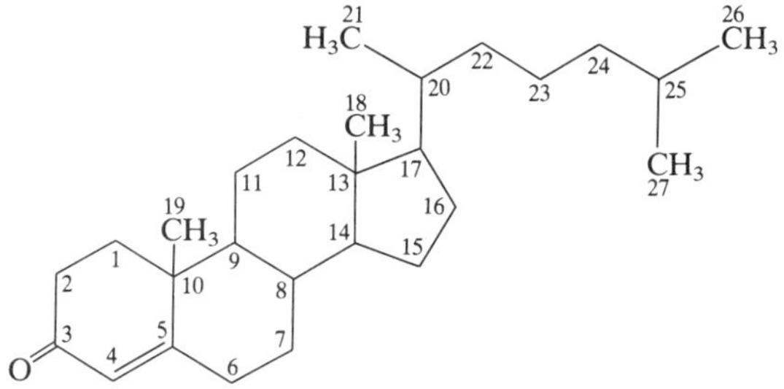

g. **^{13}C NMR**

C-1 35.7	C-10 38.6	C-19 17.4
C-2 33.9	C-11 21.0	C-20 35.7
C-3 198.9	C-12 39.4	C-21 18.7
C-4 123.6	C-13 42.4	C-22 36.1
C-5 171.0	C-14 55.9	C-23 23.8
C-6 32.9	C-15 24.1	C-24 39.6
C-7 32.1	C-16 28.1	C-25 27.9
C-8 35.7	C-17 56.1	C-26 22.5
C-9 53.8	C-18 12.0	

h. **Mass spectrum**
$C_{27}H_{44}O$, Mol. wt. 384.
m/z 384(80%)   327(5%)   271(15%)   229(40%)
    369(7%)   299(14%)   261(30%)   124(100%)

i. **IR spectrum of Δ⁴-cholesten-3-one**
(region 1300–1800 cm^{-1} in CCl$_4$; region 600–1300 cm^{-1} in CS$_2$)
1672 cm^{-1}. C=O stretching, α,β-unsaturated in a six-membered ring

1620: C=C stretching, conjugated with carbonyl
1475: unperturbed side-chain CH$_2$ groups (in C$_{27}$—C$_{29}$ steroids)
1438: unsubstituted CH$_2$ group in a six-membered ring adjacent to a double bond
1425: free methylene at C$_2$
1378: methyl groups, angular
1368: terminal gem-dimethyl group of side chain
955: out-of-plane bending of olefinic C—H bond
866: vibration of a group composed of a carbonyl at C$_3$ conjugated with C$_4$=C$_5$

j. UV spectrum of Δ^4-cholesten-3-one

λ_{max}^{EtOH} 241 mμ (logε 4.25); 312 (2.00)

The peak at 241 mμ is characteristic of a conjugated double bond system of the type C=C—C=O. The peak at 312 mμ is due to carbonyl absorption.

References

1. Windaus, A. (1906). *Ber.* **39**, 518.
2. Fieser, L. F. (1953). *J. Am. Chem. Soc.* **75**, 5421.
3. Butenand, A., and Schmidt-Thome, J., (936). *Ber.* **69**, 882.
4. Schoenheimer, R. (1935). *J. Biol. Chem.* **110**, 461.
5. Bretschneider, H., and Ajtai, M. (1943). *Monatsh. Chem.* **74**, 57.
6. Julian, P. L., Cole, W., Magnani, A., and Meyer, E. W., (1945). *J. Am. Chem. Soc.* **67**, 1728.
7. Oppenauer, R. V., (1941). *Org. Synth.* **21**, 18.
8. Kleiderer, E. C., and Kornfeld, E. C. (1948). *J. Org. Chem.* **13**, 455.
9. Djerassi, C., Engle, R. R., and Bowers, A. (1956). *J. Org. Chem.* **21**, 1547.
10. Bowden, K., Heilbron, I. M., Jones, E. R. H., and Weedon, B. C. L. (1946). *J. Chem. Soc.* **39**.
11. Fieser, L. F. (1963). *Org. Synth. Coll.* **4**, 195.

Recommended Review

1. Fieser, L. F., and Fieser, M. (1959). Investigation of Cholesterol. In "Steroids," p. 26. Reinhold, New York.

G. Preparation of Vitamin D$_2$ and Its Separation by Thin-Layer Chromatography

1. Introduction

The transformation of egosterol (provitamin D$_2$) to the corresponding vitamin D begins immediately on exposure to ultraviolet light of suitable wavelength. This process occurs in overlapping stages, with the formation of a series of products. The first stage is the photochemical conversion of ergosterol to pre-ergocalciferol, this reversible reaction providing yields of up to 85% (1). The next stage is the thermal conversion of pre-ergocalciferol to ergocalciferol (vitamin D_2 (2). During the irradiation, tachysterol$_2$ is formed

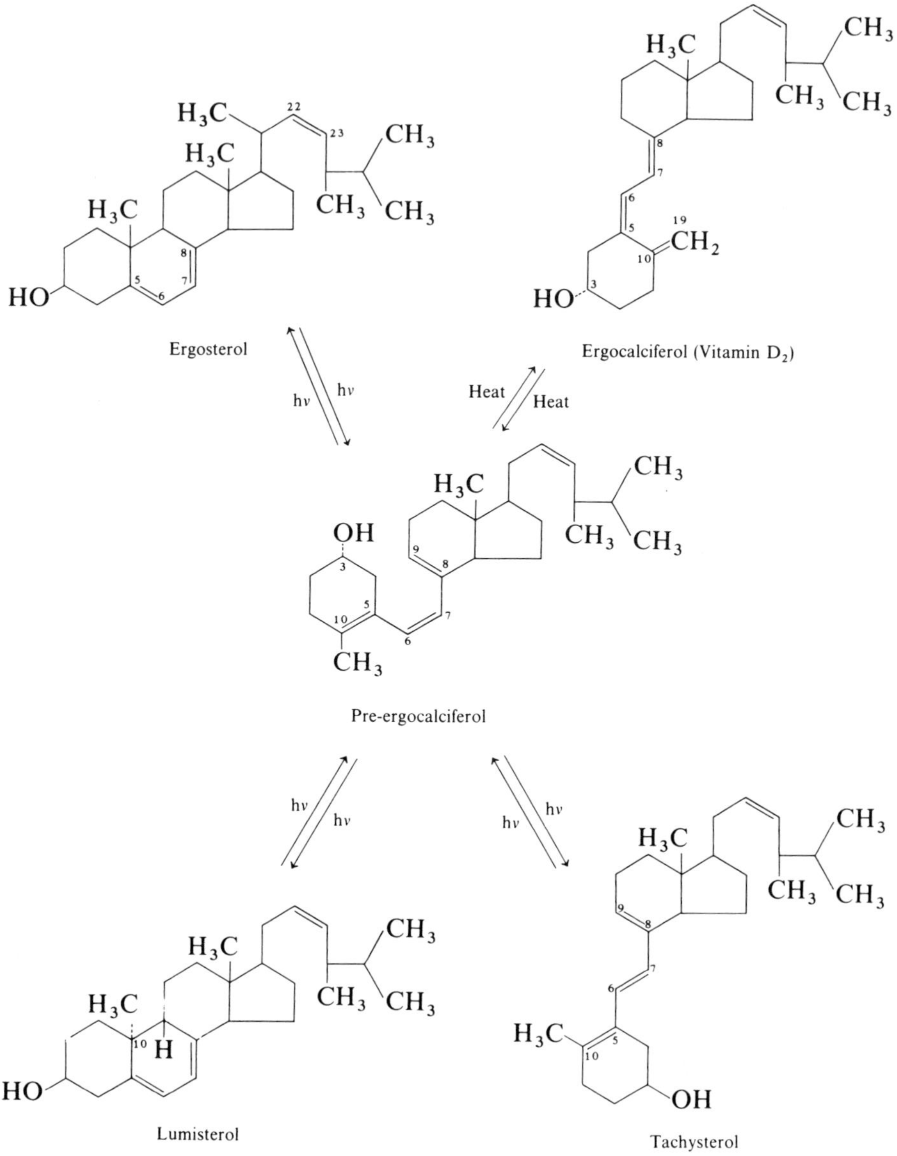

156 CHAPTER 3 Isoprenoids

via a reversible photochemical side reaction of pre-ergocalciferol. Over-irradiation of ergosterol yields biologically inactive products: toxisterol$_2$, superasterol$_2$I, and suprasterol$_2$II (3). An isomer mixture rich in lumisterol may be obtained by irradiating ergosterol with ultraviolet light of wavelength exceeding 280 mμ. It is noteworthy that irradiation of 7-dehydrocholesterol gives rise to a series of products similar to those derived from ergosterol.

A vitamin D deficiency in humans leads to rickets.

2. Principle

In the following procedure (4) ergosterol is exposed to ultraviolet rays for a few minutes, yielding tachysterol and vitamin D as the major irradiation products. They are separated on thin layers of silica gel, and their ultraviolet spectra are determined.

3. Apparatus

Spectrophotometer
TLC equipment
Ultraviolet lamp

4. Materials

Acetone	Ether	Skellysolve
Chloroform	Potassium permanganate	Sodium carbonate
Ergosterol	Silica gel	Sulfuric acid
Ethanol, 95%		

5. Time

2–3 hours

6. Procedure

a. Irradiation of ergosterol

Ergosterol 50 mg in 100 ml peroxide-free diethyl ether are placed in a 250-ml round-bottomed quartz flask connected to a reflux condenser and exposed to ultraviolet rays for 8 min. The solution is then stored at −25°C under nitrogen until such time as it is used for chromatographic analysis.

b. Preparation of plates

The suspension for five plates (20 × 20 cm) is prepared by shaking 25 g silica gel and 50 ml water for 30 sec. It is then spread on the plates with an applicator to give a layer of 0.25 mm thickness. The plates are activated at 160°C overnight.

c. Development

A solution of 10 to 20 μg of the irradiation product in 2 μl either is applied to the plate 2.5 cm from the bottom, and the plate is developed with 10% acetone in Skellysolve (solvent 1), or with chloroform (solvent 2). When the solvent has reached 10 cm above the starting line, the plate is removed from the chamber and allowed to air-dry for 5 to 10 min.

d. Detection

The plates are sprayed either with 0.2% potassium permanganate in 1.0% sodium carbonate solution or with 0.2 M sulfuric acid followed by heating at 140 to 150°C for 15 min.

e. Identification

The compounds separated are identified by the following technique: 1 to 2 mg of the ergosterol irradiation mixture is streaked in a very narrow band across the entire width of the plate at the origin. The separation is carried out as described above. After drying, the plate is covered with a sheet of paper that leaves 1-cm strips exposed at both edges. These are sprayed to locate the bands of the separated compounds. Using the bands on the edges of the plate as markers, strips of unsprayed silicic acid are scrapped off with a microscope slide. The compounds are dissolved in acetone, and silicic acid is removed by centrifugation or filtration. The acetone is evaporated with a stream of nitrogen and replaced by 95% ethanol, and the ultraviolet spectra of the fractions are determined. The R_f values are summarized in the following tables:

Identification of Compounds

Compound	R_f in Solvent	
	1	2
Vitamin D_2	0.33	0.44
Vitamin D_3	0.32	0.44
Ergosterol	0.27	0.35

Identification of Mixtures

Irradiation Mixture	R_f in Solvent	
	1	2
Spot I (tachysterol)	0.46	0.64
Spot II (vitamin D_2)	0.33	0.44
Spot III (ergosterol)	0.27	0.35

f. Vitamin D₂ (Calciferol)

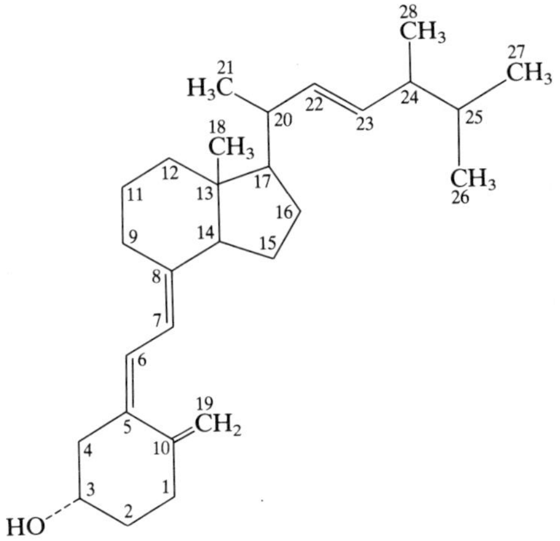

g. ¹H-NMR

Position of protons	δ (ppm)
$C_{18}(CH_3)$	0.57
$C_{22}(C\underline{H}{=}CH)$	5.20
$C_3(C\underline{H}{-}OH)$	3.92
$C_6({=}C\underline{H}{-}CH{=})$	6.05
$C_7({=}CH{-}C\underline{H}{=})$	6.05

h. ¹³C NMR

C-1 32.00	C-11 22.25	C-21 19.70
C-2 35.25	C-12 40.50	C-22 135.65
C-3 69.20	C-13 45.80	C-23 132.05
C-4 46.00	C-14 56.50	C-24 42.90
C-5 145.15	C-15 23.60	C-25 33.15
C-6 122.45	C-16 27.80	C-26 19.95
C-7 117.60	C-17 56.60	C-27 21.15
C-8 142.10	C-18 12.30	C-28 17.65
C-9 29.05	C-19 112.35	
C-10 135.20	C-20 40.35	

i. Mass Spectrum

Calciferol [Vitamin D₂] $C_{28}H_{44}O$, Mol. wt. 396.

m/z 396(46%)   119(42%)   91(32%)   69(49%)
    136(100%) 118(74%)   81(36%)   54(57%)

j. IR spectrum of vitamin D_2 (in CCl_4)
   1645, 1628, 1602 cm^{-1}: C=C stretching, triene (in chloroform)
   1458: CH_2 bending
   1374: CH_3 bending
   1345: C—H bending
   1050: C—O stretching of alcoholic OH
   995–972: out-of-plane deformation of *trans*—CH=CH—group (side-chain double bond C_{22-23})

k. UV spectrum of vitamin D_2
   λ_{max}^{EtOH} 265 mμ (logε 4.26)

The triene chromophore with its substituents gives rise to the absorption band at 265 mμ.

References

1. Velluz, L., Amiard, G., and Goffinet, B. (1955). *Bull. Soc. Chim. France* 1341.
2. Velluz, L., Amiard, G., and Pettit, A. (1949). *Bull. Soc. Chim. France* 501.
3. Windaus, A., Gaede, I., Koser, J., and Stein, G. (1930). *Ann.* **483**, 17.
4. Norman, A. W., and DeLuca, H. F. (1963). *Anal. Chem.* **35**, 1247.

Recommended Reviews

1. Bills, C. E. (1954). Vitamin D. In *"The Vitamins"* (W. H. Sebrell, and R. S. Harris eds.), Vol. 2, p. 132. Academic Press, New York.
2. Jones, H., and Rasmusson, G. H. (1980). Recent advances in the biology and chemistry of vitamin D. *Prog. Chem. Org. Nat. Prod.* **39**, 63.
3. Velluz, L., Amiard, G., and Pettit, A. (1949). *Bull. Soc. Chim. France* 501.
4. Wagner, A. F., and Folkers, K. (1964). The vitamin D group. In "Vitamins and Coenzymes," p. 330. Interscience, New York.

H. Determination of Sterols by Digitonin

1. Introduction

The ability of cholesterol to form a sparingly soluble complex with an equivalent amount of digitonin was first noted by Windaus in 1910 [1]. Some nonsteroid alcohols also form digitonides, but these differ in solubility and stability from those of the sterols. Thus, the precipitation of a sterol with digitonin is an effective method of isolation and purification of the sterols of the 3β-hydroxy series. Most of the commonly employed methods involving digitonin precipitation of cholesterol or other 3β-hydroxy sterols are lengthy, in view of the time required for the quantitative precipitation of the sterol as the digitonide complex. In 1937, Obermer and Milton (2) reported the use of aluminum hydroxide as a gathering reagent for the rapid precipitation of cholesterol digitonide. Brown, Zlatkis, Zak, and Boyle (3) have more recently compared the use of aluminum hydroxide and aluminum chloride for cholesterol digitonide precipitation from serum extracts, in combination with the ferric chloride reagent (for color development).

2. Principle

The method described here (4) involves the rapid precipitation of 3β-hydroxy sterols, e.g., cholesterol, with digitonin and aluminum chloride in the presence of dilute hydrochloric acid.

The digitonides are then cleaved to their components with hot dimethyl sulfoxide (5). When the solution is cooled to room temperature, the sterol precipitates and is extracted with hexane. Recovery of sterol is almost quantitative, and the saponin is obtained by evaporation to dryness of the dimethyl sulfoxide solution. The use of dimethyl sulfoxide has a distinct advantage of convenience and speed of isolation of the digitonin-precipitable sterols.

3. Apparatus

Centrifuge

4. Materials

Acetone	Dimethyl sulfoxide	Hydrochloric acid
Aluminum chloride	Ethanol	Methanol
Cholesterol	Ether, dry	Pyridine
Digitonin		

5. Time

3–4 hours

6. Procedure

a. Preparation of digitonides

Cholesterol (100 mg) is dissolved in 100 ml acetone–ethanol 1:1. 10 drops 3.5% hydrochloric acid (prepared by diluting 10 ml concentrated hydrochloric acid (35%) to 100 ml with distilled water), and 10 ml 0.5% digitonin in 50% aqueous ethanol (prepared by dissolving 0.5 g digitonin in 100 ml 50% ethanol at 60°C) are mixed, and added to the above solution. Aluminum chloride (10 ml, 10%) (100 g aluminum chloride hexahydrate dissolved in distilled water and made up to 1 liter) is added, and the contents are mixed and then heated at 45°C for 15 min. After cooling for 1 min, the tubes are centrifuged for 20 min at 2500 rpm, the supernatant is decanted and discarded, and the tubes are drained for 1 min by inversion. Methanol (20 ml) is then added to each tube, and the sterol digitonides are dissolved by gentle warming. Hydrochloric acid (10 drops 3.5%) and 20 ml 10% aluminum chloride are added and, following buzzing, the tubes are heated at 45°C for 20 min. The tubes are then cooled for 1 min, centrifuged at 2500 rpm for 20 min, decanted, and drained as in the previous step. To ensure complete removal of digitonin, the methanol–aluminum chloride recrystallization step is repeated. Finally, 20 ml acetone is added, and the precipitates are dispersed thoroughly by buzzing. They are centrifuged for 5 min at 2500 rpm, and the solvent is decanted.

The digitonides are dissolved in 20 ml methanol and recrystallized by addition of 2 ml 10% aluminum chloride and heating at 45°C. Quantitative recrystallization of cholesterol digitonides takes place after 15 min heating at 45°C. Complete purification is achieved by two such recrystallizations.

b. Cleavage of digitonides

The general procedure is illustrated by the following example. A mixture of 20 ml dimethyl sulfoxide and 1 g cholesteryl digitonide is heated on the steam bath for 15 min, and the resulting solution allowed to cool to room temperature, whereupon cholesterol precipitates. The mixture is transferred to a separatory funnel and extracted with 70 ml n-hexane. The dimethyl sulfoxide layer is extracted further with two 30-ml portions of n-hexane, and the combined hydrocarbon layers are allowed to stand for 20 minutes over anhydrous sodium sulfate. Evaporation of the solvent leaves 0.22 g cholesterol of mp 147–148°C. Evaporation of dimethyl sulfoxide to dryness, followed by trituration of the residue with dry ether, gives 0.6–0.7 g digitonin.

References

1. Windaus, A. (1910). *Z. Physiol. Chem.* **65**, 110.
2. Obermer, E., and Milton, R. (1937). *J. Lab. Clin. Med.* **22**, 943.
3. Brown, H. H., Zlatkis, A., Zak, B., and Boyle, A. J. (1954). *Anal. Chem.* **26**, 397.
4. Vahouny, G. V., Borja, C. R., Mayer, R. M., and Treadwell, C. R. (1960). *Anal. Biochem.* **1**, 371.
5. Issidorides, C. H., Kitagawa, I., and Mosettig, E. (1962). *J. Org. Chem.* **27**, 4693.
6. Bergmann, W. (1940). *J. Biol. Chem.* **132**, 471.

I. Gas—Liquid Chromatography of Steroids

1. Introduction

The application of gas chromatography to steroids has been restricted owing to their low volatility. The first practical demonstration of the separation of steroids was reported by Vanden Heuvel, Sweeley, and Horning (1). Before the introduction of this method, Beerthuis and Recourt (2) and others had chromatographed steroids at high temperatures on silicone oil.

A variety of liquid phases have been exploited for gas chromatography of steroids: fluorinated silicone (3), neopentyl glycol succinate (4), nitrile silicone (5), polydiethylene glycol succinate (6), and many others. Many derivatives of steroids have been prepared in order to increase their volatility, decrease the adsorption on the column, and prevent thermal breakdown. The polarity of hydroxylated steroids can be reduced by preparation of O-methyl ethers (6, 7), acetates (8, 9), trifluoroacetates (10), or trimethylsilyl ethers (11).

Considerable progress has been made in the studies of correlating the retention time of steroids with their structure. Apparently, it will be possible to predict a number of structural characteristics of a given steroid from its retention time data (12).

2. Principle

It is important that the support material should be inert with regard to the sample component. If this is not the case, interactions with the samples will take place, leading to asymmetrical peaks and partial loss of the injected amounts due to irreversible adsorption or decomposition.

In gas chromatography, diatomaceous earth-type support materials are generally used. They have the following active groups at their surface:

$$-\underset{OH}{Si} - \underset{OH}{Si}- \quad \text{and} \quad -Si-O-Si-$$

When polar substances such as steroids are analyzed, they will interact strongly with these active groups. The blocking of these groups is achieved by silanization of the support before coating, e.g., by treatment with dichlorodimethylsilane or hexamethyldisilazane. In both cases, the free OH groups become chemically bonded:

$$\begin{array}{c} Si \\ | \\ O \\ | \\ CH_3-Si-CH_3 \\ | \\ CH_3 \end{array} \quad \text{or} \quad \begin{array}{c} -Si-Si- \\ | \quad | \\ O \quad O \\ \backslash / \\ Si \\ / \backslash \\ CH_3 \quad CH_3 \end{array}$$

The reaction of a steroid with equivalent amounts of trimethylchlorosilane and hexamethyldisilazane proceeds according to the following equation:

$$3ROH + (CH_3)_3SiCl + (CH_3)_3SiNSi(CH_3)_3 \longrightarrow 3ROSi(CH_3)_3 + NH_4Cl$$

This silanization considerably reduces the support activity (13).

The ease of preparation and the excellent GLC properties are the reasons for the widespread use of trimethylsilyl ethers of steroids in GLC analysis.

In the following procedure (1, 14) the support and column are silanized with dichlorodimethylsilane, and the sterols are analyzed as methyl ethers, acetates, keto-derivatives, or trimethylsilyl ethers.

3. Apparatus

   Gas chromatograph

4. Materials

   Androstane
   Androstane-17-one
   Androstane-3,17-dione
   Cholestane
   Cholestanyl methyl ether
   Cholesteryl methyl ether
   Cholestan-3-one
   4-Cholesten-3-one
   Cholestanol
   Cholesterol
   Cholestanyl acetate
   Cholesteryl acetate
   Chromosorb W (80–100 mesh)
   Coprostane
   Dichlorodimethylsilane
   Gas Chrom P or Q
   Hexamethyldisilazane
   Silicone gum, SE–30
   β-Sitosterol
   β-Sitosteryl acetate
   Stigmastane
   Stigmasterol

5. Time

   3 hours

6. Procedure

a. Preparation of columns

A glass column 6 ft in length and 4 mm in diameter is used. The support is acid-washed Chromosorb W, inactivated by treatment with dichlorodimethylsilane. Inactivation is carried out by placing 20 g support in 80 ml 5% dichlorodimethylsilane in toluene in a round-bottomed flask fitted with a side arm. The flask is brought under reduced pressure, and the mixture is swirled for a few minutes to dislodge air bubbles from the surface of the support. After 10 min, the support is removed by filtration and washed with 100 ml toluene. The support is then washed well with methanol by resuspension in this solvent and dried for 2 hr at 80°C. It is then suspended in 100 ml 2% SE–30 in toluene, and a gentle vacuum is applied for 15 min to remove occluded air. The suspension is then poured in one lot onto a Buchner funnel, applying gentle suction, which is released as soon as the bulk of the solution has been filtered. The moist support is transferred to a filter paper and is dried in air and finally in an oven at 80°C.

Glass columns and glass wool for packing are treated with 5% dichlorodimethylsilane in toluene, washed with toluene and methanol, and dried. Columns are packed by gradual addition of the coated support and repeated tapping. The freshly packed column is heated to 300°C in a slow stream of argon or helium for 12 hr to remove volatile products.

b. Preparation of samples

The steroids are dissolved in analytical grade chloroform or tetrahydrofuran at a concentration of about 2 mg/ml, and suitable portions are mixed with cholestane for chromatography. The quantity of steroid to be injected is 0.1–1.0 μg.

c. Operating conditions

Standard conditions are column temperature, 222°C; injector maintained at 50°C above column temperature; detector, 240°C; argon flow rate, 30 ml/min at outlet; inlet pressure, 10–15 psi. The following table summarizes the retention data relative to cholestane. The retention time of cholestane is 17.6 minutes.

Retention Data of Steroids

Compound	Relative Retention Time
Androstane	0.11
Androstan-17-one	0.22
Androstan-3,17-dione	0.47
Coprostane	0.90
Cholestane	1.00
Cholestanyl methyl ether	1.78
Cholesteryl methyl ether	1.72
Cholestan-3-one	2.17
4-Cholesten-3-one	2.72
Cholestanol	1.99
Cholesterol	1.98
Cholestanyl acetate	2.84
Cholesteryl acetate	2.81
β-Sitosterol	3.26
β-Sitosteryl acetate	4.62
Stigmastane	1.65
Stigmasterol	2.84

Sterols may also be analyzed as trimethylsilyl ethers, which can be prepared by dissolving 25 mg free sterol in 1 ml anhydrous tetrahydrofuran in a small glass-stoppered test tube. Hexamethyldisilazane (0.5 ml) and 10 μl trimethylchlorosilane are added, and the tube is tightly stoppered and heated for 30 min at 55°C. The cooled mixture may be injected directly into a glass or stainless steel column filled with Gas Chrom Q (silanized Gas Chrom P) impregnated with 1 to 2% SE–30. The experimental conditions are similar to those outlined above.

References

1. Vanden Heuvel, W. J. A., Sweeley, C. C., and Horning, E. C. (1960). *J. Am. Chem. Soc.* **82,** 3481.
2. Beerthuis, R. K., and Recourt, J. H. (1960). *Nature (London)* **186,** 372.
3. Vanden Heuvel, W. J. A., Haahti, E. O. A., and Horning, E. C. (1961). *J. Am. Chem. Soc.* **83,** 1513.
4. Haahti, E. O. A., Vanden Heuvel, W. J. A. and Horning, E. C. (1961). *J. Org. Chem.* **26,** 626.
5. Vanden Heuvel, W. J. A., Creech, B. G., and Horning, E. C. (1962). *Anal. Biochem.* **4,** 191.
6. Clayton, R. B. (1962). *Biochemistry* **1,** 357.
7. Knights, B. A. (1964). *J. Gas Chromatogr.* 160.
8. Wotiz, H. H., and Martin, H. F. (1961). *J. Biol. Chem.* **236,** 1312.
9. Nishioka, I., Ikekawa, N., Yagi, A., Kawasaki, T., and Tsukamoto, T. (1965). *Chem. Pharm. Bull. (Japan)* **13,** 379.
10. Vanden Heuvel, W. J. A., Sjovall, J., and Horning, E. C. (1961). *Biochim. Biophys. Acta* **48,** 596.
11. Luukkainen, T., Vanden Heuvel, W. J. A., Haahti, E. O. A., and Horning, E. C. (1961). *Biochim. Biophys. Acta* **52,** 599.
12. Vanden Heuvel, W. J. A., and Horning, E. C. (1962). *Biochim. Biophys. Acta* **64,** 416.
13. Kabot, F. J., and Ettre, L. S. (1963). *"Gas chromatographic analysis of steroids."* The Perkin-Elmer Corporation, Norwalk, Conn.
14. Brooks, C. J. W., and Hanaineh, L. (1963). *Biochem. J.* **87,** 151.

Recommended Reviews

1. Horning, E. C., Vanden Heuvel, W. J. A., and Creech, B. G. (1963). *Separation and determination of steroids by gas chromatography. In "Methods of Biochemical Analysis,"* (D. Glick, ed.), Vol. 11, p. 69. Interscience, New York.
2. Kuksis, A. (1966). *Newer developments in determination of bile acids and steroids by gas chromatography. In "Methods of Biochemical Analysis."* (D. Glick, ed.), Vol. 14, p. 325. Interscience, New York.

J. Thin-Layer Chromatography of Sterols and Stanols on Silica Impregnated with Silver Nitrate

1. Introduction

Sterols such as cholesterol and β-sitosterol occur very often in natural sources as inseparable mixtures with the corresponding stanols. Ikan and Kashman (1) have found a mixture of β-sitosterol and β-sitostanol in Israeli peat. Similar observations have been made by McLean, Rettie, and Spring (2) on Scottish peat, and by Ives and O'Neill (3) on Canadian peat moss. It has been shown that by bromination of such critical pairs, the unchanged stanols are easily separated on thin layers from the brominated sterols. This may be accomplished either by spotting bromine in carbon tetrachloride over the mixture at the start line (4, 5), or by incorporating bromine in the mobile phase (6).

In 1937, Lucas and his coworkers (7) showed that the coordination of silver ions with unsaturated compounds is rapid and reversible. This principle has been applied by Morris (8) to cholesteryl esters, by Avigan, Goodman, and Steinberg (9) to sterols and steroids, and by Ikan (10) to tetracyclic triterpenes. All these workers used thin layers of silica gel impregnated with silver nitrate.

2. Principle

The following method is according to Ikan and Cudzinovski (11).

3. Apparatus

TLC applicator

4. Materials

Acetic acid	Chlorosulfonic acid	Silver nitrate
Campesterol	Coprostanol	β-Sitosterol
Campestanol	Ergosterol	β-Sitostanol
Chloroform	Methanol	Stigmasterol
Cholesterol	Phosphomolybdic acid	Sulfuric acid
Cholestanol	Silica gel G	

5. Time

3 hours

6. Procedure

a. Preparation of plates

The suspension for 5 plates (20 × 20 cm) is prepared by shaking 30 g silica gel G and 60 ml water for 30 sec. It is applied uniformly to a thickness of 0.25 mm with an applicator. After 30 min at room temperature, the plates are heated in an oven an 125 to 130°C for 45 min. After cooling, they are sprayed with concentrated aqueous methanolic silver nitrate solution, 5% relative to silica gel, and then activated at 120°C for 30 min. This method permits impregnation of only part of the plate, which can thus be used for comparative chromatography.

b. Development

The samples are dissolved in chloroform (1 mg/ml) and applied with micropipettes along a line 2 cm above the rim of the plate. Chloroform is used as the mobile phase; it is allowed to travel over a distance of 15 cm. The plates are then removed and dried in air.

c. Spray reagents

1. 50% sulfuric acid in water
2. 10% phosphomolybdic acid in ethanol
3. a mixture of chlorosulfonic and acetic acids, 1:1

The steroids are detected by spraying one of the three mentioned reagents, followed by heating in an oven at 150°C for 15 min. The colors of the spots are recorded before and after charring.

The following sterols are tested: campesterol, campestanol, cholesterol, cholestanol, coprostanol, ergosterol, β-sitosterol, β-sitostanol, and stigmasterol.

References

1. Ikan, R., and Kashman, J. (1963). *Israel J. Chem.* **1**, 502.
2. McLean, J., Rettie, G. H., and Spring, F. S. (1958). *Chem. Ind. (London)*, 1515.
3. Ives, D. A. J., and O'Neill, A. N. (1958). *Can. J. Chem.* **36**, 436.
4. Cargill, D. I. (1962). *Analyst* **87**, 865.
5. Ikan, R., Harel, S., Kashman, J., and Bergmann, E. D. (1964). *J. Chromatogr.* **14**, 504.
6. Copius Peereboom, J. W., and Beekes, H. W. (1962). *J. Chromatogr.* **9**, 316.
7. Alberz, W. F., Welge, H. J., Yost, D. M., and Lucas, H. J. (1937). *J. Am. Chem. Soc.* **59**, 45.
8. Morris, L. J. (1963). *J. Lipid Res.* **4**, 357.
9. Avigan, J., Goodman, de W. S., and Steinberg, D. (1963). *J. Lipid Res.* **4**, 100.
10. Ikan, R. (1965). *J. Chromatogr.* **17**, 591.
11. Ikan, R., and Cudzinovski, M. (1965). *J. Chromatogr.* **18**, 422.

Recommended Review

1. Morris, L. J. (1964). *Specific separations by chromatography on impregnated adsorbents.* In "*New Biochemical Separations*" A. T. James, and L. J. Morris, (eds.) p. 294. Van Nostrand, Princeton, NJ.

K. Questions

1. What are sterols? Write the structure for cholesterol and number the positions in the rings.
2. Point out the structural similarities and differences between the male and female sex hormones.

168 CHAPTER 3 Isoprenoids

3. Distinguish between hormones, enzymes, and vitamins.
4. Indicate a natural source of stigmasterol, vitamin D, and corticosterone.
5. Why is ergosterol called provitamin D?
6. What are the "orange juice vitamin" and the "sunshine vitamin"?
7. What are marine sterols?
8. Describe the biosynthesis of cholesterol and stigmasterol.
9. What are saponins? Give examples and describe applications.

L. Recommended Books

1. Cook R. P. (1958). *"Cholesterol."* Academic Press, New York.
2. Fieser, L. F., and Fieser, M. (1959). *"Steroids."* Reinhold, New York.
3. Klyne, W. (1957). *"The Chemistry of Steroids."* John Wiley, New York.
4. Kritchevsky, D. (1958). *"Cholesterol."* John Wiley, New York.
5. Richards, J. H., and Hendrickson, J. B. (1964). *"The Biosynthesis of Steroids, Terpenes, and Acetogenins."* Benjamin, W. A.
6. Shoppee, C. W. (1958). *"Chemistry of the Steroids."* Academic Press, New York.

III. TERPENOIDS

A. Introduction

Terpenoids are widely distributed in nature, mostly in the plant kingdom. They may be regarded as derivatives of oligomers of isoprene, $CH_2=C-CH=CH_2$, usually joined head to tail. Terpene hydrocarbons
$\quad\quad\;|$
$\quad\quad CH_2$
are classified as follows:

$\quad\quad$ Monoterpenes, $\quad C_{10}H_{16}$
$\quad\quad$ Sesquiterpenes, $\;\, C_{15}H_{24}$
$\quad\quad$ Diterpenes, $\quad\quad\; C_{20}H_{32}$
$\quad\quad$ Triterpenes, $\quad\quad C_{30}H_{48}$
$\quad\quad$ Tetraterpenes, $\;\, C_{40}H_{64}$
$\quad\quad$ Polyterpenes, $\quad (C_5H_8)_n$

Abundant sources of terpenoids are the essential oils. They consist of a complex mixture of terpenes or sesquiterpenes, alcohols, aldehydes, ketones, acids, and esters. There are four general methods for the extraction of essential oils: (1) expression; (2) steam distillation; (3) extraction with volatile solvents; (4) resorption in purified fats. The separation of individual components is accomplished by vacuum fractionation and by chromatographic methods. The unsaturated hydrocarbons are conveniently separated as their crystalline addition products with hydrochloric acid, hydrobromic acid, or nitrosyl chloride.

1. Monoterpenes

The monoterpenes are subdivided into three groups: acyclic, monocyclic, and bicyclic.

a. Acyclic monoterpenes

Among the important hydrocarbons are ocimene and myrcene.

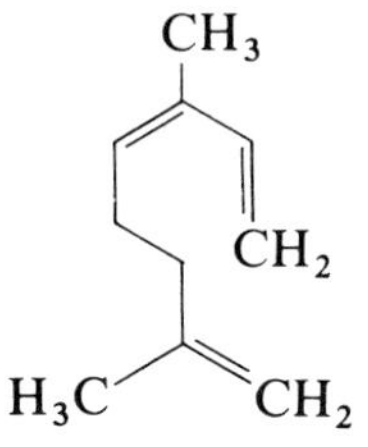

Ocimene

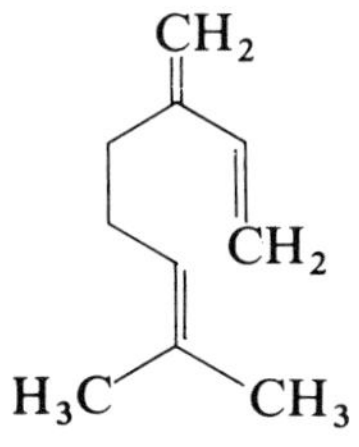

Myrcene

Aldehydes

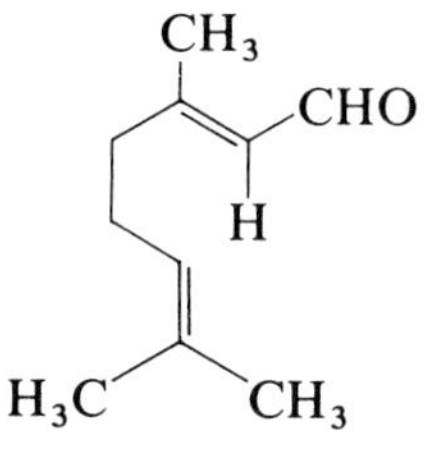

Geranial

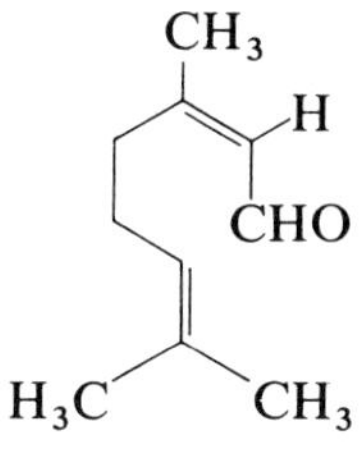

Neral

Alcohols

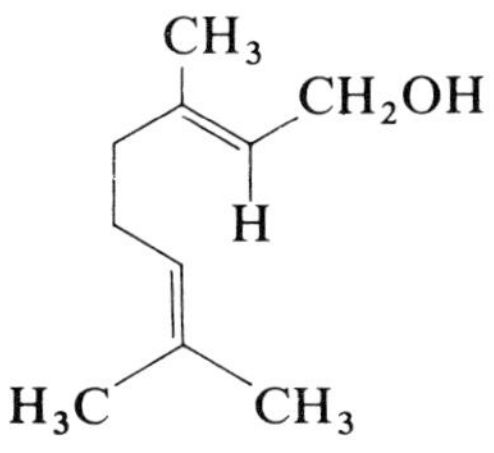

Geraniol

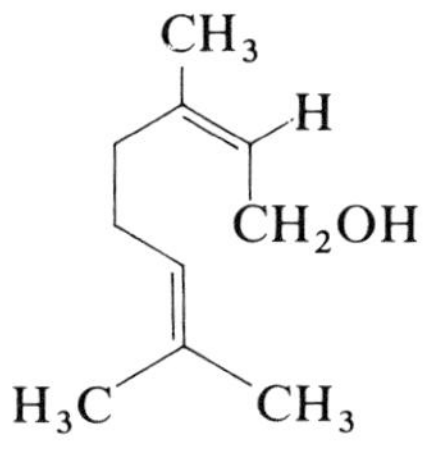

Nerol

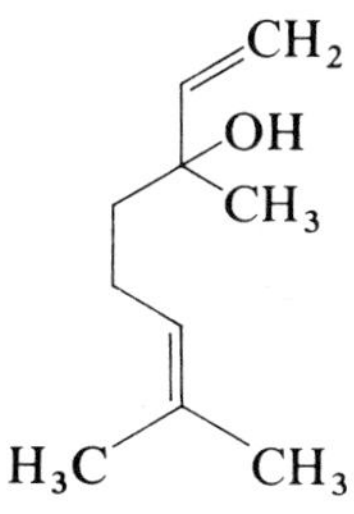

Linalool

Numerous examples of the cyclization of acyclic monoterpenes to alicyclic or aromatic compounds are known, e.g., the conversion of citral to α- and β-cyclocitral.

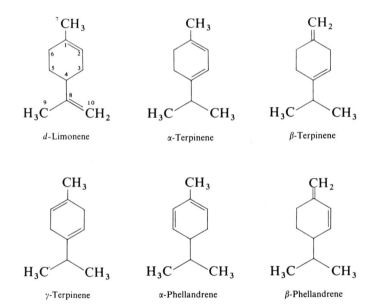

b. Monocyclic monoterpenes
Important hydrocarbons are:

d-Limonene   α-Terpinene   β-Terpinene

γ-Terpinene   α-Phellandrene   β-Phellandrene

Alcohols

α-Terpineol   Menthol   Piperitol   Carveol

III. Terpenoids

Aldehydes

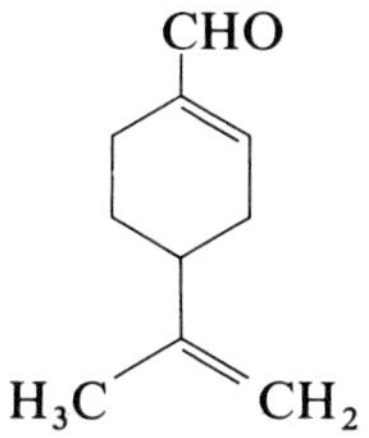

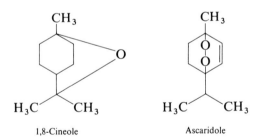

Perillaldehyde 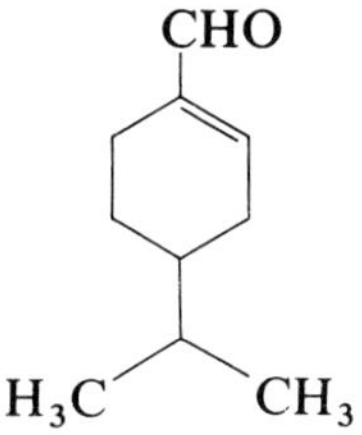   Phellandral

Ketones

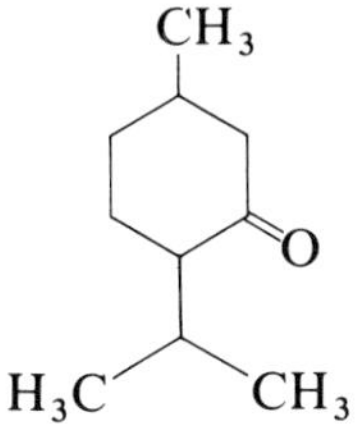

Menthone 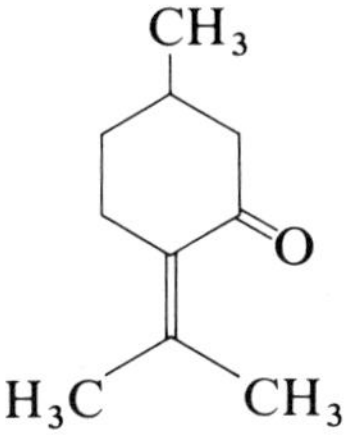   Pulegone 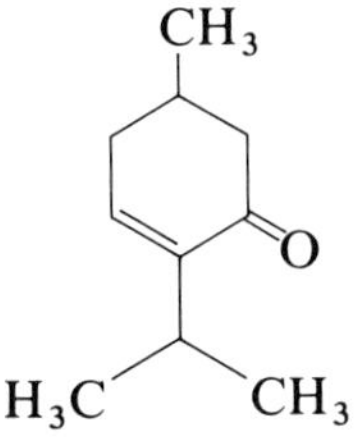   Piperitone 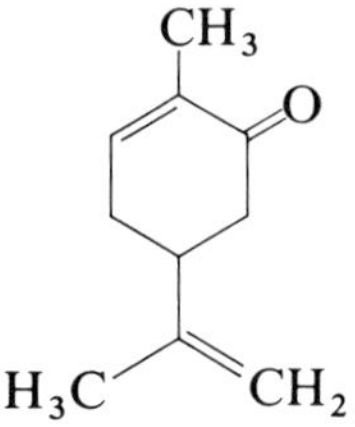   Carvone

Oxides

1,8-Cineole   Ascaridole

c. **Bicyclic monoterpenes**

The bicyclic monoterpenes may be divided into five groups.

Thujane group

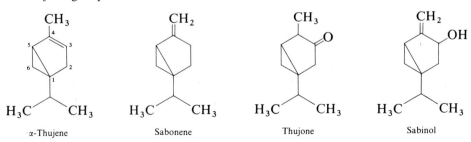

α-Thujene   Sabonene   Thujone   Sabinol

Carane group

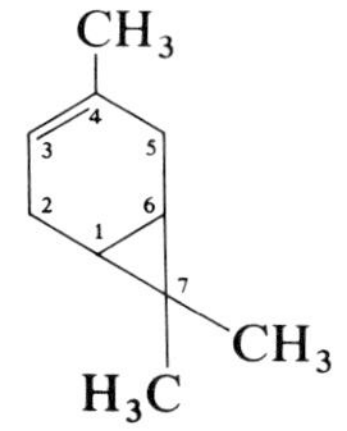

Car-3-ene 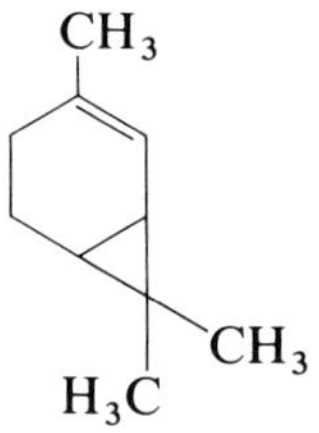   Car-4-ene 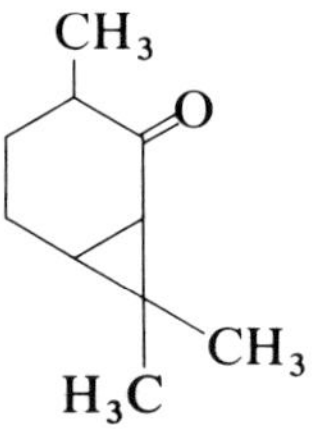   Carone

Pinane group

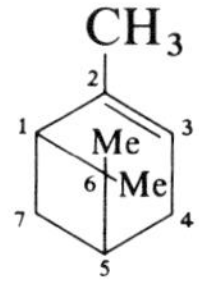

α-Pinene 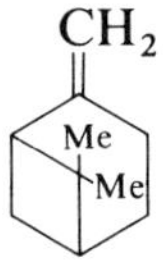   β-Pinene 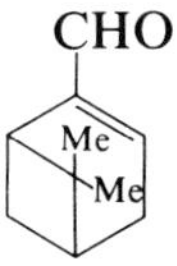   Myrtenal 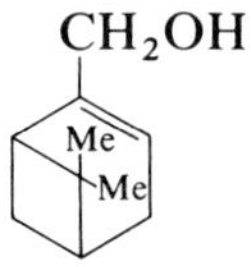   Myrtenol

Camphane group

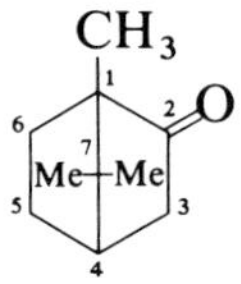

Camphor 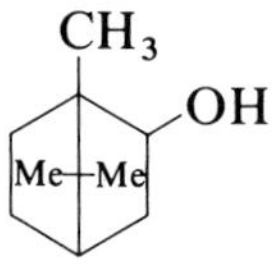   Borneol 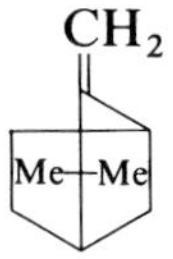   Camphene

Fenchane group

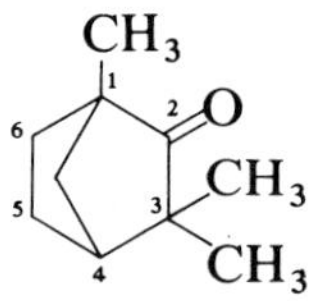

Fenchone 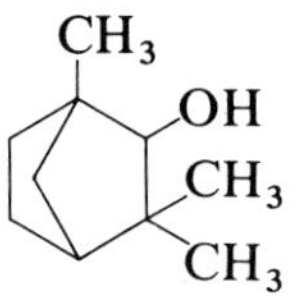   Fenchyl alcohol 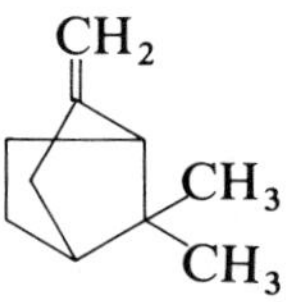   α-Fenchene

2. Sesquiterpenes

The sesquiterpenes form the higher-boiling fraction of the essential oils. They are formed by the union of three isoprene units. Sesquiterpenes are unsaturated compounds and may be acyclic, monocyclic, bicyclic, and tricyclic.

III. Terpenoids   173

a. Acyclic sesquiterpenes

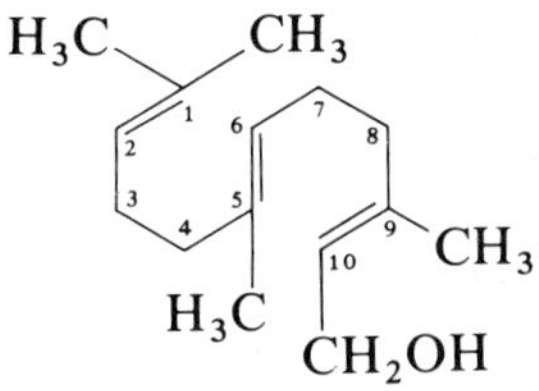

Farnesol

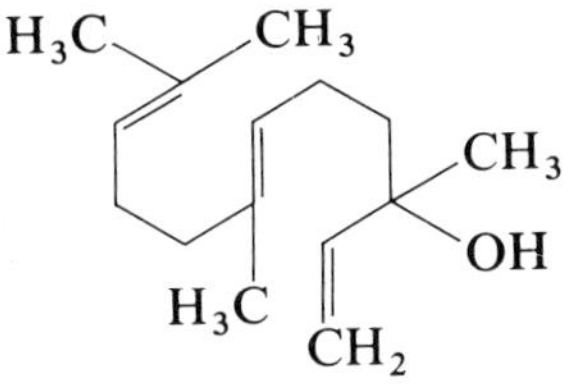

Nerolidol

b. Monocyclic sesquiterpenes

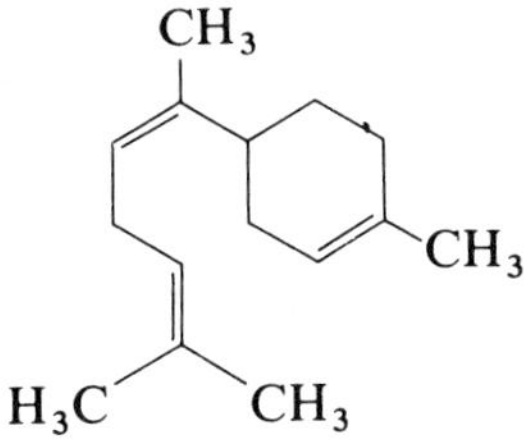

α-Bisabolene

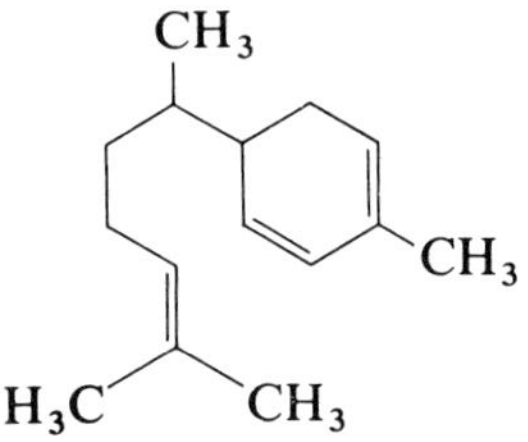

Zingiberene

c. Bicyclic sesquiterpenes

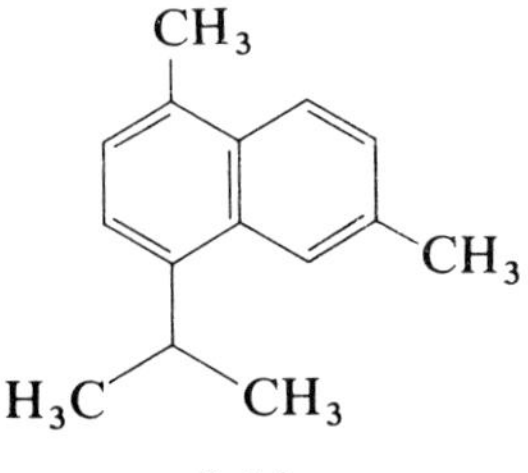

Cadalene

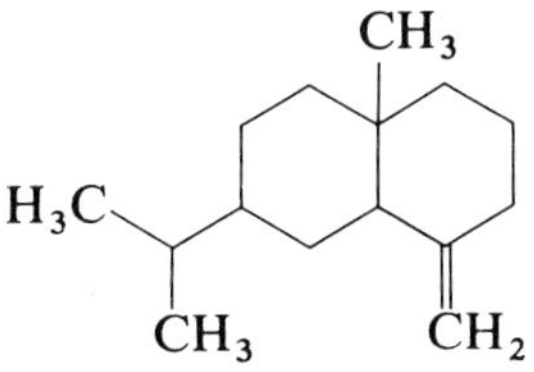

β-Selinene

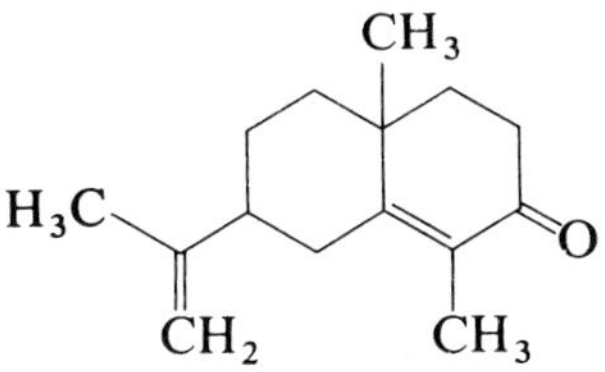

α-Cyperone

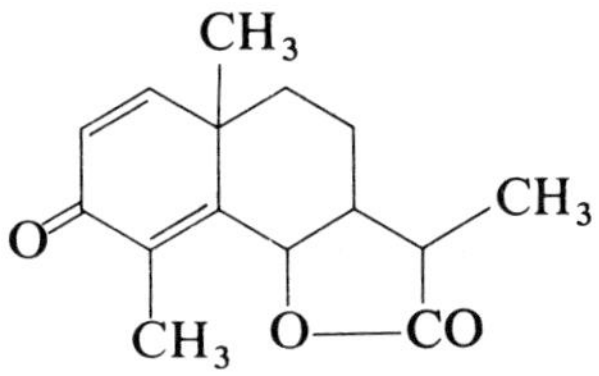

Santonin

d. Azulenes

Many essential oils have a blue or violet color, or may take on such colors after dehydrogenation with sulfur, selenium, or palladium.

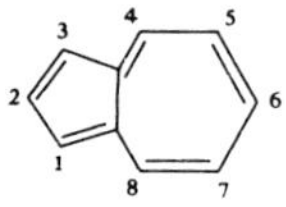

Azulene

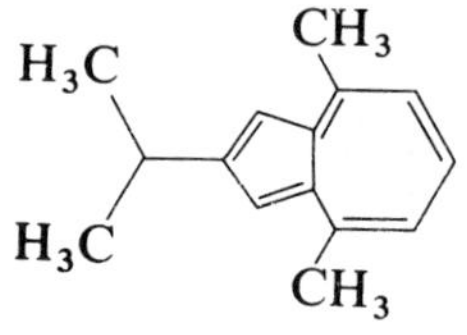

Vetivazulene

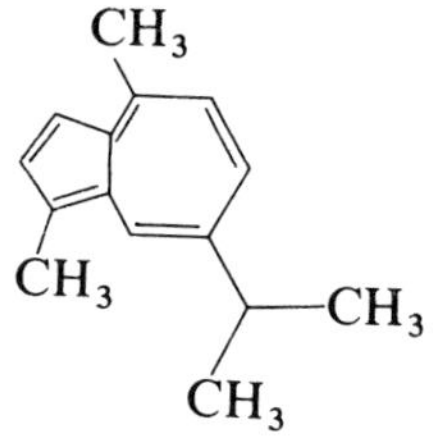

Guaiazulene

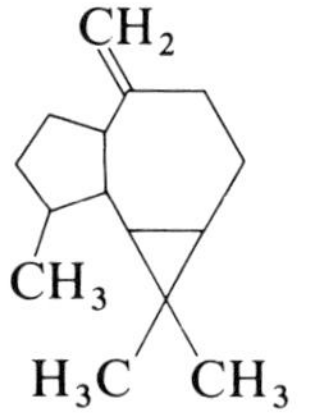

Aromadendrene

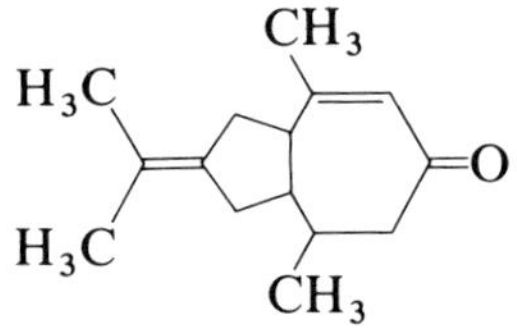

β-Vetivone

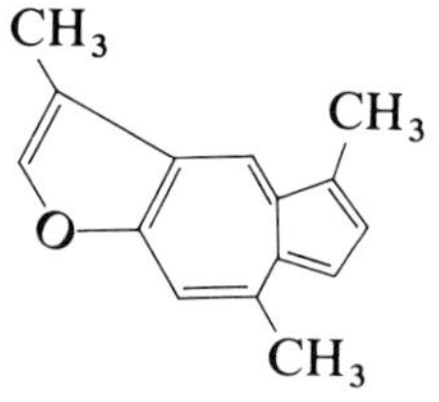

Linderazulene

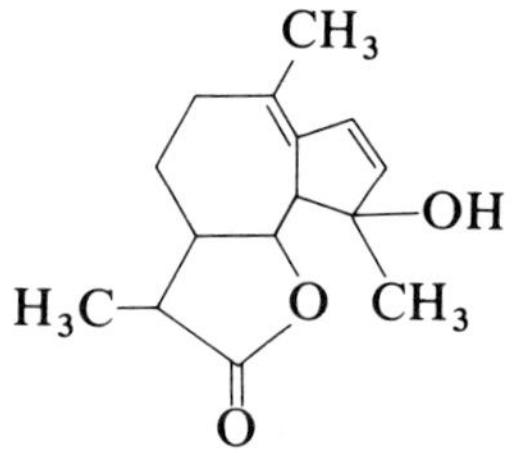

Artabsin

Some sesquiterpenes show greater structural complexity, e.g., caryophyllene.

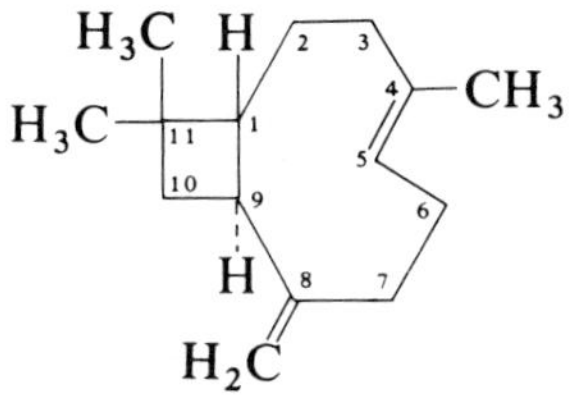

Caryophyllene

3. Diterpenes

The diterpenes are formed by the union of four isoprene units and are found mainly in the resins of plants.

III. Terpenoids

a. Acyclic diterpene

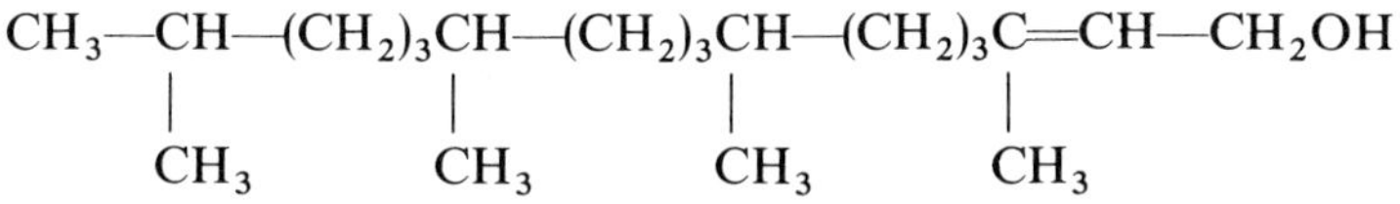

$$CH_3-CH-(CH_2)_3CH-(CH_2)_3CH-(CH_2)_3C=CH-CH_2OH$$
$$\quad\;\;|\qquad\qquad|\qquad\qquad|\qquad\qquad|$$
$$\quad\;CH_3\qquad\;\;CH_3\qquad\;\;CH_3\qquad\;\;CH_3$$

Phytol

b. Monocyclic diterpene

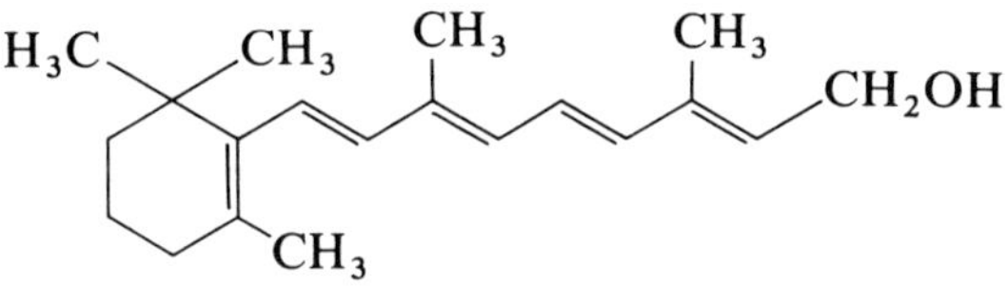

Vitamin A$_1$

c. Bicyclic diterpenes

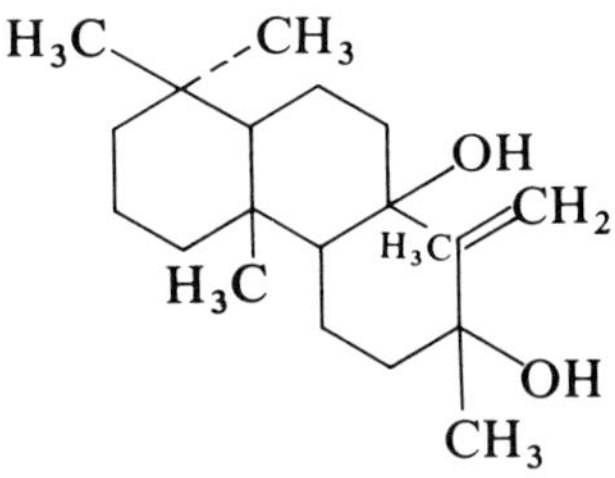

Sclareol

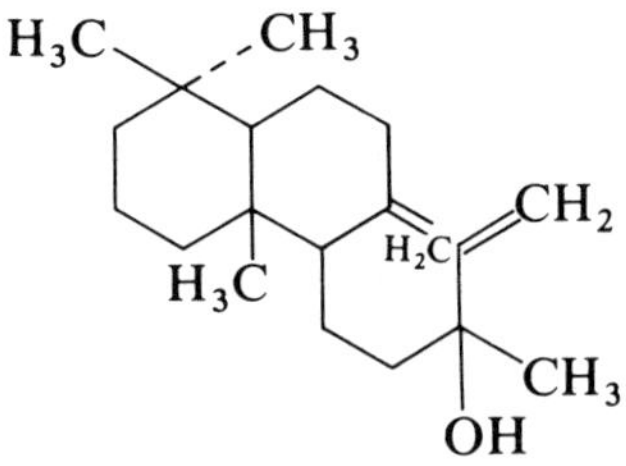

Manool

d. Tricyclic diterpenes

The resin acids form the major nonvolatile part of many natural resins, especially those of coniferous trees.

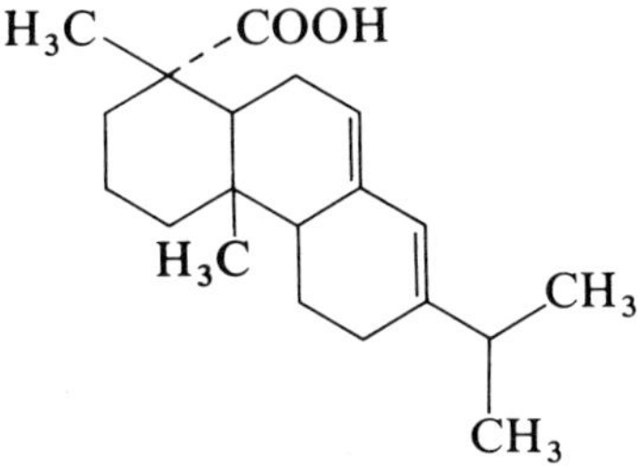

Abietic acid

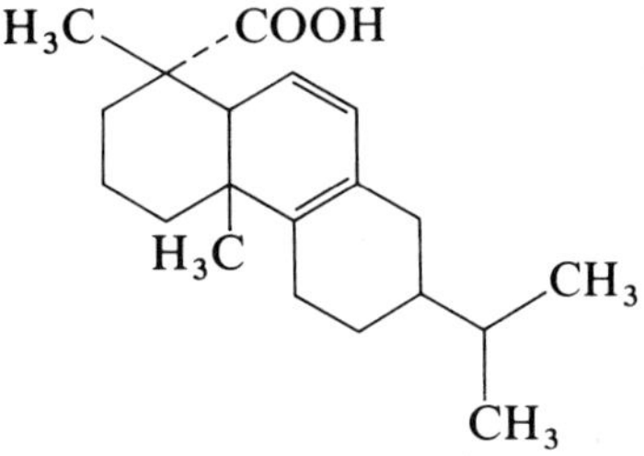

Levopimaric acid

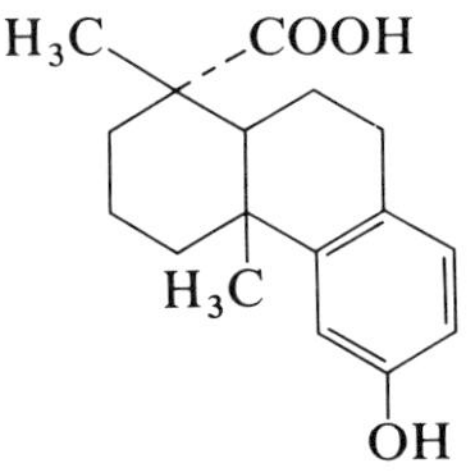

176 CHAPTER 3 Isoprenoids

Podocarpic acid

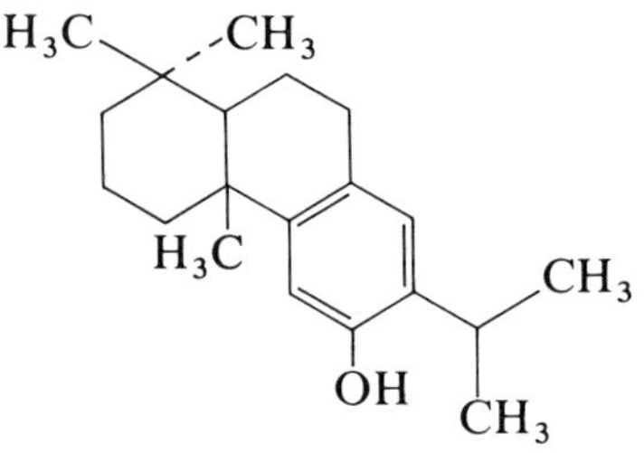

Ferruginol

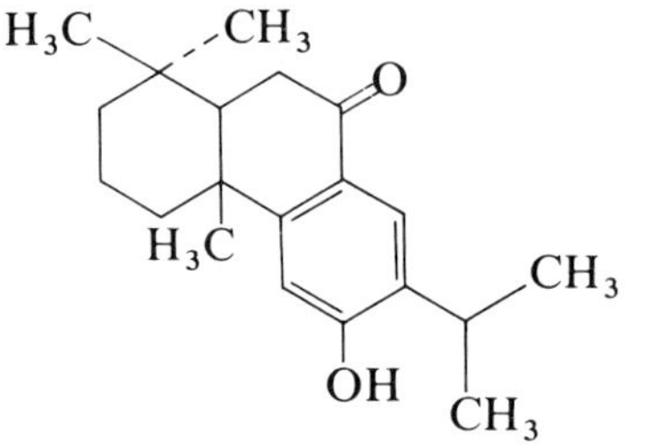

Sugiol

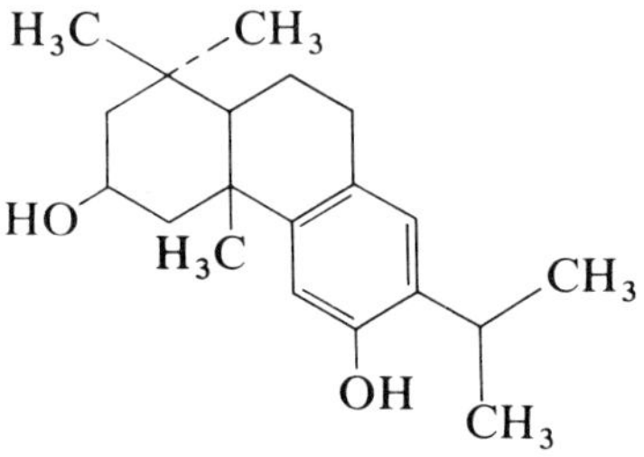

Hinokiol

4. Triterpenes

The triterpenes are widely distributed in the plant and animal kingdoms, where they occur either in the free state, as esters, or as glycosides. They may be classified into three groups: (1) acyclic; (2) tetracyclic; (3) pentacyclic.

Many conversions from one group to another have been accomplished. Some of the transformations involved the change of a —COOH to the corresponding —CH$_3$ or to CH$_2$OH.

$$\begin{array}{c} \text{RCOOH} \longrightarrow \text{RCOCl} \longrightarrow \text{RCHO} \longrightarrow \text{RCH}_3 \\ \downarrow \\ \text{RCH}_2\text{OH} \longrightarrow \text{RCHO} \longrightarrow \text{RCH}_3 \\ \downarrow \\ \text{RCH}_2\text{OTs} \longrightarrow \text{RCH}_3 \end{array}$$

An important reaction is the detection of a gem-dimethyl group at the C$_4$ position.

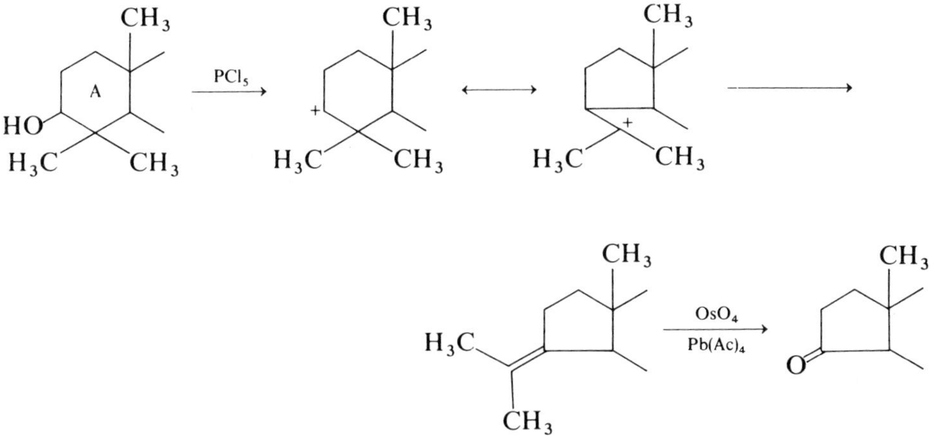

Valuable information on the carbon skeletons of triterpenoids can be obtained by dehydrogenation, which yields the following products:

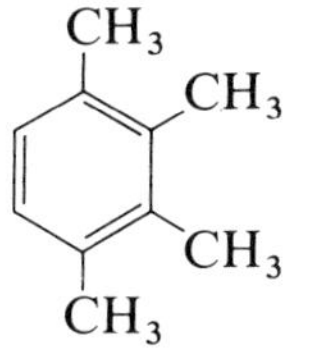

1,2,3,4-Tetramethylbenzene

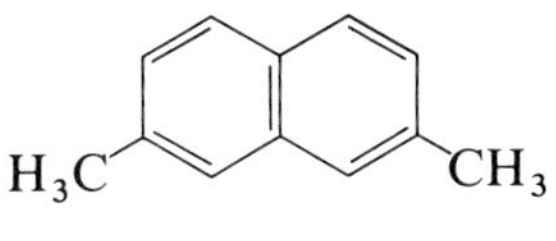

2,7-Dimethylnaphthalene

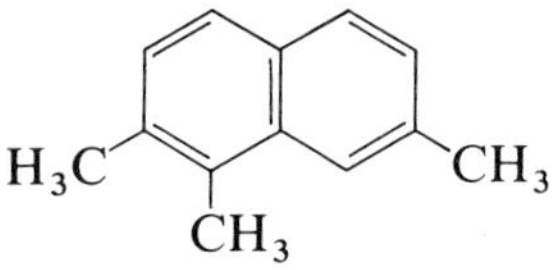

1,2,7-Trimethylnaphthalene

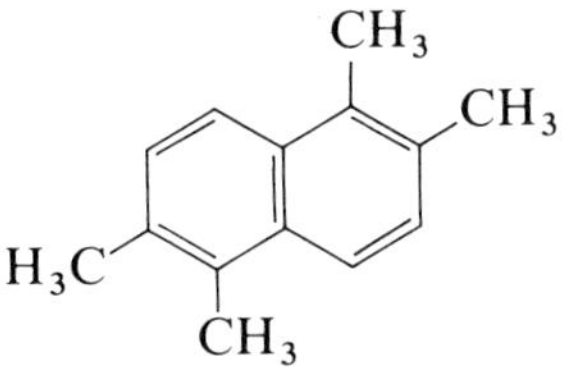

1,2,5,6-Tetramethylnaphthalene

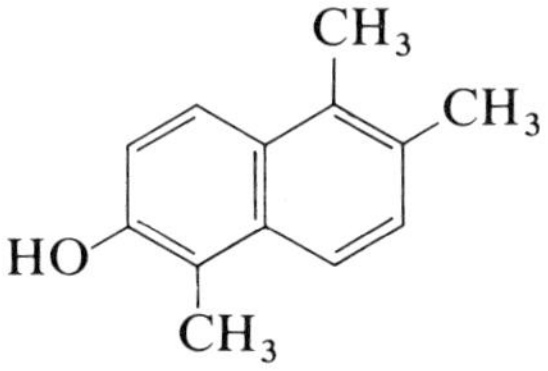

1,5,6-Trimethyl-2-naphthol

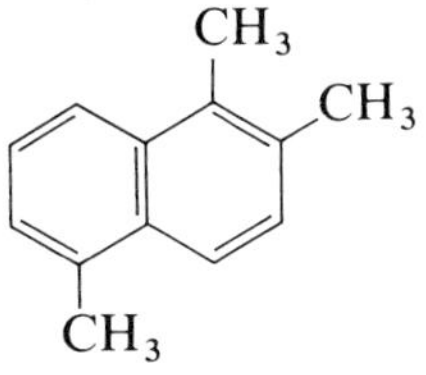

1,5,6-Trimethylnaphthalene

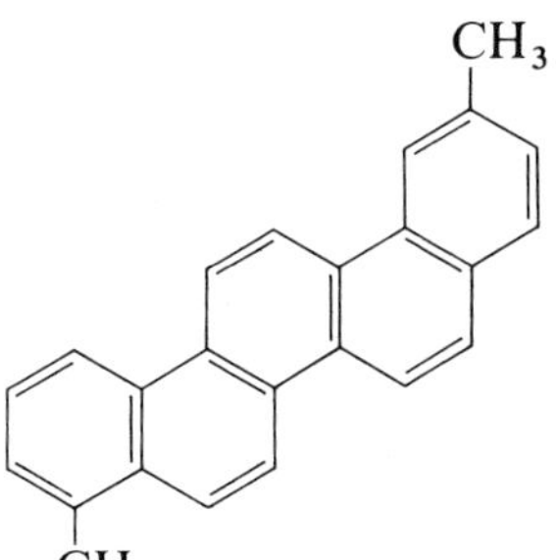

1,8-Dimethylpicene

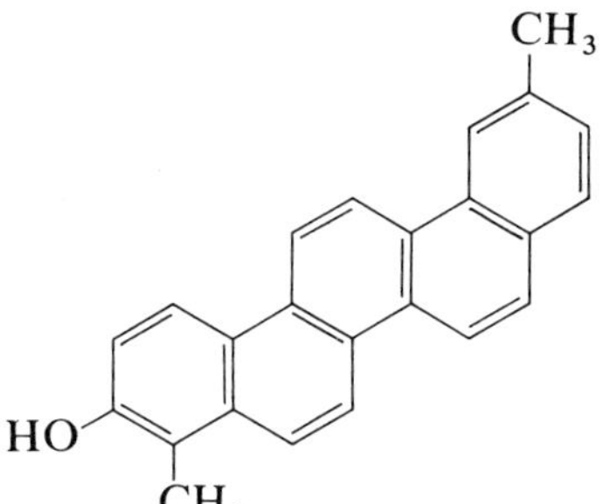

2-Hydroxy-1,8-dimethylpicene

The main dehydrogenation product of tetracyclic triterpenes is

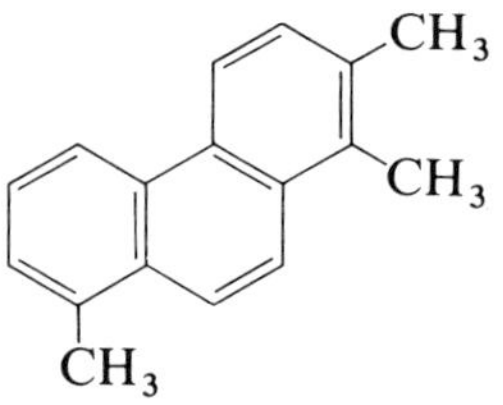

1,2,8-Trimethylphenanthrene

a. Acyclic triterpenes

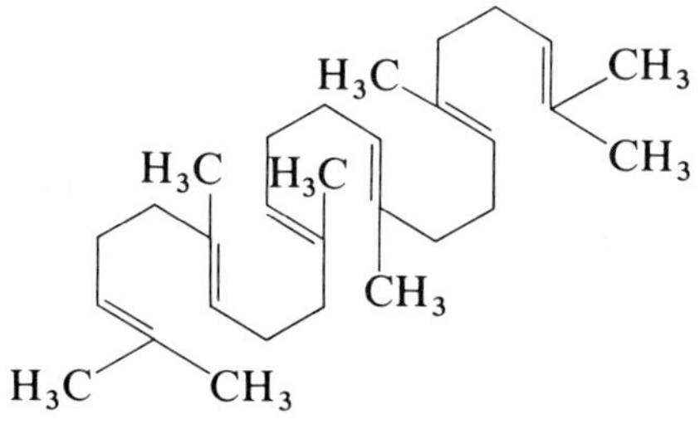

Squalene

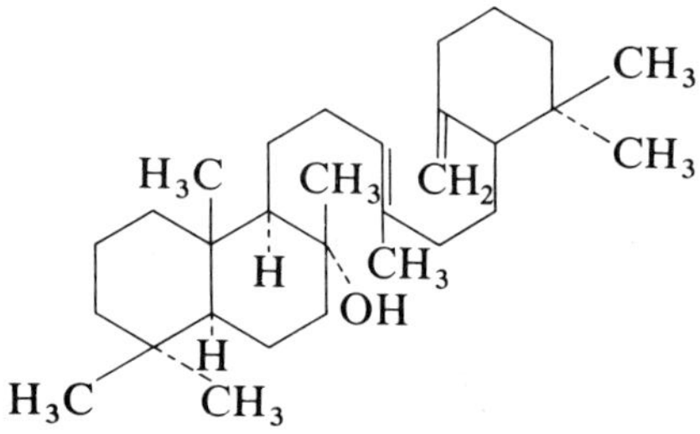

Ambrein

b. Tetracyclic triterpenes

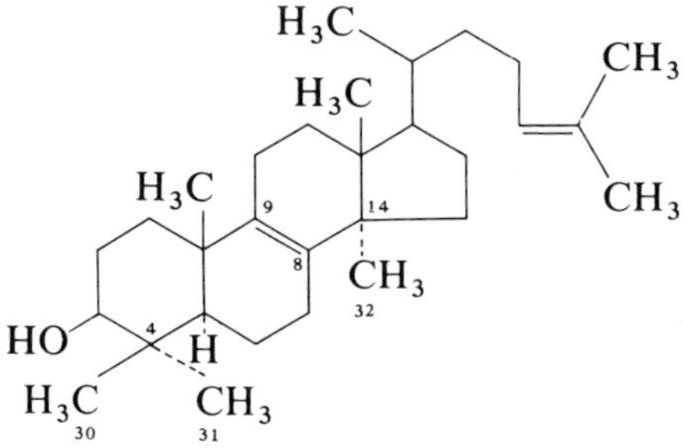

Lanosterol

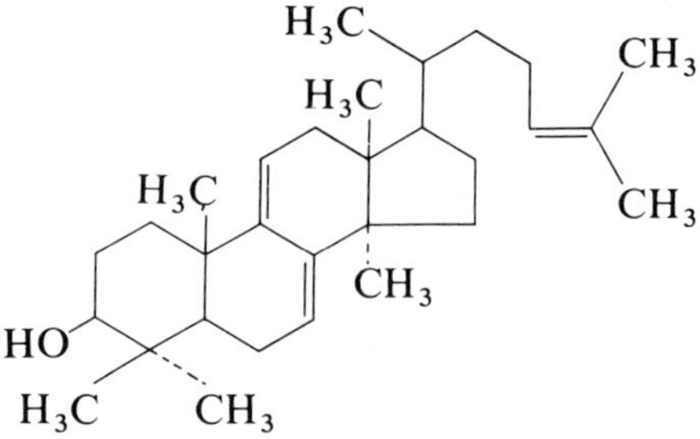

Agnosterol

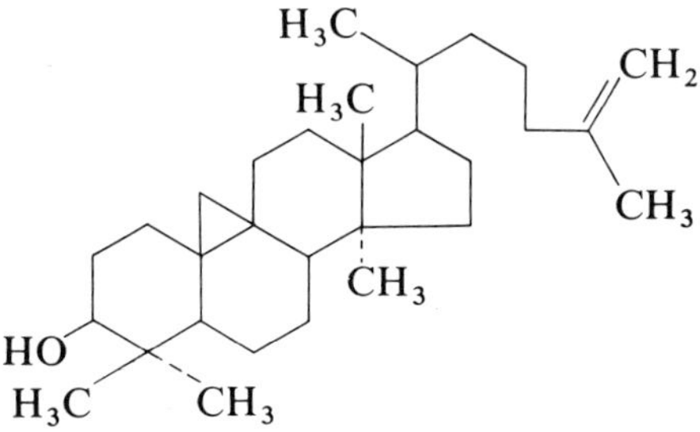

Cyclolaudinol

III. Terpenoids   179

c. Pentacyclic triterpenes

These compounds are subdivided into three groups based on established interrelationships with one of three standards, α-amyrin, β-amyrin, and lupeol.

α-Amyrin group

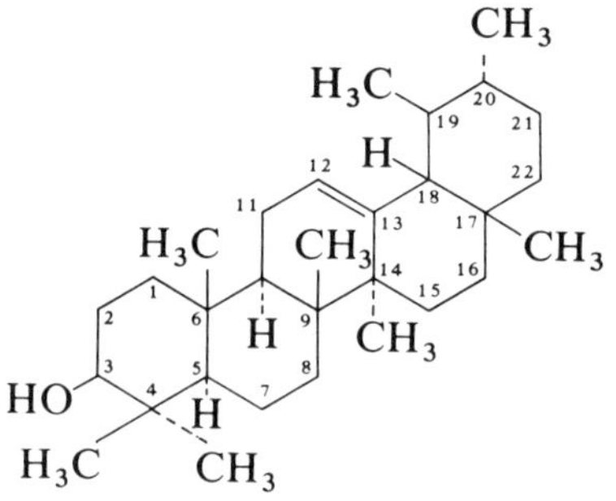

α-Amyrin

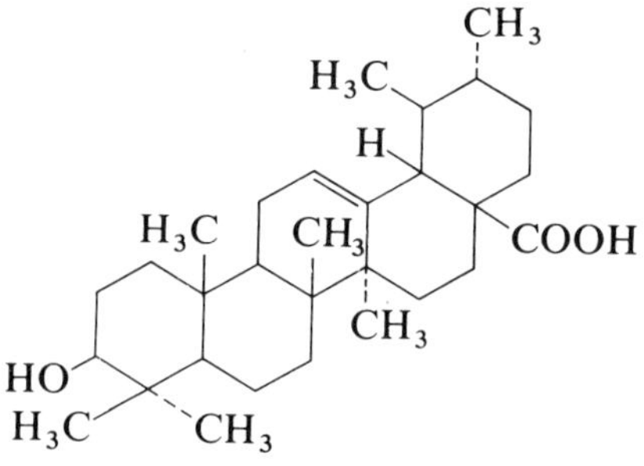

Ursolic acid

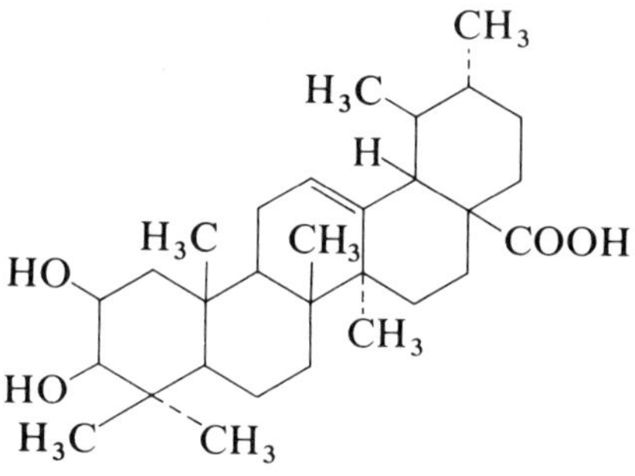

Asiatic acid

β-Amyrin group

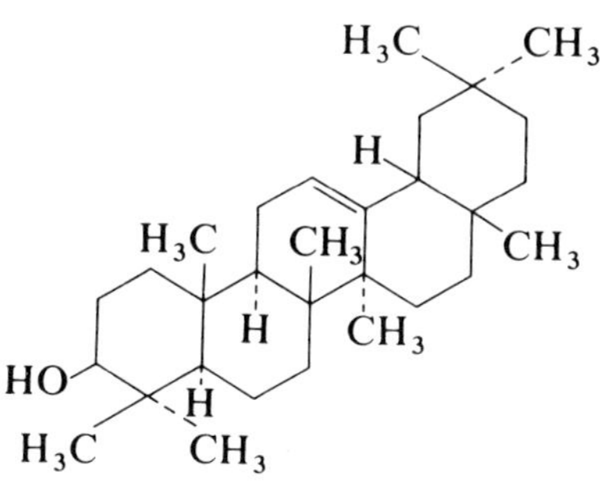

β-Amyrin

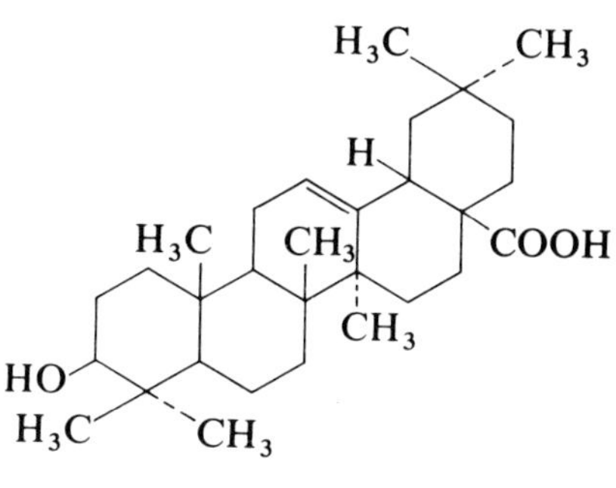

Oleanolic acid

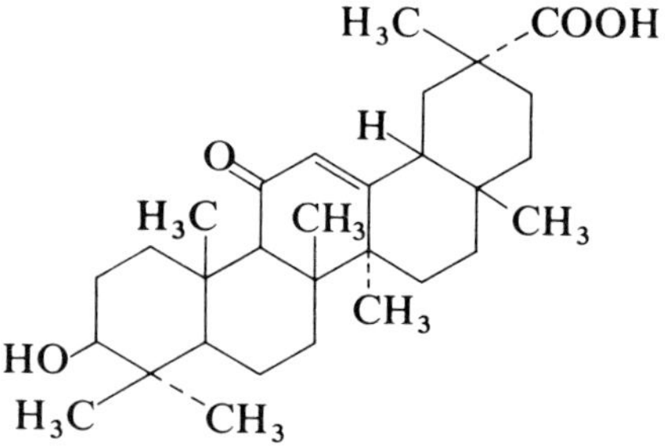

180 CHAPTER 3 Isoprenoids

Glycyrrhetic acid

Lupeol group

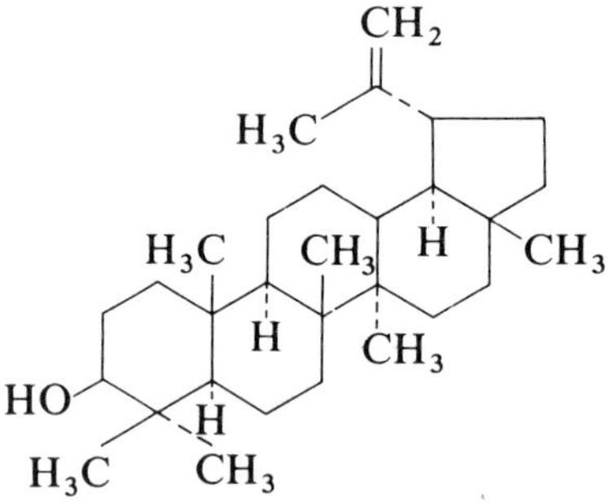

Lupeol

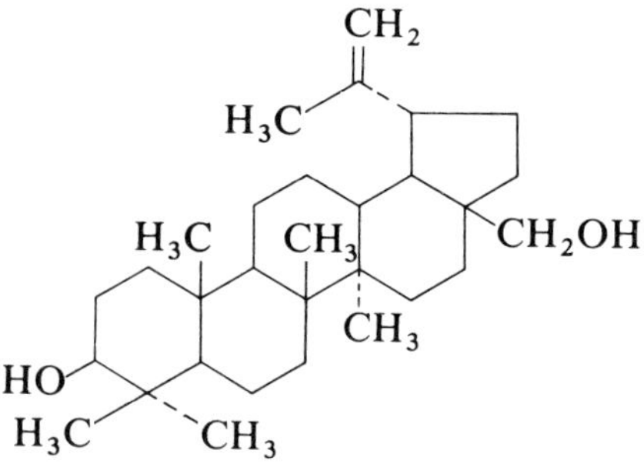

Betulin

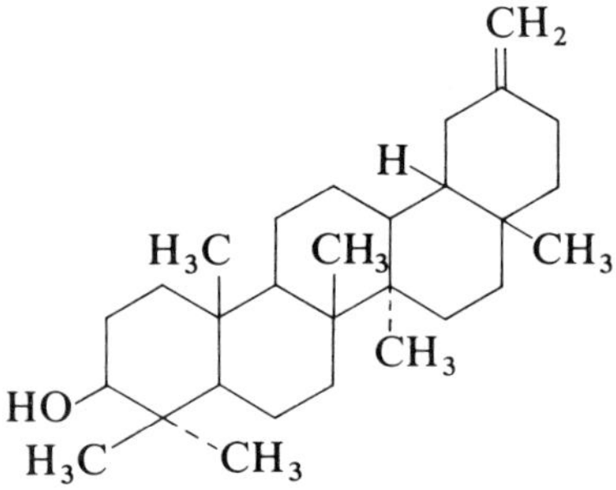

Taraxasterol

It is important to note that squalene and lanosterol-like products are intermediates in the biosynthesis of cholesterol.

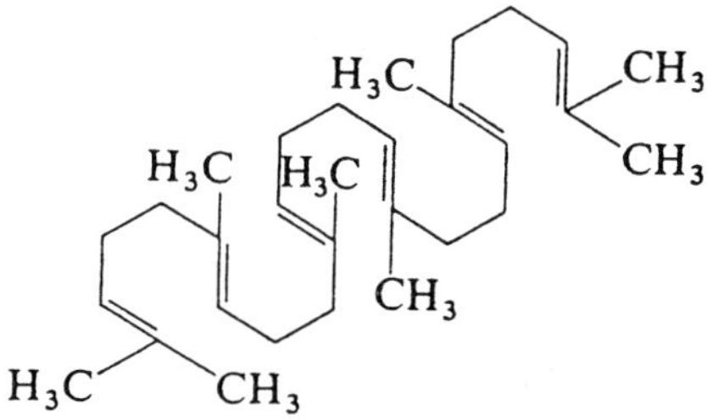

III. Terpenoids 181

Squalene

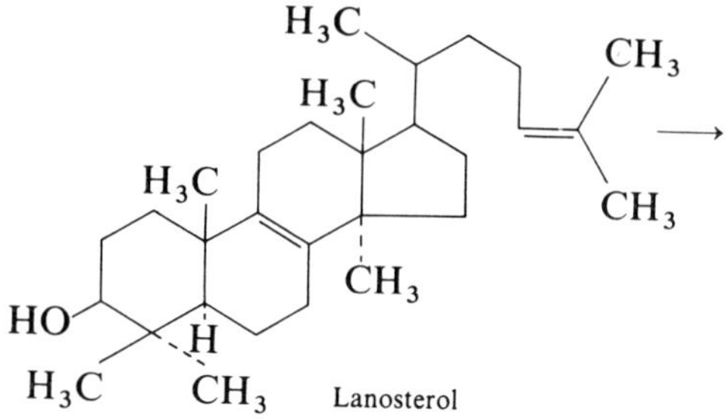

Lanosterol

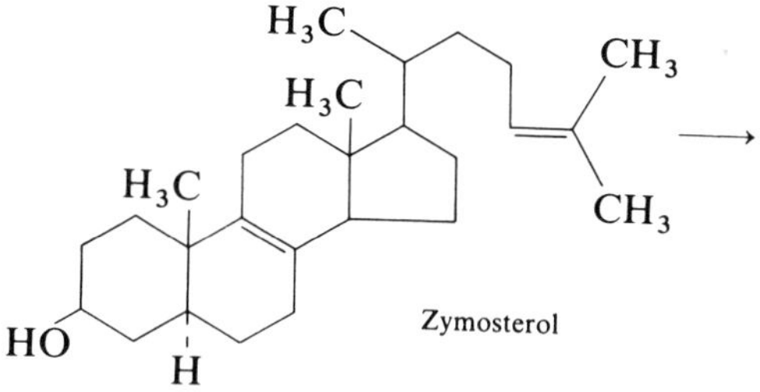

Zymosterol

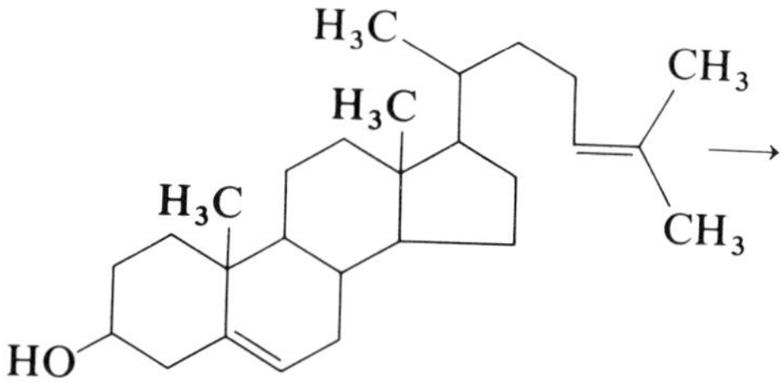

Desmosterol

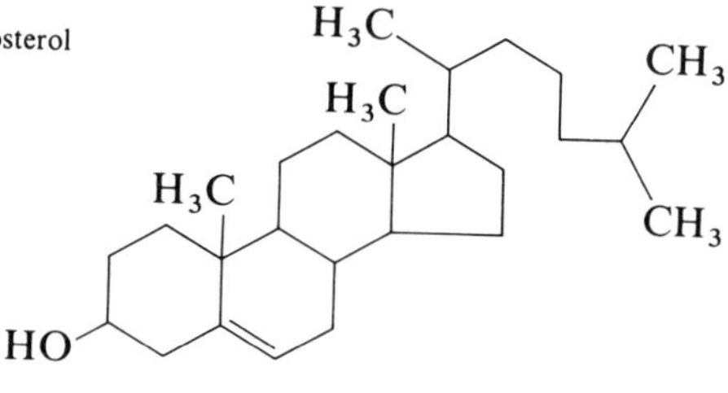

Cholesterol

B. Conversion of D-Limonene to L-Carvone and Carvacrol

1. Introduction

The terpene hydrocarbon, D-limonene, is the major constituent of all citrus peel oils. Sweet orange contains about 95% D-limonene, 1–2% n-octanal and n-decanal, 0.05–1.5% n-nonanal and D-linalool, and traces of other hydrocarbons, carbonyl compounds, alcohols, esters, and acids. Bernhard (1) applied gas-partition chromatography to the examination of lemon oil and identified about 30 constituents.

D-Limonene is structurally related to L-carvone and can be converted into it via the sequence of reactions outlined below. The conversion of

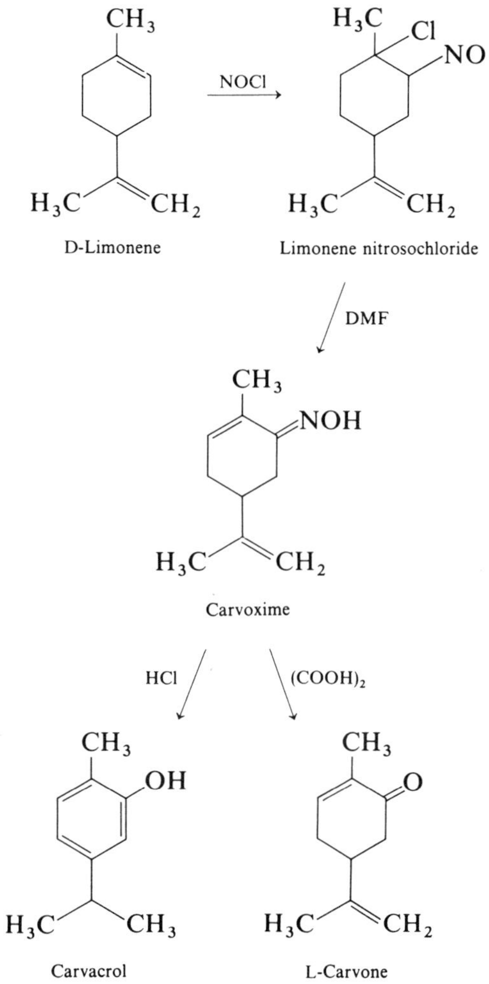

D-limonene to the nitrosochloride has been effected by the action of gaseous nitrosyl chloride (2), ethyl nitrite (3), amyl nitrite (4), and nitrogen trioxide (5). Dehydrohalogenation was effected by the following reagents: alcohol (6), alcoholic solutions of sodium hydroxide and potassium hydroxide (7), sodium methylate (8), and pyridine (3, 9). Hydrolysis of L-carvoxime to L-carvone without racemization or isomerization is readily effected by refluxing with 5% aqueous oxalic acid. The use of dilute (5N) mineral acids leads to the formation of carvacrol as the major hydrolysis product.

2. Principle

According to the following method (10, 11), limonene nitrosochloride is formed *in situ* by simultaneous addition of sodium nitrite and hydrochloric acid to limonene. It is then converted into carvoxime by boiling with dimethylformamide. Hydrolysis of L-carvoxime to L-carvone is effected with hot oxalic acid. In order to minimize the time of contact of carvoxime and carvone with the acidic media, carvone is removed from the reaction mixture by steam distillation as rapidly as it is formed.

Treatment of carvoxime with mineral acid leads to the formation of carvacrol, which may be characterized through the formation of 2-methyl-5-isopropylphenoxy-acetic acid.

3. Apparatus

Stirrer

4. Materials

Chloroform	Isopropanol
Chloroacetic acid	Methanol
Dimethylformamide	Oxalic acid
2,4-Dinitrophenylhydrazine	Semicarbazide hydrochloride
Essential oil of oranges	Sodium carbonate
Ether	Sodium hydroxide
Hydrochloric acid	Sodium nitrite

5. Time

5–6 hours

6. Procedure

a. Isolation of D-limonene from the essential oil of oranges

Crude essential oil of oranges (100 g) is fractionally distilled under reduced pressure. D-Limonene distills at 75°C/27 mm as a colorless oil weighing 75 g.

b. NMR spectrum of limonene

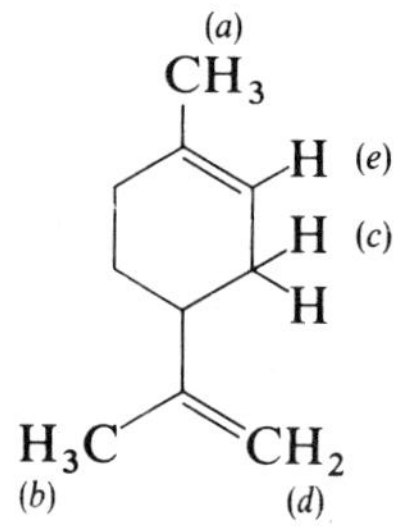

Position	δ (ppm)
a	1.64
b	1.72
c	1.97
d	4.66
e	5.35

c. Limonene nitrosochloride

A solution of 40.8 g D-limonene in 40 ml isopropanol is cooled below 10°C. To this solution are added simultaneously, through separate dropping funnels, a solution of 120 ml concentrated hydrochloric acid in 80 ml isopropanol and a concentrated aqueous solution of 20.7 g sodium nitrite. The addition is controlled so as to maintain the temperature below 10°C. The mixture is stirred for an additional 15 min and is allowed to stand in a refrigerator for 1 hr. The solid is filtered on a Buchner funnel and washed with cold ethanol; yield 50 g. It is crystallized from a mixture of chloroform–methanol, 1:3; mp 103°C.

d. Carvoxime

Limonene nitrosochloride (40 g) (crude product is permissible) and 20 ml dimethylformamide are boiled for 30 min under reflux in 125 ml isopropanol. The solution is then poured into 500 ml cracked ice and water, stirred vigorously, and filtered after the ice has melted. The solid is washed three times with 50 ml cold water and once with 15 ml cold isopropanol. The dry product weighs 27 g; mp 66–69°C. If the product is an oil, it is extracted with dichloromethane, which is then evaporated.

e. L-Carvone

A mixture of 10 g L-carvoxime and 100 ml 5% aqueous oxalic acid is refluxed for 2 hr. At the end of this period, the reaction mixture is steam distilled, and the distillate is extracted with ether. The ethereal solution is dried over sodium sulfate and fractionally distilled to give 7 g L-carvone, bp 88–90°C at 4 mm. Semicarbazone is prepared in methanol; it melts at 162°C. 2,4-Dinitrophenylhydrazone melts at 189°C.

f. Carvone

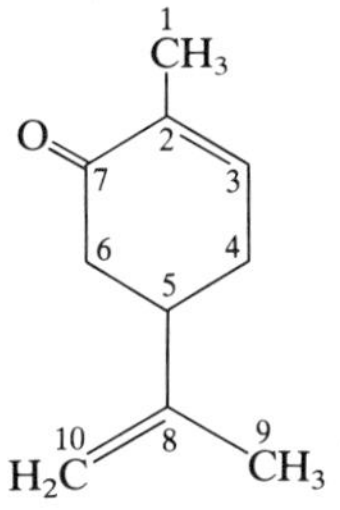

g. ¹H NMR
δ_H 1.76(6H, 2 × CH$_3$) 4.79(2H, C=CH$_2$)
2.00–2.90(5H, 2 × CH$_2$, CH) 6.74(1H, C=CH)

h. ¹³C NMR
C-1 δ 15.4   C-6   43.1
C-2   135.2   C-7   197.7
C-3   143.8   C-8   147.2
C-4   31.3   C-9   20.2
C-5   42.7   C-10 110.3

i. Mass spectrum
Carvone, C$_{10}$H$_{14}$O, Mol. wt. 150
m/z 150(1%)   82(100%)   54(67%)   41(32%)
    108(18%)   79(14%)   53(23%)   39(58%)
    93(25%)

j. IR spectrum of carvone (sandwich cell)
3320 cm^{-1}: overtone of C=O
2955, 2865: stretching of CH$_3$
2910, 2840: stretching of CH$_2$
1675: C=O stretching, α,β-unsaturated
1645: C=C stretching of terminal methylene
1450: CH$_2$ scissor
1435, 1370: CH$_2$ and CH$_3$ bending
1055: aliphatic ketone
890: out-of-plane bending of terminal methylene
802: out-of-plane bending of trisubstituted double bond

k. UV

λ_{max}^{EtOH} 235 mμ (logε 3.93); 318(1.62)

The 235 mμ peak is characteristic of a conjugated double bond system of the type C=C—C=O.

The 318 mμ peak is due to the excitation of one of the lone pair electrons of the oxygen (carbonyl absorption). Note its lower intensity.

l. Carvacrol

A mixture of 10 g L-carvoxime and 50 ml 5N hydrochloric acid is refluxed for 15 min. The mixture is steam distilled, the distillate is extracted with ether, and the ethereal solution is dried over anhydrous sodium sulfate. Removal of the ether leaves 6.5 g oil (which is soluble in 5% aqueous sodium hydroxide).

The oil is characterized as carvacrol through the formation of 2-methyl-5-isopropylphenoxyacetic acid (mp 152°C) which is prepared as follows: to a mixture of 1 g of the phenol and 3.5 ml 33% sodium hydroxide solution is added 2.5 ml 50% chloroacetic acid solution; if necessary, a little water is added to dissolve the sodium salt of the phenol. The test tube containing the solution is stoppered loosely and heated for 1 hr in a gently boiling water bath. The solution is cooled, diluted, acidified to Congo red with a mineral acid, and extracted once with ether. The ether extract is washed once with a little water, and the aryloxyacetic acid is removed by washing with dilute sodium carbonate. Acidification of this extract gives the free acid, which is recrystallized from water.

m. Thin-layer chromatography of carvone and carvacrol

A drop each of carvone and carvacrol (5% in chloroform) is spotted on a thin-layer plate (Silica Gel GF$_{254}$) and developed with chloroform–benzene 3:1.

Detection: (1) UV$_{254}$—dark spots. (2) Spraying with phosphomolybdic acid reagent, prepared by dissolving phosphomolybdic acid (20 g) in ethanol (100 g). After heating at 110°C for 10 min, blue spots on a yellow background are observed. Carvone, R_f 0.55.

References

1. Bernhard, R. A. (1958). *J. Food Res.* **23**, 213.
2. Tilden, W. A., and Shenstone, W. A. (1877). *J. Chem. Soc.* **31**, 554.
3. Royals, E. E., and Horne, S. E. (1951). *J. Am. Chem. Soc.* **73**, 5856.
4. Wallach, O. (1889). *Ann.* **252**, 109.
5. Rupe, H. (1921). *Helv. Chim. Acta.* **4**, 149.
6. Goldschmidt, H., and Zurrer, R. (1885). *Ber.* **18**, 2220.
7. Wallach, O. (1892). *Ann.* **270**, 175.
8. Deussen, E., and Hahn, A. (1909). *Ann.* **369**, 60.
9. Wallach, O. (1918). *Ann.* **414**, 257.
10. Reitsema, R. H. (1958). *J. Org. Chem.* **23**, 2038.
11. Earl, E. E., and Horne, S. E. (1951). *J. Am. Chem. Soc.* **73**, 5856.

Recommended Reviews

1. Guenther, E. (1949). *Essential oils of the genus Citrus. In* "The Essential Oils," Vol. 3, p. 5. Van Nostrand, New York.

2. Kefford, J. F. (1959). *The chemical constituents of citrus fruits. In "Advances in Food Research."* (C. O. Chichester, E. M. Mrak, and G. F. Stewart, eds.), Vol. 9, p. 286. Academic Press, New York.
3. Kirchner, J. G. (1961). *Oils of peel, juice sac, and seed. In "The Orange"* (W. B. Sinclair, ed.), p. 265. University of California Press, Berkeley, California.
4. Shaw, P. E., and Wilson, C. W. (1980). Importance of selected volatile components to natural orange, grapefruit, tangerine, and mandarin flavors. *In* "Citrus Nutrition and Quality." (S. Nagy, and J. A. Attaway, eds.), p. 167, ACS Symposium Series 143.

C. Isolation and Determination of Optically Pure Carvone Enantiomers from Caraway (*Carum carvi* L.) and Spearmint (*Mentha spicata* L.)

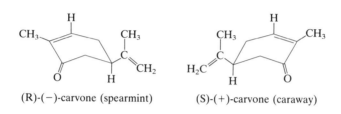

(R)-(−)-carvone (spearmint)   (S)-(+)-carvone (caraway)

1. Introduction

Optical purity of enantiomers is of great importance in the chemistry of natural products, especially in the fields of flavors, fragrances, and pheromones. Odor receptors in mammals and in insects can discriminate between enantiomers, so it is important to describe the enantiomeric composition of chiral molecules that impinge on these receptors.

The human olfactory organ is capable of distinguishing between chiral compounds. Odor quality and potency of enantiomeric compounds may show considerable differences. Thus, a distinct differentiation in odor perception could be observed in pairs of enantiomeric monoterpenoid and sesquiterpenoid odorants. The intensity of the odor of enantiomeric compounds exhibits as great a variability as does the quality. In theory, chiral receptor sites should acknowledge the spatial arrangements of enantiomers; careful experiments have shown that some insects are capable of making the distinction.

Methods for the determination of enantiomeric composition are as follows:

1. Polarimetry;
2. NMR (^{1}H and ^{13}C): chiral shift reagent; chiral derivatizing agent; chiral solvating agent;
3. Achiral chromatography (GC or HPLC): separation of diasteriomeric derivatives;

4. Chiral chromatography (GC or HPLC): separation of enantiomers; and
5. X-ray crystallography of a diasteriomeric derivative.

The classical procedure for determination of enantiomeric composition (or optical purity) is polarimetry. The basic requirements are a relatively large sample, a relatively high specific rotation, and a knowledge of the specific rotation of the pure enantiomer under similar conditions. High-performance methods of enantiomer analysis are urgently needed in order to keep pace with the development of highly enantioselective reactions and to cope with the requirements of mechanistic investigations.

The simple and effective method for determination of enantiomeric composition is NMR-polarimetry—NMR spectroscopy using chiral Lanthanide shift reagents (LSR). LSR are the europium and praseodym salts of β-diketones:

Ln = Eu   Eu(tfc)$_3$
Ln = Pr   Pr(tfc)$_3$

Eu(dcm)$_3$

Ln = Eu   Eu(hfc)$_3$
Ln = Pr   Pr(hfc)$_3$

These chiral LSR are used to determine the enantiomeric purity of optically active compounds (1). LSR interacts diastereomerically with a pair of enantiomers forming a chelate; this leads to separated NMR signals for corresponding nuclei in the two enantiomers. Europium chelates generally cause downfield shifts, whereas the praseodym analogues effect upfield shifts. The LSR reagents induce a considerable spread of the chemical shifts without significantly broadening the lines (2). The method is applicable to a much broader range of compounds than some of the other techniques.

McCreary and colleagues (3) reported the first use of chiral LSR with ^{1}H-NMR spectrometry to determine the enantiomeric composition of the monoterpenes camphor, menthol, and menthyl acetate. Plummer and associates (4) and Borden and colleagues (5) determined the enantiomeric purity of several pheromones, including the monoterpene alcohols *trans*-verbenol, ipsdienol and sulcatol using a chiral shift reagent. In the following experiment the determination of the enantiomeric composition of natural carvones isolated from various essential oils, using a chiral LSR, tris[3-heptafluoropropyl-hydroxymethylene)-(+)-camphorato europium] [Eu(hfc)$_3$], is described.

Carvone is the major constituent in the essential oils of the ripe fruits from caraway (51.2%) and of the fresh leaves from *Menta spicata* (69.9%). It exists in two enantiomeric forms and has different physical and physiological properties. S(+) carvone has a caraway odour, and is detectable at rather high threshold. It is used as a starting material for the synthesis of a pheromone component of the female California red scale (6). R(−) carvone has a spearmint odor, is detectable at a lower threshold than S(+) and is used as a starting material in the preparation of picrotoxinin (7).

The enantiomeric purity of S(+) and R(−) carvones, isolated from natural oils, is easily determined by NMR-polarimetry using LSR.

2. Apparatus

Flash chromatography assembly
Gas chromatograph
NMR instrument

3. Materials

Caraway seeds	Eu(hfc)$_3$	Silica gel 60Å
Chromosorb W	Hexane	Spearmint seeds
Ethyl acetate		

4. Procedure

Each of the commercial variety of *Mentha spicata* (received from a nursery) and ripe fruit of caraway are steam-distilled for 1 hr. The essential oils are dried over anhydrous sodium sulfate. Carvone is isolated from the oil by flash chromatography (as described in the experiment Flash Chromatography of Essential Oil Constituents, p. 200) on silica gel Davisil 60Å, 200–425 mesh, and eluted by 1 to 5% solution of ethyl acetate in hexane. The chemical purities of natural and commercial samples of carvone are detected by gas chromatography on a packed glass column (3m × 4mm inside diameter [i.d.]) with 5% Carbowax 20M on acid-washed silanized Chromosorb W (80–100 mesh). Operating conditions: temperature programming 70–200°C (5°C/min), carrier gas nitrogen, flow 30 ml/min. Optical rotations are measured (in ethanol) using a polarimeter.

Before the determination of the enantiomeric composition of natural carvone, it is essential to carry out the NMR measurements of a mixture of the two synthetic enantiomers of carvone in the presence of the chiral shift reagent Eu(hfc)$_3$. (The pure synthetic enantiomers can be obtained from certain chemical companies). These measurements provide an important information regarding the L:S (lanthanide to substrate) ratio, which leads to a good separation of the two synthetic enantiomers and hence for the natural enantiomeric mixture (such as (S)-(+) and (R)-(−)-carvones).

Equal volumes (5 µl) of each of the S and R-carvones (1:1 ratio), are placed in a NMR tube, 300 µl of deuterochloroform (CDCl$_3$) and 50µl of 1% Tetramethyl silane (TMS) in deuterochloroform are added. The NMR spectrum shows no difference of the two enantiomers (there is no separation in NMR signals of the nuclei near asymmetric carbon). The Eu(hfc)$_3$ shift reagent is then added portionwise to the NMR tube (the first portion is 5 mg, ≈3.7 µmole) and the NMR spectra are recorded until the splitting of the olefinic C$_7$ methyl protons singlet is observed. In order to determine which

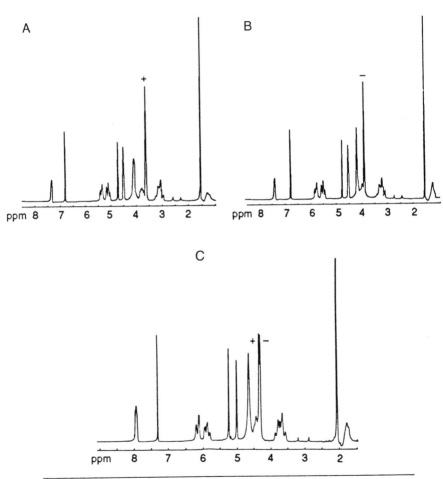

Figure 3.1 A: ^{1}H-NMR spectra of (S)-(+)-carvone with Eu(hfc)$_3$ at [L]/[S] = 0.7–0.8; B: ^{1}H-NMR spectra of (R)-(−)-carvone with Eu(hfc)$_3$ at [L]/[S] = 0.7–0.8; C: ^{1}H-NMR spectra of (RS)-carvone with Eu(hfc)$_3$ at [L]/[S] = 0.7–0.8.

peak corresponds to the S and the R enantiomer, the ratio of the two enantiomers is increased to 1:2.

Having found the right conditions and the L:S ratio for the satisfactory separation of the synthetic enantiomers, the natural product is measured using the same L:S ratio. The fact that there is no splitting of the singlet of the allylic C-7 methyl protons in the NMR spectrum confirms the enantiomeric purity of the natural product.

In Figs. 3-1 (a–c), the ^{1}H-NMR spectra of R, S, and RS-carvones with Eu(hfc)$_3$ reagent at [L]/[S] = 0.7–0.8 is presented.

References

1. Sullivan, G. R. (1978). *Topics in Stereochemistry* **10**, 287.
2. Martin, M. L., Delpuech, J. J., and Martin, G. J. (1980). "Practical NMR Spectroscopy." Heyden and Sons, London.
3. McCreary, M. D., Lewis, D. W., Warnick, D. L., and Whitesides, G. M. (1974). *J. Amer. Chem. Soc.* **96**, 1038.
4. Plummer, E. L., Stewart, T. E., Pyrne, K., Pierce, C. T., and Silverstein, R. M. (1976). *J. Chem. Ecol.* **2**, 307.
5. Borden, J. H., Handley, J. R., McLean, J. A., Silverstein, R. M., Chang, L., Slesser, K. N., Johnson, B. D., and Shuler, H. R. (1980). *J. Chem. Eco.* **6**, 445.
6. Roelfs, W., Gieselmann, M., Carde, A., Tashiro, H., Moreno, D. S., Henrick, C. A., and Anderson, R. J. (1978). *J. Chem. Ecol.* **4**, 211.
7. Corey, E. J., and Pearce, H. L. (1979). *J. Amer. Chem. Soc.* **101**, 5841.
8. Ravid, U., Bassat, M., Putievsky, E., Weinstein, V., and Ikan, R. (1987). *Flavour and Fragrance Journal* **2**, 95.

Recommended Reviews

1. Pruthi, J. S. (1980). Academic Press, "Spices and Condiments: Chemistry, Microbiology, Technology."
2. Ravid, U., Putievsky, E., Bassat, M., Ikan, R., and Weinstein, V. (1988). Determination of the enantiomeric composition of natural and nature-identical semiochemicals by ^{1}H-NMR "polarimetry." In "Bioflavour '87" (P. Schreier, (ed.), p. 97. Walter de Gruyter and Co.
3. Ravid, U., Putievsky, E., Bassat, M., Ikan, R., and Weinstein, V. (1988). The use of chiral lanthanide shift reagent to determine the enantiomeric purity of some essential oil constituents. *Flavour and Fragrance Journal* **3**, 117.

D. Transformation of α-Pinene to Camphor

1. Introduction

α-Pinene is widely distributed in nature. It is present in the majority of the essential oils of conifers and is the principal constituent of turpentine oil. The name pinene was selected to indicate its association with the various species of *Pinus*. By a series of transformations, it is possible to convert α-pinene to camphor. Pinene hydrochloride is stable only at low temperatures and readily undergoes rearrangement to form bornyl chloride. Furthermore, the dehydrochlorination of bornyl chloride with a base does not yield the expected bornylene, but camphene. On treatment with acetic anhydride, camphene forms isobornyl acetate, which is treated with a base to form isoborneol. Upon oxidation with chromic acid, isoborneol yields camphor.

Camphor is used as a counterirritant, anesthetic, and mild antiseptic.

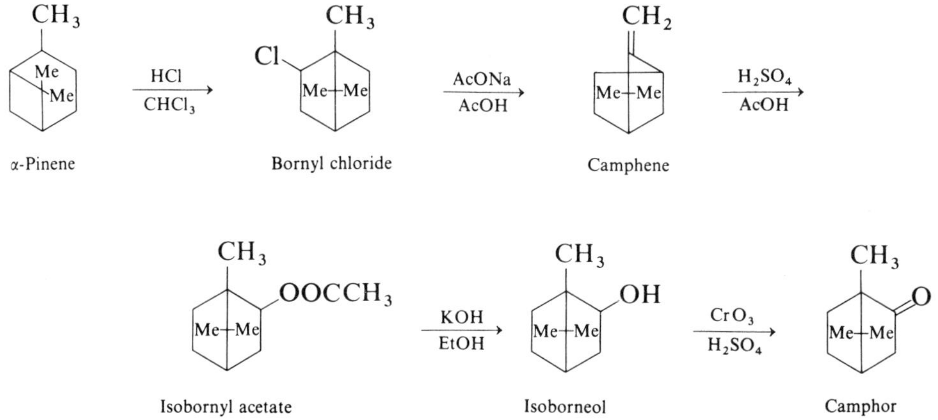

$$\alpha\text{-Pinene} \xrightarrow[CHCl_3]{HCl} \text{Bornyl chloride} \xrightarrow[AcOH]{AcONa} \text{Camphene} \xrightarrow[AcOH]{H_2SO_4}$$

$$\text{Isobornyl acetate} \xrightarrow[EtOH]{KOH} \text{Isoborneol} \xrightarrow[H_2SO_4]{CrO_3} \text{Camphor}$$

2. Principle

Many rearrangements of terpenes are of the Wagner-Meerwein type, which take place via the formation of a carbonium ion. They may be divided into two groups—eliminative and noneliminative changes.

The first stage of the transformation of α-pinene to camphor, the transformation of pinene hydrochloride into bornyl chloride, probably proceeds by the following mechanism:

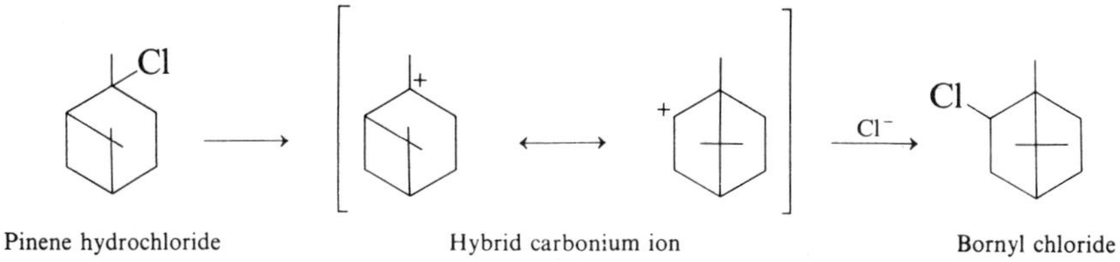

Pinene hydrochloride → [Hybrid carbonium ion] $\xrightarrow{Cl^-}$ Bornyl chloride

The formation of camphene from bornyl chloride, a typical elimination reaction proceeds via a similar mechanism:

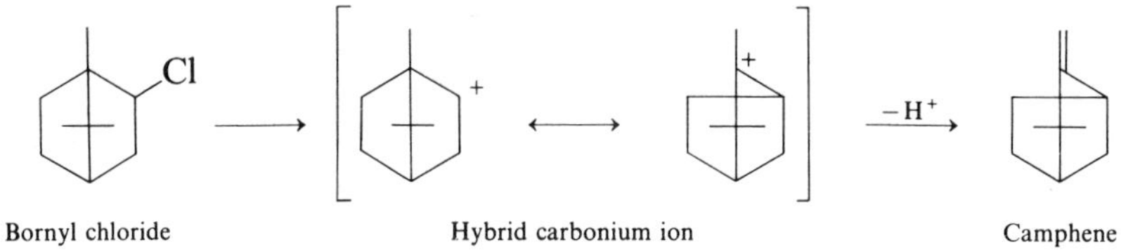

Bornyl chloride → [Hybrid carbonium ion] $\xrightarrow{-H^+}$ Camphene

The stereospecificity of the rearrangement of camphene hydrochloride to isobornyl chloride requires a special explanation. Representing camphene hydrochloride by the stereochemical formula 1, we see that the positively charged organic moiety resulting from chloride anion extraction can be represented by resonance hybrid forms 2a, 2b, and 2c, or by cation 2. Such a species, in which the positive charge is delocalized by means of a multicenter molecular orbital formed by sigma overlap of atomic orbitals, is called a non-classical carbonium ion.

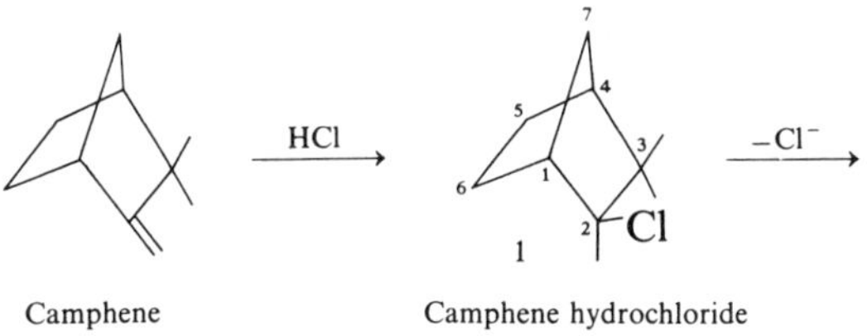

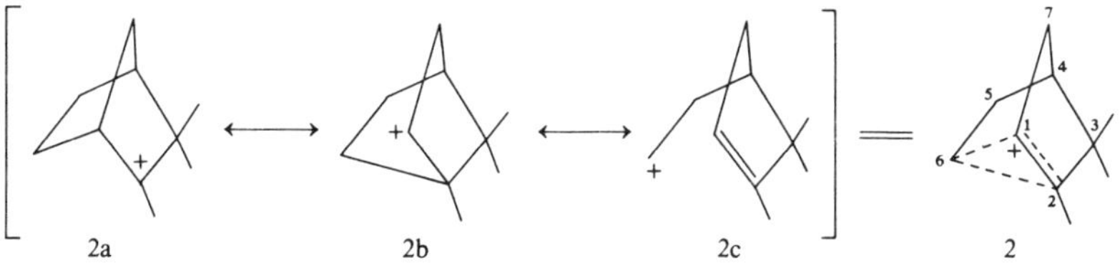

Alkaline hydrolysis of isobornyl acetate forms isoborneol, which is oxidized to camphor by means of chromic acid.

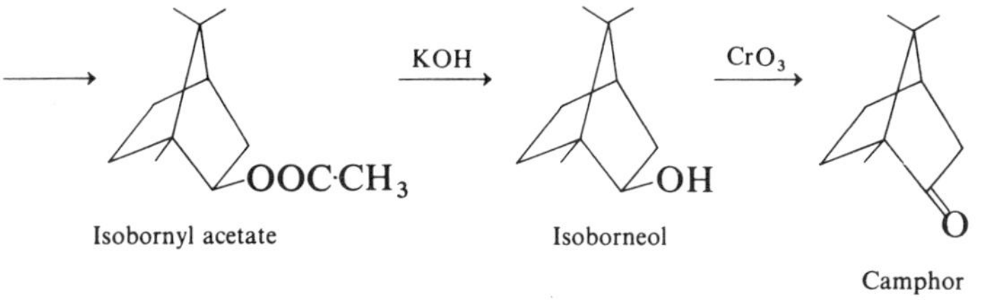

3. Apparatus

Steam generator

4. Materials

Acetic acid	Potassium hydroxide
Acetone	Sodium acetate
Chloroform	Sodium bicarbonate
Chromium trioxide	Sodium carbonate
Ethanol	Sulfuric acid
Hydrochloric acid, dry	Turpentine

194 CHAPTER 3 Isoprenoids

5. Time

10 hours

6. Procedure

a. Isolation of α-pinene from turpentine oil
A reasonable amount of commercial turpentine oil is dried with sodium sulfate and fractionally distilled. α-Pinene distills at 154 to 155°C and β-pinene, at 163 to 164°C.

b. Pinene hydrochloride
Pure pinene is mixed in a flask with an equal volume of chloroform, the flask is cooled below 0°C, and dry hydrochloric acid (gas) is bubbled through the solution until it becomes saturated. An equal volume of water is added, and excess acid is neutralized with sodium bicarbonate. The mixture is immediately subjected to steam distillation; the product solidifies and is recrystallized from ethanol; mp 130–131°C.

c. Camphene
One part pinene hydrochloride is mixed with one part sodium acetate (freshly fused) and two parts glacial acetic acid, and heated at 200°C for 3 to 4 hr. Steam distillation gives pure DL-camphene; mp 50°C, bp 159–160°C.

d. NMR spectrum of camphene

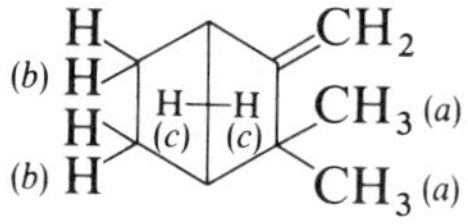

Position	δ (ppm)
a	1.01
b	1.52
c	1.52

e. Isobornyl acetate
Camphene (10 g) in 30 ml acetic acid and 2 ml 50% sulfuric acid are heated on a water bath for 20 min. Twenty ml water is added, and the mixture is allowed to cool. The ester is separated and washed with dilute sodium carbonate and water, and dried. It is then distilled under reduced pressure; bp 105°C/15 mm; yield, 10 g.

f. Isoborneol
Ten g isobornyl acetate and 5 g potassium hydroxide in 30 ml ethanol and 10 ml water are refluxed for 1 hr on a water bath. Cold water is added, and

the solidified isoborneol is filtered off on a Buchner funnel and washed with water; yield, 8 g.

g. Camphor

To 5 g isoborneol in 10 ml acetone is added 7 ml of a solution prepared by dissolving 18 g chromium trioxide in 16 ml sulfuric acid. 30 ml water is then added, and the mixture is left for 1 hr at room temperature. It is then subjected to steam distillation, and the solid camphor is filtered and washed with water; yield, 2 g. Sublimation yields pure camphor, mp 176°C.

h. UV Spectrum of camphor

λ_{max}^{EtOH} 289mμ (logε 1.51)

The peak is due to carbonyl absorption.

i. NMR Spectrum of camphor

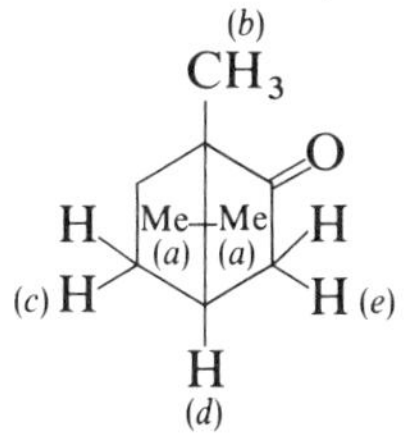

Position	δ (ppm)
a	0.86
b	0.97
c	1.49
d	1.71
e	2.04

Recommended Reviews

1. Berson, J. A. (1964). *Carbonium ion rearrangements in bridged bicyclic systems.* In "*Molecular Rearrangements*" (P. de Mayo, ed.), Vol. 1, p. 111. Interscience, New York.
2. Sargent, G. D. (1966). *Bridged, non-classical carbonium ions.* Quart. Revs. **20**, 301.

E. Photoprotonation of Limonene to *p*-Menth-8-en-1-yl Methyl Ether

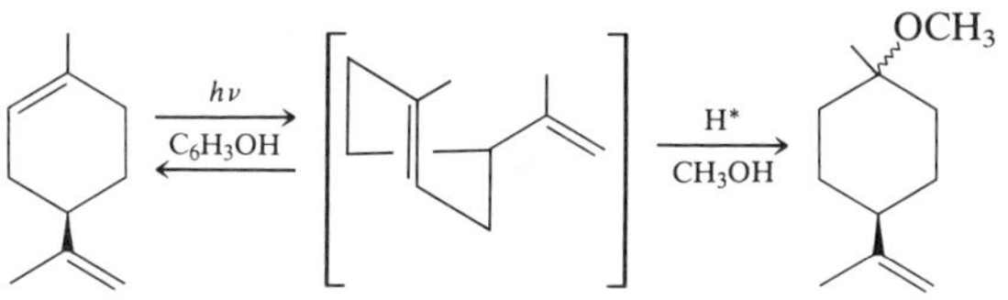

Limonene

p-Menth-8-en-1-yl Methyl Ether

196 CHAPTER 3 Isoprenoids

1. Introduction

Acid-catalyzed ground state additions to limonene generally afford a mixture of products resulting from competing protonation of both double bonds. The photoprotonation of cycloalkenes, described in this procedure, is believed to proceed via initial light-induced cis → trans isomerization of the alkene. The resulting highly strained trans isomer undergoes facile protonation.

2. Principle

The following (1) procedure permits the protonation of cyclohexenes and cycloheptenes under neutral and mildly acidic conditions. In addition to facilitating protonation of cycloalkenes, the procedure affords a means of selectively protonating a double bond located in a six- or seven-membered ring, in the presence of another double bond contained in a acyclic, exocyclic, or larger-ring cyclic environment.

3. Apparatus

Photochemical reactor
Gas chromatograph equipped with a stainless steel column (3m × 3.2m) packed with 20% SF-96 on Chromosorb W.

4. Materials

(+)-Limonene   Phenol
Methanol   Sodium hydroxide

5. Time

48 hours for complete conversion

6. Procedure

A 250-ml photochemical reactor (Fig. 3.2) (equipped with a 450 watt Hanovia mercury lamp, a water-cooled condenser, and a trap filled with methanol) is flushed with nitrogen and charged with a solution of 20 g (+)-limonene, 5 g phenol, and 5 drops concentrated sulfuric acid in 210 ml anhydrous methanol. Water flow through the condenser is started, and the nitrogen flow is adjusted to provide agitation of the contents of the vessel. After 15 min, irradiation is started and the reaction followed by GLC. A 3 m × 3.2 mm i.d. stainless steel column packed with 20% SF-96 on Chromosorb W (60–80 mesh) is used. The flow rate of helium is 60 ml/min. Temperature programming 10°C/min (from 50 to 200°C), the retention time of the cis and the trans isomers are 17.9 and 18.7 min, respectively. About 48 hr is the approximate time needed for essentially complete conversion. The solution is then poured into 900 ml 5% aqueous sodium hydroxide solution containing 125 g sodium chloride, and the

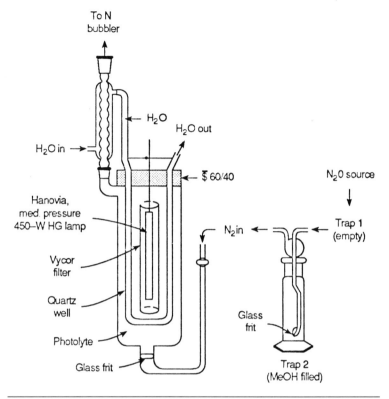

Figure 3.2 A 250-ml photochemical reactor with a 450-watt Hanovia mercury lamp, a water-cooled condenser, and a trap filled with methanol.

mixture is extracted with two 100-ml portions of ether. The ether layers are combined, washed with 50 ml saturated sodium chloride solution, and dried over anhydrous sodium sulfate. The drying agent is removed by filtration, and the filtrate is concentrated under reduced pressure. The liquid is distilled at reduced pressure through a short column. The material that distills at 90 to 95°C (10 mm) consists of a mixture of *cis*- and *trans*-p-menth-8-en-1-yl methyl ether; yield, 13 g.

The products can be checked by GLC using 2.4 m × 3.2 mm i.d. stainless steel column packed with 7% SE-30 on Chromosorb W (60–80 mesh) at 160°C. The retention time of the trans isomer is 1.6 min, and that of cis is 1.8 min; flow rate of helium is 55 ml/min.

a. IR (neat) cm^{-1}
3080 (=C—H); 2964, 2939, 2860, 2825 (C—H), 1645 (C=C); 1464, 1453 and 1442 (overlapping peaks): 1370, 1124 and 1082 (C—OC); 885 (=CH).

b. ¹H NMR (CDCl₂)

δ 1.10 [s, 3H, CH₃ (trans)], 1.19 [s, 3H, CH₃ (cis)], 1.30–2.00 (8H, CH₂—), 1.71 [s, 3H, CH₃ (cis/trans)], 3.14 [s, 3H, OCH₃ (trans)], 3.21 [s, 3H, OCH₃, (cis)], 4.69 [s, 2H, =CH₂ (cis/trans)].

Reference

1. Tise, F. P., and Kropp, P. J., (1983). *Org. Synth.*, **61**, 112.

Recommended Reviews

1. Kropp, P. J. (1978). Photochemistry of alkenes in solution. *Molecular Photochemistry* **9**, 39.
2. Schuster, D. I. (1987). Organic photochemistry. *In* "Encyclopedia of Physical Science and Technology," Vol. 10, p. 375. Academic Press, New York.

F. Flash Chromatography of Essential-Oil Constituents

1. Introduction

Essential oils are complex and volatile mixtures of terpenoids and phenylpropane derivatives. The number of components in any single oil may vary from a few to over 100. These multicomponent mixtures make the analysis of the oils by GC rather difficult because of the overlap of peaks and similarity of retention times. Thus, column chromatographic prefractionation of the oils before the GC analysis is of prime importance. The conventional long column chromatography for separating complex mixtures of organic compounds is time consuming and frequently gives poor recovery owing to band tailing.

A substantially faster technique for the separation of complex mixtures is flash chromatography. It is basically an air pressure–driven hybrid of medium pressure and short column chromatography, which has been optimized for rapid separations. The apparatus required for this technique consists of a chromatography column, nitrogen cylinder, and a flow-controller valve. The flash chromatographic technique can be summarized as follows:

1. A solvent is chosen that gives a good separation and moves the desired component to $R_f \sim 0.3$ on analytical TLC;
2. The column of appropriate diameter is selected. The ratio of the essential oil and the adsorbent (such as silica gel (40–60 μm) is about 1:30;
3. The column is filled with nonpolar solvent, and pressure is used on the top to rapidly push all the air from the silica gel; and
4. The sample is applied, and the column is refilled with the solvent and eluted at a flow rate of 2 in/min.

Fractions are monitored by TLC (using UV-detection lamp or sprayed with an appropriate chromogenic reagent such as 2,4-DNP). Increasing amounts of ethyl acetate in hexane or petroleum ether appeared to be the most generally satisfactory eluant. The order of elution of essential oil–oxygenated constituents with an increasing gradient of ethyl acetate in hexane is as follows: oxides, esters, ketones, and alcohols.

2. Principle

Carvone is a natural terpenoid ketone found in the essential oils of caraway seeds and the spearmint plant. The fact that caraway and spearmint exhibit distinctly different odors suggests that the carvones occurring in these oils are stereoisomers. The relatively polar carvone is adsorbed more strongly by silica gel than are the terpenoid hydrocarbons (and other components), which can be eluted from the column by a nonpolar solvent such as hexane. The polarity can then be increased by adding increasing amounts of ethyl acetate to elute the carvone. The optical rotations and the infrared spectra of the enantiomeric carvones are then measured by a polarimeter.

3. Apparatus

Glass column for flash chromatography

4. Materials

Clean sand
Hexane
Ethyl acetate
Essential oils of caraway, spearmint, thyme, nutmeg, clary sage, lavender, citrus
Silica gel (40–60 mesh)
Microscope TLC plates.

5. Procedure

Having chosen a column of the appropriate diameter, a small plug of glass wool (A) is placed in the tube connecting the teflon stopcock and the column body, or preferably a sintered glass disc is fused here. Dried 40–60 mesh silica gel is poured into the column in a single portion to give a depth of 5.5 to 6 in. The column is then gently tapped vertically on the bench top to pack the gel. Next, a 1/8-inch layer of sand is carefully placed on the flat top of the dry gel bed, and the column is clamped for pressure packing and elution. The solvent chosen is then poured carefully over the sand to fill the column completely. The stopcock of the flow controller (Fig. 3.3) is opened all the way; this will cause the pressure of nitrogen (or air) above the adsorbent layer to climb rapidly and compress the silica gel, as solvent is rapidly forced through the column. It is important to maintain the pressure until the air is expelled. The

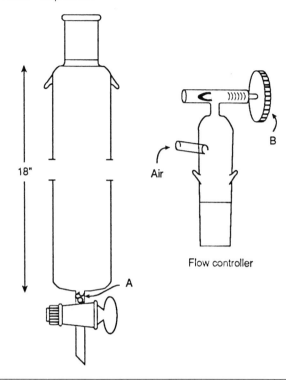

Figure 3.3 Flash chromatography column with (A) small plug of glass wool, and (B) stopcock of flow controller.

pressure is then released and excess eluant is forced out of the column. The essential oil (of caraway or spearmint) is applied by a pipette as a 20% solution in the eluant to the top of the adsorbed bed, and the pressure is applied to push all the sample into the silica gel (for each gram of oil, 30 g of silica gel is used). The flow of nitrogen is then adjusted to cause the surface of the solvent in the column to fall 2.0 in/min. The fractions eluted are checked on TLC plates as described in the introduction. The fractions containing various series of compounds (hydrocarbons, acetates, ethers, ketones, alcohols) are then run on GC columns and finally identified by GC/MS, IR, NMR and other techniques. **Note: Never let the upper part of the adsorbent in the column get dry!**

Recommended Reviews

1. Clark-Still, W., Kahn, M., and Mitra, A. (1978). Rapid chromatographic technique for preparative separations with moderate resolution. *J. Org. Chem.* **43,** 4923.

2. Crane, L. J., Zief, M., and Howarth, J. (1981). Low-pressure preparative liquid chromatography. *American Laboratory.*
3. Rogers, J. A. (1981). Essential oils. *In* "Kirk-Othmer Encyclopedia of Chemical Technology," Vol. 16, p. 307. John Whiley.

G. Isolation and Determination of Grape and Wine Aroma Constituents

1. Introduction

The aromatic compounds of some grape varieties are present in the berry either in a free state or bound to sugars in the form of glycosides. (1–3) The free forms are present in different quantities in different varieties; in muscats in particular, they form the major part of the aroma and consist essentially of terpenols in various states of oxidation and of terpenoid polyols (Fig. 3.4).

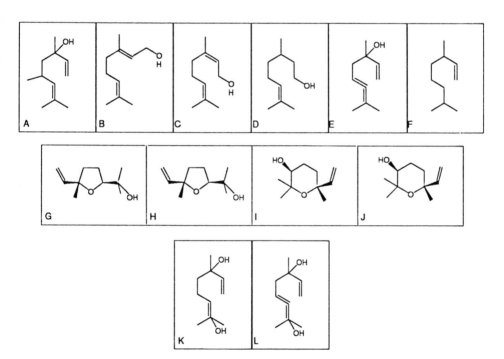

Figure 3.4 Aromatic compounds of grape varieties, including (A) linalool; (B) geraniol; (C) nerol; (D) citronellol; (E) hotrienol; (F) *p*-menth-1-ene-9-ol; (G) *cis*-furan-linalool oxide; (H) *trans*-furan-linalool oxide; (I) *cis*-pyran-linalool oxide; (J) trans-pyran-linalool oxide; (K) 3,7-dimethylocta-1-ene-3,7-diol; and (L) 3,7-dimethylocta-1,5-diene-3,7-diol (and also benzyl- and 2-phenylethyl alcohols).

The nonvolatile, nonaromatic bound fraction consists of disaccharide glycosides, namely α, L-arabinofuranosyl-β-D-glucopyranosides or α, L-rhamnofuranosyl-β, D-glucopyranosides, the aglycone of which is a terpenol, such as geraniol, linalool and nerol, a terpene diol, 2-phenylethanol, and benzyl alcohol.

2. Principle

In the following procedure (4), the free and the bound aromas of grape juice or wine are separated on XAD-2 column. The bound fraction (terpene glycosides) is cleaved by an enzyme (Pectinol) into free terpenes and sugars. The free terpenes of both fractions are determined by GC on capillary column.

3. Apparatus

Centrifuge   Glass capillary column
Gas chromatograph   Mixer

4. Materials

Amberlite XAD-2 resin   Geraniol   2-Octanol
Benzyl alcohol   Grapes   Pectinol
Citrate–phosphate buffer   Hydrogen (cylinder)   Pentane
Diethyl ether   Linalool   Phenylethyl
Ethanol   Methanol   alcohol
Ethyl acetate   Nerol   Sorbic acid
        Wine

5. Time

15 hours

6. Procedure

a. Plant material

Fresh grapes or wines can be used.

b. Juices and wines

Fresh grapes are cooled to 1°C immediately after picking, then crushed and pressed, and the juice is centrifuged (9000 rpm, 15 min). All these operations should be carried out at 1°C. Sorbic acid (200 ppm) is then added to the clarified juice. When the analysis is to be carried out some time after the picking, immediately after harvesting, the grapes are deep-frozen at $-20°C$ and stored at this temperature. They are thawed to 1°C just before analysis and then crushed in a whirlmixer; the pulp is filtered through gauze and centrifuged at the same temperature. The clear juice is stabilized with sorbic acid. Wines are diluted in an equal volume of water before analysis.

c. Column preparation

Amberlite XAD-2 resin suspended in methanol is poured into a glass column (35×1 cm i.d.) whose lower part is fitted with a Teflon stopcock and a glass-wool layer. The packed column should contain about 10 cm resin. Several 25 ml volumes of methanol and then of diethyl ether are passed through it, followed finally by 50 ml distilled water. The column is now ready for use.

d. Fractionation and determination of free and bound fractions of the aroma

A sample of 50 to 100 ml of juice or wine is passed through the column with 10 μl of 0.1% solution of 2-octanol in ethanol (added as internal standard). The flow rate is about 2 to 2.5 ml/min. The column is then rinsed with 50 ml of water to eliminate sugars, acids, and other water-soluble constituents.

e. Fractionation of the free fraction

The free fraction of the aroma fixed on the column is eluted using 50 ml of pentane at a flow rate of 2 to 2.5 ml. The pentane extract is dried over anhydrous sodium sulfate, filtered, and concentrated to a final volume of 50 μl by distillation at 40°C. The concentrate is used for GC analysis.

f. Fractionation of the bound fraction

After the elution of the free fraction, the bound fraction is eluted with 50 ml of ethyl acetate. It is then dried over anhydrous sodium sulfate, filtered and concentrated, first to 1 ml under vacuum at 40°C, and then to dryness at 45°C, using a stream of nitrogen.

g. Hydrolysis of the bound fraction

To the above dry glycoside extract is added 0.1 ml of $2 \times 10^{-1} M$, citrate–phosphate buffer, pH 5.0. The solution is washed four times with 0.1 ml pentane to eliminate possible traces of the free fraction, using strong agitation in a whirlmixer. Pectinol solution (0.4 mg of Pectinol in 0.1 ml of $2 \times 10^{-1} M$, citrate–phosphate buffer, pH 5) is then added to the glycoside solution. After agitation, the mixture is poured into an ampule, hermetically sealed, and placed in a water bath for 12 hr at 40°C. The medium is then extracted five times with 0.1 ml pentane. After addition of 10 μl 0.1% solution of 2-octanol in ethanol, the solvent is eliminated by distillation at 40°C to give a final volume of 40 μl. The extract is then used for GC analysis.

h. Gas-chromatographic analysis of aroma constituents

The terpenoid alcohols (geraniol, nerol linalool) and benzyl- and 2-phenylethyl alcohols are determined on GC equipped with a glass capillary column (70 m × 0.5 mm i.d.) coated with 0.3 µm of free fatty acid phase and connected to an integrator. The fractionation is carried out first at a constant temperature of 70°C for 5 min, then with a temperature gradient from 70° to 180°C at a rate of 2°C/min, and finally at a constant temperature of 180°C for 30 min. The carrier gas (hydrogen) flow rate is 25 ml/min. The volumes of pentane extracts injected are between 2 and 3 µl.

i. Notes

1. The choice of pentane for free terpenol elution is the result of a compromise: a more polar solvent like dichloromethane gives a better elution of these compounds, but also removes part of the glycosides. Pentane does not elute glycosides but does not enable full recovery of free terpenols. Ethyl acetate elutes glycosides as well as methanol, but the former is more selective.
2. The glycosidic part of the bound fraction usually consists of glucose, rhamnose, and arabinose. The identity of the aglycones as determined by GC-MS are the following terpenols: linalool, α-terpineol, citronellol, nerol, and geraniol, together with linalool pyranic oxide, 3, 7-dimethyl-1, 5-octadiene-3, 7-diol, and two aromatic alcohols, benzyl- and 2-phenylethyl alcohols.
3. Enzymatic hydrolysis of the glycosides is chosen as the best method; an acid hydrolysis may induce considerable modification of the terpenol composition (rearrangements of the nonterpenoid alcohols).
4. The bound terpenols (terpene glycosides) are the hidden aromatic potential in grapes. Certain technological applications, such as enzymatic cleavage of the bound forms and release of free terpenol into medium, will increase the aromatic potential of wines.

References

1. Williams, P. J., Strauss, C. R., Wilson, B., and Massy-Westropp, R. A., (1982). *J. Agric. Food Chem* **30,** 1219.
2. Williams, P. J., Strauss, C. R., Wilson, B., and Massy-Westropp, R. A. (1982). *J. Chromatogr.* **235,** 471.
3. Shoseyov, O., Bravdo, B. A., Ikan, R., and Chet, I. (1988). *Phytochem.* **27,** 1973.
4. Gunata, Y. Z., Bayonove, C. L., Baumes, R. L., and Cordonnier, R. E. (1985). *J. Chromatogr.* **331,** 83.

Recommended Reviews

1. Rapp, A., Mandery, H., and Guntert, M., (1984). Terpene compounds in wine. *In* "Proc. Alko Symp. on Flavor Research of Alcoholic Beverages." (L. Nykanen, and P. Lehtonen, eds.), Vol. 3, p. 255.
2. Strauss, C. R., Wilson, B., Gooley, P. R., and Williams, P. J. (1986). Role of monoterpenes in grape and wine flavor. *In* "Biogeneration of Aromas." Amer. Chem. Soc., Symp. Series, Vol. 317, p. 222.

3. Montedoro, G., and Bertuccioli, M., (1986). The flavour of wines, vermouth, and fortified wines. *In* "Food Flavours, Part B." (I. D. Morton and A. J. Macleod, eds.), p. 171. Elsevier, Amsterdam, the Netherlands.
4. Merritt, C., and Robertson, D. H. (1982). Techniques of analysis of flavours: Gas-chromatography and mass spectrometry. *In* "Food Flavours, Part A." (I. D. Morten and A. J. MacLeod, eds.), p. 49. Elsevier, Amsterdam, the Netherlands.

H. Qualitative and Sensory Evaluation of Aromatic Herb Constituents by Direct Headspace—Gas Chromatography

1. Introduction

Gas chromatographic methods in which the volatile components in the gas space around liquid or solid samples are determined indirectly are known under the general name of headspace-GC methods.

A sample from the equilibrated gas phase above the solid or liquid phase is transferred either manually or automatically to the gas chromatograph for analysis. The gas sample is either introduced directly into the GC column, or concentrated using liquid or solid traps, and then transferred into the column by means of thermal desorption or liquid extraction. HSGC methods have been widely applied in the fields of environmental analysis, clinical research, industrial processing, and quality control, as well as in microbiological and food chemistry studies. HSGC is a rapid, simple, precise, sensitive, and nondestructive technique. It has been used for the quality control of aromatic herbs (1–3) and of wine (4).

Sensory analysis of GC effluent (GC-sniff) has been used by several investigators to evaluate the aroma significance of separated components in coffee (5), bilberries (6), peaches (7), and wine (8). In this way it is possible to examine the substances responsible for the specific olfactory sensations excited by a particular product by evaluating the constituents of the odor.

2. Principle

Herb sample is ground in a glass grinder. After conditioning in a gas-tight glass vial, the gaseous phase is transferred to a syringe and injected into a GC coupled with a mass spectrometer. A series of volatile markers may be injected into GC for comparative purposes.

A gas syringe is the most popular device for transferring a gaseous sample of volatiles from the headspace vial into a gas chromatograph (Method 1). Apart from the contamination and adsorption problems with such a device, there is a more serious drawback: a syringe is an open system during the sample transfer. Any pressure that has been generated by the increase of the partial vapor pressures of all the volatiles in the sample by heating the vial, also extends into the syringe.

During sample transfer, however, the syringe needle is open to the atmosphere, and an undefined amount of gas sample will be lost by expansion of the headspace gas through the needle.

In order to get reproducible results, the volume and the pressure in the sample must be held constant. This is achievable by using a sampling device as described in method 2 (Fig. 3.5) in the procedure (9).

Both methods, the syringe and the sampling device, may be coupled with a splitter for mass spectroscopic analysis.

3. Apparatus

Gas chromatograph Headspace apparatus
GC column 50 m coated with QV-1 Thermostat bath

4. Materials

Herbs such as bitter orange, clary sage, caraway, coriander, lemon grass,
   Mentha, peppermint, *Salvia dominica*, spearmint, sweet basil.
Monoterpenes (hydrocarbons, ketones, alcohols, etc.)
Syringe
Vials

5. Time

8 hours

6. Experimental

a. Method 1

A suitable sample of an herb such as Clary sage (*Salvia sclarea* L.) and *Salvia dominica* L.; lavender (*Lavandula angustifolia* Mill.), sweet basil (*Ocimum basilicum*), coriander (*Coriandrum sativum*), sweet marjoram (*Origanum majorana* L.), caraway (*Carum carvi* L.), spearmint (*Mentha spicata* L.) and *Mentha longifolia* (2), yarrow (*Achillea millefolium*), bitter orange (*Citrus aurantium*), lemon tree (*Citrus limonium*), and peppermint (*Mentha piperita*), is finely milled in a glass grinder, avoiding overheating. A weighed amount of the ground herb (1 g) is then put in a glass vial, sealed with a rubber septum. After conditioning in a thermostat bath for 7 hr at 60°C, a sample of the gaseous headspace is injected into a gas chromatograph by means of a syringe. The following operating conditions may be used: column, 50 m, OV-1 0.15 μm stationary phase thickness; carrier gas, hydrogen, flow rate 2 ml/min; splitting ratio, 1:4. Injector and flame ionization detector temperatures, 200°C; oven temperature, from 20°C to 150°C at a rate of 3.5°C/min.

In order to identify the volatile constituents, it is desirable to couple the GC with a mass spectrometer. A separate injection should be used for the sniffing of the GC effluent.

b. Method 2

The sampling device (Perkin-Elmer HS-100 Automatic Headspace Sampler) comprises a heated movable sampling needle with two vents and a solenoid valve, V_1, in the carrier gas supply line. This needle moves up and down in a heated cylinder and is sealed by three O-rings. In the standby position, the lower needle vent is placed between the two lower O-rings and thus sealed against atmosphere, while the carrier gas flows through valve V_1 to the column. A small cross-flow purges the cylinder and is vented via valve V_2.

At the end of the selected thermostating period, the sampling needle descends, pierces the septum cap into the vial, and pressurizes it either up to the head pressure of the column (as shown in Fig. 3.5) or to any other higher pressure value, which is provided by an additional pressure regulator and a solenoid valve. The latter arrangement is recommended for columns with a small pressure drop, such as a wide-base capillary column. At elapse of the preselected injection time (a few seconds), both valves open again. Carrier gas streams to the column and branches in order to pressurize the sample vial again. This immediately stops the injection, and the residual sample

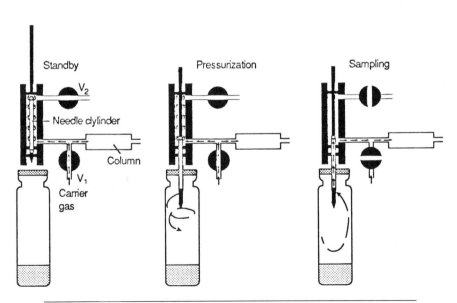

Figure 3.5 Balanced pressure sampling with the Perkin-Elmer HS-100 automatic headspace sampler.

vapors in the needle are flushed back into the vial, while the sample moves through the GC column.

c. Quantitative analysis of the aroma composition

When a state of equilibrium is achieved, the phase distribution of a volatile solute is determined by its partition coefficient K according to Equation 1, and the peak size A in the resulting headspace chromatogram is thus proportional to the gas phase concentration Cg (Equation 2), while Cl is the concentration of the liquid sample.

[1] $\quad K = Cl/Cg$

[2] $\quad A \sim Cg = Cl/K$

The application of external standards for comparative purposes is recommended.

References

1. Gabri, S., and Chialva, F. J. (1981). *High Resol. Chromatogr. and Chromatogr. Commun.* **4**, 215.
2. Chialva F., Doglia, G., Gabri, G., and Ulian, F. (1983). *J. Chromatogr.* **279**, 333.
3. Hiltunene, R., Vuorela, H., and Laakso, L. (1985). *In* "Essential Oils and Aromatic Plants." (B. Svendsen and J. J. C. Sheffer (eds.), p. 23. Martinus Nijhoff and W. Junk, Dordrecth, (1980). The Netherlands.
4. Noble, A. C., Flalth, R. A., and Forrey, R. R. (1980). *Agric. Food Chem.* **28**, 346.
5. Tassan, C., and Russell, G. F. (1974). *J. Food Sci.* **39**, 64.
6. Von Sydow, E., Anderson, J., Anjon, K., and Karlsson, G. (1970). *Lebensm.-Wiss. Technol.* **3**, 11.
7. Spencer, M., Panghorn, R. M., and Jennings, W. G. (1978). *J. Agric. Food Chem.* **26**, 725.
8. Rapp, A., Hastrich, H., Enel, L., and Knipser, W. (1978). "Flavor in Foods and Beverages." Academic Press, New York.
9. Kolb, B. (1985). *In* Essential Oils and Aromatic Plants. (A. B. Svendsen and J. J. C. Scheffer eds.), p. 3. Martinus Nijhoff/Dr. W. Junk. Boston, Massachusetts.

Recommended Reviews

1. Bauer, K., and Garbe, D. (1985). "Common Fragrance and Flavor Materials." VCH Publishers, Weinheim, FRG.
2. Cagen, R. H., and Kare, M. R. (1981). "Biochemistry of Taste and Olfaction." Academic Press, New York.
3. Charalambous, G., (ed.) (1978). "Analysis of Foods and Beverages. Headspace Techniques." Academic Press, New York.
4. Lawrence, B. M., Mookherjee, B. D., and Willis, B. J., (eds.). (1988). "Flavors and Fragrances: A World Perspective." Elsevier, Amsterdam, The Netherlands.
5. Nunez, A. J., and Gonzalez, L. F. (1984). Preconcentration of headspace volatiles for trace organic analysis by gas chromatography. *J. Chromatogr,* **127**, 300.

I. Preparation of Abietic Acid from Wood Rosin and Its Dehydrogenation to Retene

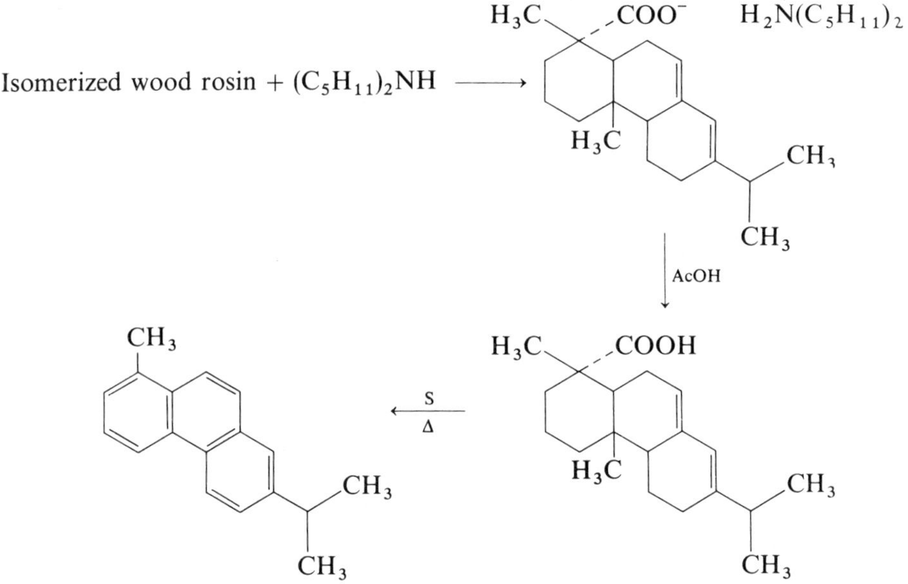

1. Introduction

The oleoresinous exudate of conifers, mainly pines, can be separated into two components—rosin and turpentine. Turpentine is generally removed from oleoresin by steam distillation, and rosin remains in the still as a nonvolatile residue. Rosin is known as naval stores, as it was used for caulking the hulls of ships.

Rosin consists mainly of isomeric diterpenoid acids having an empirical formula $C_{20}H_{30}O_2$. Abietic acid is the most abundant resin acid; it can also be prepared by acid-catalyzed isomerization of other natural resin acids (1). Abietic acid has been purified through its diamylamine, piperidine, and brucine salts (2).

The nature of the ring system of abietic acid was established by dehydrogenation with sulfur to retene (3), and was identified as 1-methyl-7-isopropylphenanthrene by oxidative degradation and by synthesis.

2. Principle

Rosin is isomerized by refluxing in a solution of ethanol–hydrochloric acid for 2 hr (4).

Abietic acid is then isolated from the isomerized rosin by formation of the diamylamine salt; its crystallization and acidification with acetic acid affords pure abietic acid. On dehydrogenation with sulfur, abietic acid forms retene, identified as a picrate.

3. Materials

Acetic acid	Ethanol, 95%	Sodium hydroxide
Acetone	Ether	Sulfur
Carbon dioxide	Hydrochloric acid	Wood rosin
Diamylamine		

4. Time

5–6 hours

5. Procedure

a. Preparation of abietic acid

Into a 2-liter round-bottomed flask fitted with a 35-cm reflux condenser are introduced 250 g wood rosin, 740 ml 95% ethanol, and 42 ml hydrochloric acid. A stream of carbon dioxide is passed over the surface of the solution through a glass tube during this reaction. The mixture is boiled under reflux for 2 hr. At the end of this period, the ethanol and acid are removed by steam distillation, and the water is decanted. The residue is cooled to room temperature and dissolved in 1 liter ether. The ether solution is extracted with water and dried over anhydrous sodium sulfate. Evaporation of ether *in vacuo* yields 245 g isomerized rosin, which is placed in a 1-liter Erlenmeyer flask and dissolved in 375 ml acetone by heating the mixture on a steam bath. To this solution at incipient boiling are added slowly, with vigorous agitation, 127 g diamylamine. Upon cooling to room temperature, crystals appear in the form of rosettes. The mass in agitated, cooled well in an ice bath, and filtered by suction. The crystalline salt is washed on a Buchner funnel with 150 ml acetone and dried in a vacuum oven at 50°C for 1 hr. The solid is recrystallized four times from acetone. Each time a sufficient quantity (20 ml per gram) of acetone is used to ensure complete solution, and the solvent is evaporated until incipient precipitation of the salt; yield, 118 g; $[\alpha]_D^{24} -60°$.

The amine salt is placed in a 4-liter Erlenmeyer flask and dissolved in 1 liter 95% ethanol by heating the mixture on a steam bath. To the solution, which has been cooled to room temperature, is added 36 ml glacial acetic acid. The resulting solution is stirred, 900 ml water is added, at first cautiously and with vigorous agitation, until crystals of abietic acid begin to appear; the remainder of the water is then added more rapidly. The abietic acid is collected on a Buchner funnel and washed with water until the acetic acid has been removed completely, as indicated by tests with indicator paper. Recrystallization can be effected by dissolving the crude product in 700 ml 95% ethanol, adding 600 ml water as described above, and cooling the solution. The yield of abietic acid is 98 g (40% based on the weight of isomerized rosin); $[\alpha]_D^{24} - 106°$.

b. ^{1}H-NMR

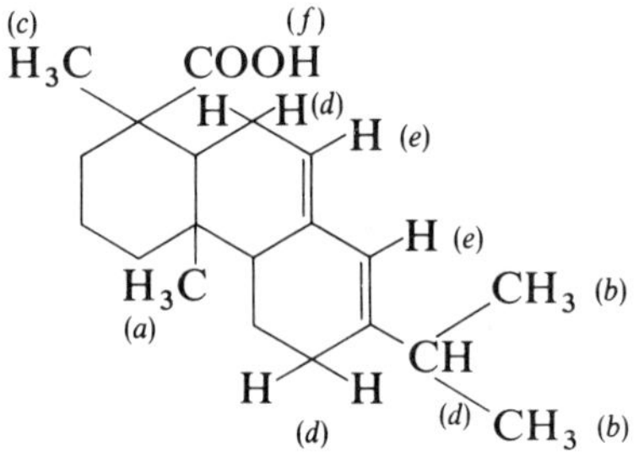

Position	δ (ppm)
a	0.67
b	0.83 (doublet)
c	1.18
d	1.96
e	5.68
f	12.45

c. Abietic Acid

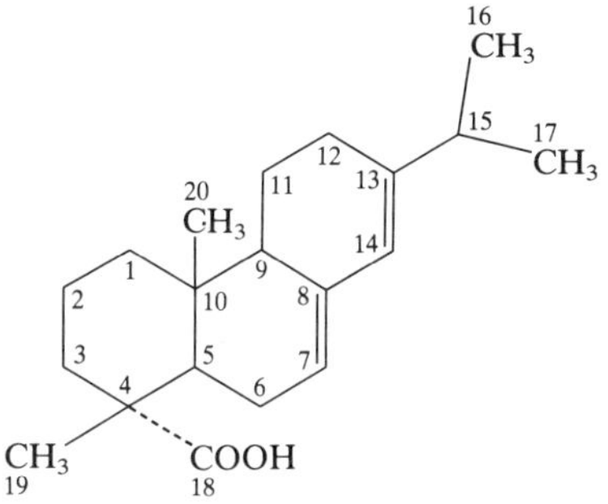

d. ^{13}C NMR

C-1	38.3	C-8	135.5	C-15	34.8
C-2	18.1	C-9	51.0	C-16	20.9
C-3	37.2	C-10	34.5	C-17	21.4
C-4	46.3	C-11	22.5	C-18	185.4
C-5	44.9	C-12	27.5	C-19	16.7
C-6	25.6	C-13	145.1	C-20	14.0
C-7	120.5	C-14	122.5		

e. Mass Spectrum
Abietic Acid $C_{20}H_{30}O_2$, Mol. wt. 302.
m/z 302(62%)   105(78%)   91(100%)   77(56%)
    121(49%)   93(49%)   79(55%)   67(45%)

f. UV Spectrum of abietic acid

λ_{max}^{EtOH} 237 mμ (logε 4.2)

Abietic acid is a heteroannular conjugated diene in which the two double bonds are conjugated, but located in two adjoining rings; the parent homoannular diene has an absorption maximum at 253 mμ.

g. Dehydrogenation of abietic acid to retene

Abietic acid, 50 g, and 25 g sulfur are heated at 200°C (in a hood) until hydrogen sulfide begins to evolve more slowly; heating at 250°C is then continued for a short period. The crude product is distilled under reduced pressure; bp 260–270°C/20 mm; yield, 11–12 g.

The solid distillate is extracted with ether, and the ethereal solution is washed with alkaline solution. Crystallization of the residue (after removal of ether) from ethanol yields 3 g crystals of mp 98–99°C; the picrate melts at 127°C.

References

1. Ruzicka, L., and Meyer, J. (1922). *Helv. Chim. Acta* **5**, 315.
2. Lombard, R., and Frey, J. M. (1948). *Bull. Soc. Chim. France* 1194.
3. Vesterberg, A. (1903). *Ber.* **36**, 4200.
4. Wheeler, T. S. (1963). *Org. Synth. Coll.* **4**, 478.

Recommended Reviews

1. Barton, D. H. R. (1949). *The chemistry of the diterpenoids. Quart. Revs.* **3**, 36.
2. McCrindle, R., and Overtone, K. H. (1965). *The chemistry of the cyclic diterpenoids.* In "Advances in Organic Chemistry" (R. A. Raphael, E. C. Taylor, and H. Wynberg, eds.), Vol. 5, p. 472. Interscience, New York.
3. Weaver, J. C. (1982). Natural resins. In "Kirk-Othmer Encyclopedia of Chemical Technology" Vol. 20, p. 197. Wiley, New York.

J. Conversion of Betulin to Allo- and Oxyallobetulin

1. Introduction

Betulin is one of the more plentiful triterpenes and was first isolated from the bark of the birch tree in 1836 (1). It constitutes up to 25% of the extractable part of the outer bark of the white birch (*Betula alba*) (2), and as much as 35% in the case of Manchurian white birch (*Betula platyphylla*) (3). Betulin also occurs in the bark of *Corylus avellana* (4) and *Lophopetalum toxicum* (5), in rose hips (6), in brown coal (7), in lignite (8), and in the cactus *Lemaireocereus griseus* (9).

Betulin may be converted to the lactone, oxyallobetulin, by the action of formic acid (10) or hydrobromic acid (11), followed by CrO_3 oxidation (12). This sequence of reactions connects the lupeol and β-amyrin series of the terpenoids.

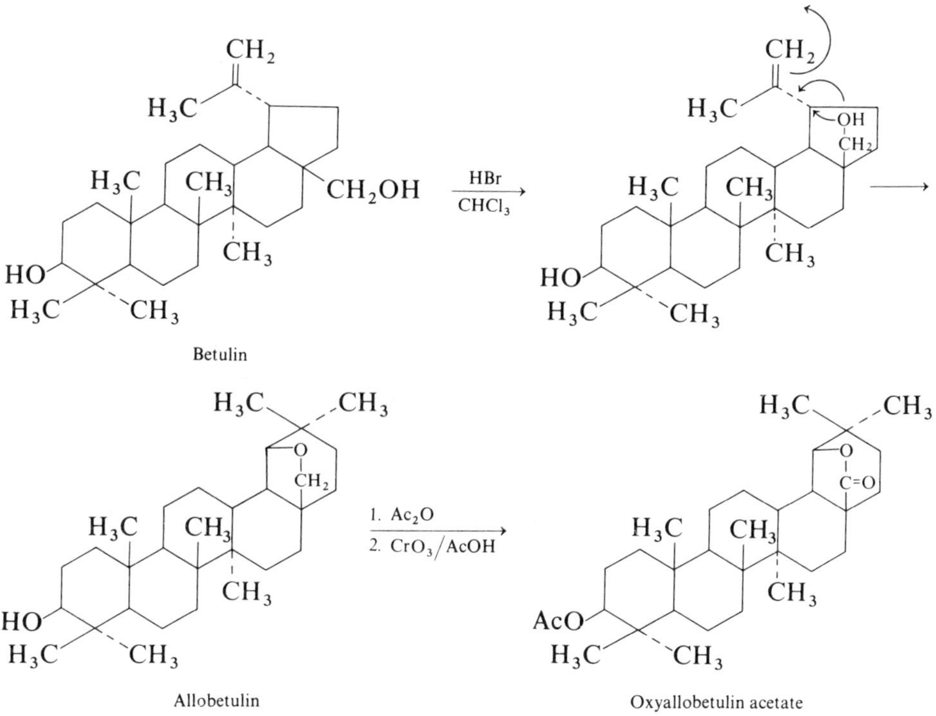

2. Principle

Birch bark is extracted with a hot solution of sodium carbonate to remove any acidic constituents. Betulin is then extracted with methanol. The betulin may be isomerized to allobetulin by treatment with hydrobromic acid or with formic acid followed by alkali. Acetylation of allobetulin, followed by oxidation with chromium trioxide and hydrolysis, gives the lactone, oxyallobetullin.

3. Apparatus

Chromatography column

4. Materials

Alumina	Acetic anhydride	Benzene
Acetic acid	Bark of birch tree	Chloroform

214 CHAPTER 3 Isoprenoids

Chromium trioxide	Formic acid	Potassium hydroxide
Ethanol	Hydrobromic acid	Sodium carbonate
Ether	Methanol	

5. Time

    7 hours

6. Procedure

a. Isolation of betulin

Birch bark (150 g) is cut into small pieces and boiled in an Erlenmeyer flask with 2% sodium carbonate solution. The contents are filtered, and the residue is boiled twice with 500-ml portions of hot water, and filtered on a Buchner funnel. The solid residue is dried in air, extracted with hot methanol for 3 hr, and filtered. On concentration of the filtrate, crude betulin separates as white crystals. The esters of betulin, remaining in the mother liquor, can be saponified by adding 25 volumes 1% ethanolic potassium hydroxide and leaving overnight at room temperature. Crude betulin is filtered, dried, and crystallized once from benzene and once from ethanol; needles, mp 251–252°C.

b. 3β-Betulin [Lupendiol]

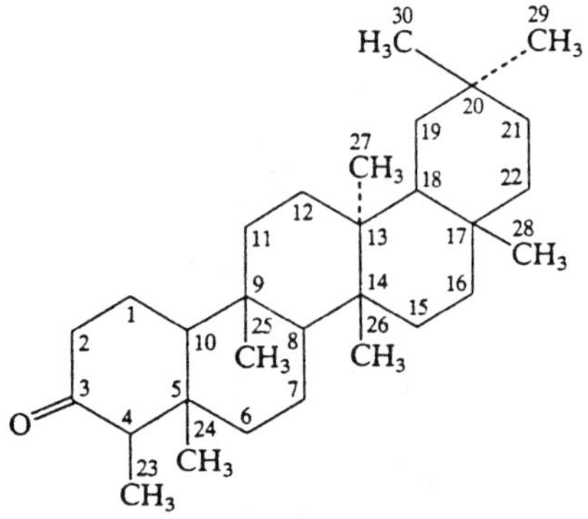

c. ^{13}C NMR

C-1 38.7	C-8 41.0	C-15 27.1
C-2 27.4	C-9 50.5	C-16 29.2
C-3 79.1	C-10 37.3	C-17 47.9
C-4 38.8	C-11 20.9	C-18 47.9
C-5 55.3	C-12 25.2	C-19 48.8
C-6 18.3	C-13 37.2	C-20 150.5
C-7 34.3	C-14 42.8	C-21 29.8

| C-22 | 34.1 | C-24 | 15.4 | C-26 | 16.1 | C-28 | 60.6 | C-30 | 109.7 |
| C-23 | 28.1 | C-25 | 16.1 | C-27 | 14.8 | C-29 | 19.1 | | |

d. Mass Spectrum

3β-Betulin [Lupendiol] $C_{30}H_{50}O_2$, Mol. wt. 442.

m/z 442(38.46%) 203(71%) 135(69%) 95(78%)
    411(64%)   189(100%) 121(51%) 81(53%)
    207(75%)

e. IR Spectrum of betulin (in KBr)

3420 cm^{-1}: O—H stretching
2925, 2855: stretching of CH_2
1645: C=C stretching of terminal methylene
1450, 1380: asym. and sym. CH_3
1075: C—O stretching of secondary OH
1030: C—O stretching of primary OH
880: out-of-plane bending of terminal methylene

f. Preparation of allobetulin (method A)

1. *Allobetulin formate*

   1 part betulin and 8 parts formic acid (90–95%) are refluxed for 45 min; the solution turns violet. After cooling, it is filtered, and the solid is extracted with ethanol and then crystallized from benzene; rhombic crystals, mp 311–312°C.

2. *Allobetulin*

   Allobetulin formate (4.5 g), 30 ml 1N ethanolic potassium hydroxide, and 5 ml benzene are refluxed for 30 min. Removal of the solvents *in vacuo* leaves allobetulin, which is filtered and washed with a little ethanol and water. Recrystallization from ethanol yields crystals of mp 260–261°C; yield, 3 g.

g. Preparation of allobetulin (method B)

Two g betulin in 10 ml chloroform and 5 ml hydrobromic acid (sp gr 1.39) are refluxed for 1 hr. After distilling off the chloroform, the dark residue is dissolved in ethanol, and water is added until the onset of turbidity. Recrystallization from ethanol yields material of mp 260–261°C. $[\alpha]_D^{15} + 48.25°$ (in chloroform).

h. IR Spectrum of allobetulin (in KBr)

3425 cm^{-1}: O—H stretching
2920, 2850: stretching of CH_2

1465, 1450: CH_2 and CH_3 bending
1380, 1370: gem-dimethyl
1140, 1040: C—O stretching of ring C—O—C

i. Allobetulin acetate

One g allobetulin and 10 g acetic anhydride are left overnight at room temperature. The crystals are then filtered off and crystallized from ethanol as plates of mp 275–276°. $[\alpha]_D^{15}$ + 54.16° (in chloroform).

j. Oxyallobetulin acetate

To a boiling solution of 500 mg allobetulin acetate in 50 ml acetic acid is slowly added 650 mg chromium trioxide in 45 ml acetic acid. After boiling under reflux for 15 min, the mixture is cooled, diluted with water, and extracted with chloroform. Distillation of chloroform leaves about 350 mg which is dissolved in benzene, adsorbed on a column of 30 g alumina, and eluted with benzene and benzene-ether. Oxyallobetulin acetate crystallizes as plates of mp 360°C.

k. Oxyallobetulin

Oxyallobetulin acetate (300 mg) dissolved in 100 ml benzene, and 100 ml 5% ethanolic (95%) potassium hydroxide solution are refluxed for 1 hr. The solution is concentrated *in vacuo* to half its volume, diluted with water, and extracted with chloroform. The residue is crystallized as needles from ethanol. Mp 336–337°C; yield, 150 mg.

l. IR Spectrum of oxyallobetulin (in KBr)
3425 cm^{-1}: O—H stretching
2920, 2850: CH_2 stretching
1765: C=O stretching of γ-lactone
1705: C=O stretching of δ-lactone
1465, 1450: CH_2 and CH_3 bending
1380, 1370: gem-dimethyl
1215, 1190: two methyl groups attached to quaternary carbon atom
1255, 1120: C—O—C stretching of lactone

References

1. Hunefeld, J. (1836). *Prakt. Chem.* **7**, 53.
2. Sosa, A. (1940). *Ann. Chem.* **14**, 5.
3. Hirota, K., Takano, T., Taniguichi, K., and Iguchi, K. (1944). *J. Soc. Chem. Ind. Japan* **47**, 922.
4. Brunner, O., and Wohrl, R. (1934). *Monatsh. Chem.* **64**, 21.
5. Dieterle, H., Leonhardt, H., and Dorner, K. (1933). *Arch. Pharm.* **271**, 264.
6. Zimmermann, J. (1944). *Helv. Chim. Acta* **27**, 332.

7. Ruhemann, S., and Raud, H. (1932). *Brennstoff Chem.* **13**, 341.
8. Ikan, R., and McLean, J. (1960). *J. Chem. Soc.* 893.
9. Djerassi, C., Bowers, A., Burstein, S., Estrada, H., Grossman, J., Herran, J., Lemin, A. J., Manjarrez, A., and Pakrashi, S. C. (1956). *J. Am. Chem. Soc.* **78**, 2312.
10. Schulze, H., and Pieroh, K. (1922). *Ber.* **55**, 2332.
11. Dischendorfer, O. (1923). *Monatsh. Chem.* **44**, 123.
12. Davy, G. S., Halsall. T. G., Jones, E. R. H., and Meakins, G. D. (1951). *J. Chem. Soc.* 2702.

Recommended Reviews

1. Barton, D. H. R. (1953). *The chemistry of triterpenoids.* In *"Progress in Organic Chemistry"* J. W. Cook, (ed.), Vol. 2, p. 67. Butterworth, London.
2. Steiner, M., and Holtzem, H. (1955). *Triterpene und triterpene saponine.* In *"Modern Methods of Plant Analysis"* (K. Paech, and M. V. Tracey, eds.), Vol. 3, p. 58. Springer-Verlag, New York.
3. De Mayo, P. (1963). *Molecular rearrangements.* In *"Techniques of Organic Chemistry"* (K. W. Bentley, ed.), Vol. 11, part 2, p. 1087. Interscience, New York.

K. Isolation of Cerin and Friedelin from Cork

1. Introduction

Chevreul was the first to describe the cork alcohols cerin and friedelin (1). The important fact that the sparingly soluble, crystalline material from cork consists principally of at least two substances, which can be separated by a series of recrystallizations from chloroform, was demonstrated by Istrati and Ostrogovich (2). In honor of Friedel, Istrati named the more soluble substance friedelin.

Drake and Jacobsen (3) have isolated about 1.3% friedelin and 0.1–0.15% cerin from cork. Friedelin was also isolated by Jefferies (4) from *Ceratopetalum apetalum*, by Arthur, Lee, and Ma (5) from leaves of *Rhododendron westlandii*, and by Weizmann, Meisels, and Mazur (6) from *Citrus paradisi*.

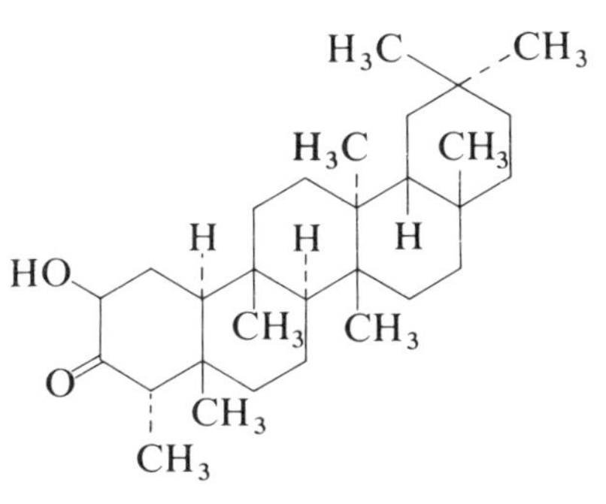

Cerin

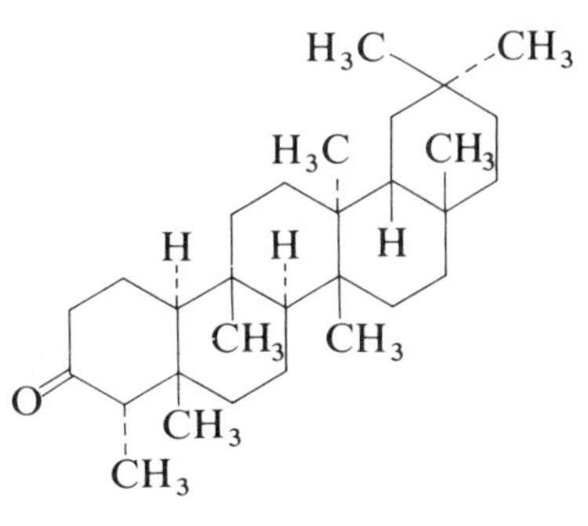

Friedelin

2. Principle

Ground cork is extracted with hot ethyl acetate. The filtrate contains cerin, friedelin, and other constituents. On crystallization of cerin from chloroform, friedelin is left in the mother liquor, which is concentrated and subjected to column chromatography. Crystallization of friedelin from ethyl acetate gives white needles. Both terpenes yield 2,4-DNP (dinitrophenylhydrazine) derivatives.

3. Apparatus

Chromatography column

4. Materials

Acetone	Chloroform	Ethyl acetate
Alumina	Cork, ground	Hydrochloric acid
Benzene	2,4-Dinitrophenylhydrazine	Methyl cellosolve

5. Time

4–5 hours

6. Procedure

a. Isolation of cerin

To 75 g ground cork (20–30 mesh) in a 1-liter round-bottomed flask is added 500 ml ethyl acetate. The contents of the flask are heated on a water bath under reflux for 3 hr, filtered on a Buchner funnel, and washed with 200 ml hot ethyl acetate. The brown-colored filtrate is evaporated to dryness under reduced pressure, and the residue obtained is crystallized twice from 3 to 5 ml chloroform. Cerin is obtained in the form of silky needles, mp 255–256°C, whereas friedelin is left in the mother liquor; yield, 50 mg.

b. IR Spectrum of cerin (in KBr)

3475 cm^{-1}: O—H stretching of secondary hydroxyl and intramolecular hydrogen bond with C=O
2920, 2850: stretching of CH_2
1710: C=O stretching, free
1650: bonded carbonyl
1460, 1445: CH_2 and CH_3 bending
1380, 1360: gem-dimethyl group
1200, 1185: two methyl groups attached to quaternary carbon atom
1050: C—O stretching of alcoholic OH

III. Terpenoids

c. Cerin 2,4-dinitrophenylhydrazone

Cerin (50 mg) is dissolved in 5 ml hot methyl cellosolve, and 30 mg 2,4-dinitrophenylhydrazine is added. The mixture is warmed until homogeneous. One drop of concentrated hydrochloric acid is added, and the mixture is boiled for 5 to 10 min. On cooling, the solution deposits orange-colored, lath-like crystals, which are filtered off and recrystallized from a benzene–alcohol mixture in the form of orange laths, decomposition point 253–255°C.

d. Isolation of friedelin

The chloroform filtrates from the first two recrystallizations of cerin are evaporated under reduced pressure until solid separates from the hot solution. An equal volume of acetone is added, and the crude crystalline friedelin is chromatographed over 5 g neutral alumina. Elution is started with a mixture of benzene–chloroform, 9:1, and the percentage of chloroform is gradually increased. Recrystallization from ethyl acetate gives white needles of friedelin melting at 262 to 263°C.

e. Friedelin

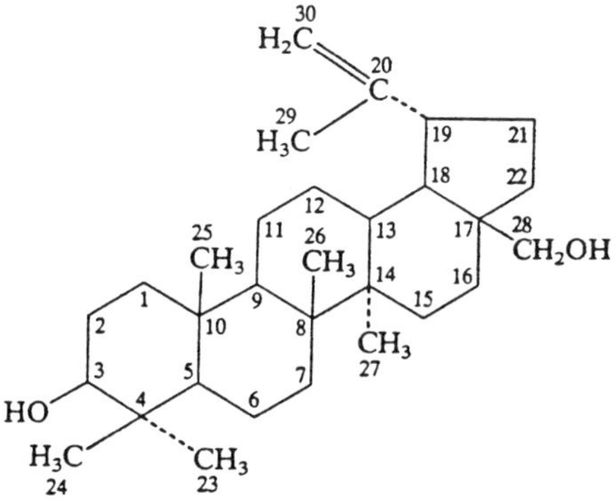

f. ^{13}C NMR

C-1	41.3	C-11 32.4	C-21 35.0
C-2	41.5	C-12 36.0	C-22 35.3
C-3	213.0	C-13 38.3	C-23 6.8
C-4	59.5	C-14 39.7	C-24 22.3
C-5	42.1	C-15 31.8	C-25 14.6
C-6	35.6	C-16 32.8	C-26 17.9
C-7	30.5	C-17 30.0	C-27 18.2
C-8	53.1	C-18 42.8	C-28 20.3
C-9	37.4	C-19 39.2	C-29 32.1
C-10	58.2	C-20 28.1	C-30 18.6

220 CHAPTER 3 Isoprenoids

g. Mass Spectrum
Friedelin $C_{30}H_{50}O$, Mol. wt. 426.

m/z	426(90%)	218(26%)	193(32%)	163(62%)
	411(63%)	205(42%)	191(42%)	109(100%)
	341(26%)	207(16%)		

h. IR (in KBr)
3425 cm^{-1}: O—H stretching
2920, 2860: stretching of CH_2
1715: C=O stretching, six-membered ketone
1460, 1440: CH_2 and CH_3 bending
1385, 1370: gem-dimethyl
1200, 1185: two methyl groups attached to quaternary carbon atom
1110: aliphatic ketone

i. Friedelin 2,4-dinitrophenylhydrazone
Friedelin (50 mg) is dissolved in 5 ml hot methyl cellosolve, and 30 mg 2,4-dinitrophenylhydrazine is added. The mixture is warmed until homogeneous. One drop concentrated hydrochloric acid is added, and the mixture is boiled for 5 to 10 min. On cooling, the solution deposits orange-colored, lath-like crystals which are recrystallized from benzene. The product melts with decomposition at 297 to 299°C.

References

1. Chevreul. (1807). *Ann. Chim.* **62**, 323.
2. Istrati, C., and Ostrogovich, . (1899). *Compt. Rend.* **128**, 1581.
3. Drake, N. L., and Jacobsen, R. P. (1935). *J. Am Chem. Soc.* **57**, 1570.
4. Jefferies, P. R. (1954). *J. Chem. Soc.* 473.
5. Arthur, H. R., Lee, C. M., and Ma, C. N. (1956). *J. Chem. Soc.* 1461.
6. Weizmann, A., Meisels, A., and Mazur, Y. (1955). *J. Org. Chem.* **20**, 1173.

L. Removal of Bitter Components from Citrus Juices with β-Cyclodextrin Polymer

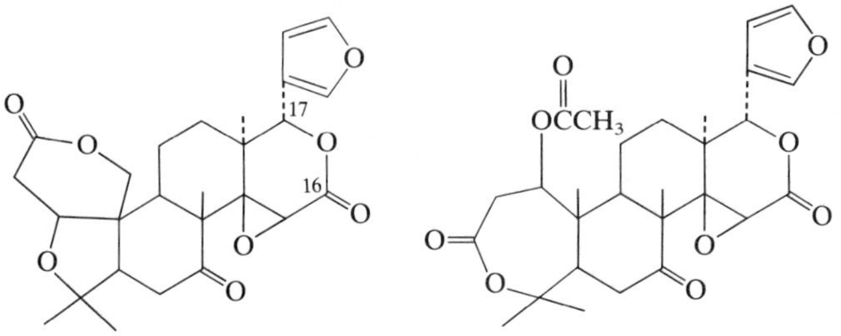

Limonin   Nomilin

Naringin

1. Introduction

Limonoids are a group of chemically related triterpene derivatives found in Rutaceae and Meliaceae families.

Limonin and nomilin are the major causes of juice bitterness, along with the bitter-flavored naringin. In general, citrus juices contain limonoids at levels below the bitterness threshold of 6 ppm. Intact fruits normally contain a nonbitter precursor, limonoate A-ring lactone, which differs from limonin only in having a free carboxyl group at C-16 and a free hydroxyl group at C-17. When the juice is extracted, this precursor lactonizes to the intensely bitter dilactone, limonin, upon standing for a few hours under the juice's acidic conditions (pH 3.5 to 4.5) or upon heating. The conversion is accelerated by the enzyme limonin D-ring lactone hydrolase present in citrus.

It has been recently reported (1) that citrus tissues and juices contain very high concentrations of limonoid glucosides, which are mildly bitter compared to their limonoid aglycones.

Excessive bitterness stemming from the bitter principles such as limonin and naringin in navel orange and grapefruit juice is an undesirable flavor quality. Many attempts have been made to remove bitter components from navel orange juice (2). Treatment of grapefruit juice with enzymes, such as naringinaze, to hydrolyze the bitter glucoside naringin to the nonbitter aglycone, naringenin, has been reported by Kefford and Chandler (3).

Recently ion-exchange resins were used to remove titratable acids, as well as limonin and naringin, from grapefruit juice. (4).

β-Cyclodextrin monomer, which is soluble in aqueous solution, was shown to decrease the bitter taste of limonin and naringin in citrus juice (5).

2. Principle

The following procedure (7) uses α- or β-cyclodextrin polymer with commercial citrus juices in a continuous fluid-bed or batch process to reduce levels of limonin, nomilin, and naringin in the juices. The debittered juices are evaluated by chromatographic (HPLC, TLC) and by taste panel techniques.

It has been found (6)) that β-cyclodextrin polymer reduced the levels of limonin and naringin in citrus juice and that water or organic solvents regenerated the polymer for use in debittering additional juice samples.

3. Apparatus

   High-pressure liquid chromatograph
   Silica gel TLC plates

4. Materials

Acetic acid	Ethanol
Acetone	Ethyl ether
Benzene	Grapefruit and orange concentrated juices
Chloroform	Sodium borohydride
Cyclodextrin, α- or β-	Sodium hydroxide
Epichlorohydrin	

5. Time
   a. Overnight
   b. 7 hours

6. Procedure

a. Preparation of Cyclodextrin polymers.

Navel oranges and grapefruit and frozen concentrated juices are purchased at a local market. A mixture of α- or β-cyclodextrin and 11 ml of water is treated with a freshly mixed solution (80°C) of 13.5 g sodium hydroxide in 13.5 ml water. To the magnetically stirred mixture, 50 mg sodium borohydride is added, followed by rapid, dropwise addition of 24 ml epichlorohydrin. The resulting pasty mixture undergoes a rapid, exothermic reaction is 10 to 20 min with brief, spontaneous refluxing, and a frothy, hard, glassy reaction product is formed. The mixture is allowed to stand at room temperature overnight and then heated to 50°C for 5 hr. If the spontaneous reflux fails to occur, the mixture is heated to 60°C for 30 min. The glassy polymeric reaction product is easily broken up and washed with acetone, 10×75 ml water (until neutral), and 3×75 ml ethanol to remove water, and dried in air overnight to yield 20–30 g polymer. The ethanol wash yields a dry polymer, easier to handle than when the last wash is with water.

b. Batch process procedure for debittering

500 ml juice is stirred magnetically for 60 min with 10 g 20–60 mesh β-cyclodextrin polymer. The polymer is removed by filtration through a 60-

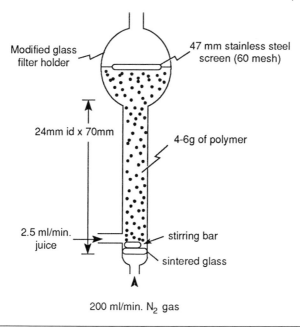

Figure 3.6 Fluid-bed apparatus for debittering citrus juices.

mesh screen (Fig. 3.6) and regenerated by stirring the polymer for 30 min each time with two or three 200 ml portions of 2% sodium hydroxide solution or absolute ethanol. Polymer, regenerated with dilute sodium hydroxide solution, is then washed with 5 × 20 ml water and 2 × 50 ml ethanol and dried in air before reuse.

c. Analytical Procedures
1. High-performance liquid chromatography (HPLC): Separations for quantitation of naringin and naringenin 7β-rutinoside in grapefruit juice are carried out on a high-speed 5 μm C-18 column, 12.5 cm long. The eluting solvent is 20% acetonitrile to 80% water at 1.5 ml/min.
2. Thin-layer chromatography (TLC) is carried out on silica gel plates. The developing solvent is benzene–chloroform–ether–acetic acid (100:50:50:3).
   The R_f value of limonin is 0.08 and of nomilin, 0.10.

d. Flavor tests
A taste panel may be used to compare the flavors of debittered navel orange and grapefruit juices to those of the starting juices. Debittered grapefruit juice may still have some of the bitterness expected of grapefruit juice, but at a greatly reduced level, compared to the starting juice.

The debittering process with β-cyclodextrin polymer does not adversely affect the flavor of navel orange and grapefruit juices, nor does it decrease the levels of major desirable components—sugars, acids, and ascorbic acid. The decrease in the essential oil content due to the polymer treatment can be corrected by addition of oil back to the juice.

References

1. Hasegawa, S. Bennett, D., Herman, 2., Fong C. H., and Ou, P. (1989). *Phytochem.* **23**, 1717.
2. Hasegawa, S., Patel, M. N., and Snyder, R. C. (1982). *J. Agric. Food Chem.* **30**, 509.
3. Kefford, J. F., and Chandler, B. V. (1970). "The Chemical Constituents of Citrus Fruits," Academic Press, New York.
4. Johnson, R. L., and Chandler, B. V. (1982). *J. Sci. Food Agric.* **33**, 287.
5. Konno, A., Misaki, M., Toda, J., Wada, T, and Yasumatsu, K. (1982). *Agric. Biol. Chem.* **46**, 2203.
6. Shaw, P. E., and Wilson, C. W. *J. Food Sci.* **48**, 646.
7. Shaw, P. E., Tatum, J. H., and Wilson, C. W. (1984). *J. Agric. Food Chem.* **32**, 832.

Recommended Review

1. Shaw, P. E. (1990). Cyclodextrin polymers in the removal of bitter compounds in citrus juices. *In* "Bitterness in Foods and Beverages." (R. Rouseff, ed.), pp. 309–324. Elsevier, Amsterdam, The Netherlands.

M. Questions

1. How are terpenes classified? Give an example of each class.
2. Write equations for the reactions you could expect to take place when limonene is oxidized, reduced catalytically, treated with HBr.
3. What reagents and experimental conditions are required for the reduction of thymol to menthol? How many stereoisomers can you expect?
4. How can carvone be isolated from oil of caraway, and how can it be synthesized?
5. What is the relationship of geraniol to nerol?
6. What products are formed by the action of ozone on ocimene, limonene, and α-pinene?
7. Explain why dehydration of borneol with aqueous acid gives camphene.
8. Show how β-amyrin can be formed by cyclization of squalene.
9. Discuss the medicinal and industrial uses of terpenes.
10. Draw the structures of the following compounds and indicate the isoprene units they contain: menthol, α-pinene, camphor, farnesol, caryophyllene, abietic acid, and β-amyrin.

N. Recommended Books

1. Guenther, E. *"The Essential Oils."* Van Nostrand, New York.
2. Goodwin, T. W. (1971). *"Aspects of Terpenoid Chemistry and Biochemistry."* Academic Press, New York.
3. De Mayo, P. (1959). *"Mono- and Sesquiterpenoids."* Interscience, New York.
4. De Mayo, P. (1959). *"The Higher Terpenoids."* Interscience, New York.
5. Pinder, A. R. (1960). *"The Chemistry of the Terpenes."* Chapman and Hall, London.
6. Simonsen, J. L., and Owen, L. N. (1931). *"The Terpenes, Vol. 1, The Simpler Acyclic and Monocyclic Terpenes and Their Derivatives."* Cambridge University Press, New York.
7. Simonsen, J. L. (1932). Vol. 2, *"The Dicyclic Terpenes, Sesquiterpenes, and Their Derivatives."* Cambridge University Press, New York.
8. Simonsen, J. L. and Barton, D. H. R. (1952). Vol. 3, *The Sesquiterpenes, Diterpenes, and Their Derivatives."* Cambridge University Press, New York.
9. Simonsen, J. L., and Ross, W. S. J. (1957). *"Vols. 4 and 5, The Triterpenes and Their Derivatives."* Cambridge University Press, New York.

NITROGENOUS COMPOUNDS 4

I. ALKALOIDS

A. Introduction

Alkaloids are nitrogenous compounds occurring in plants. Most of them are optically active, and nearly all of them are of basic nature. Accordingly, they very often form salts with plant acids such as quinic or meconic acid. Some alkaloids are present in plants in combination with sugars (solanine), while others occur as acid amides (piperine) or esters (cocaine, atropine). There are certain exceptions to these general statements; thus, ricinine and colchicine are almost neutral compounds. Some alkaloids are quaternary salts while others are tertiary amine oxides and therefore not basic.

The indole alkaloid, bufotenine, is found not only in plants (*Piptadenia pergrina*), but also in toads (*Bufo vulgaris*) and fungi (*Amanita mappa*). In general terms, the nitrogenous compounds found in fungi or other microorganisms should also be regarded as alkaloids, but this is not customary. Other such compounds are: gliotoxin (fungus *Trichoderma viride*), pyocyanine (bacterium *Pseudomonas aeruginosa*), and erythromycin (strains of *Streptomyces*).

The alkaloids found in the seeds, roots, and bark of plants are isolated by extraction with dilute acids (hydrochloric, sulfuric, acetic) or alcohol. If they are present as salts, they are liberated by treatment with calcium hydroxide before extraction. The individual alkaloids are usually separated from the very complex mixtures in which they occur by chromatographic methods.

Most alkaloids are colorless, crystalline compounds; a few, e.g., coniine, nicotine, and hygrine, are liquids, and some are colored, e.g., berberine, which is yellow. Many alkaloids have a bitter taste and quite a number of them possess curative properties. For example, morphine has a narcotic action, reserpine is a tranquilizer, atropine has an antispasmodic action, cocaine is a local anesthetic, and strychnine is a nerve stimulant.

The function of alkaloids in plants is still a matter of controversy; they are regarded as byproducts of plant metabolism, as reserve materials for protein synthesis, as protective substances discouraging animal or insect attacks, as plant stimulants or regulators similar to hormones, or simply as detoxication products.

The presence of alkaloids is detected by either precipitants or color reagents. The more important precipitants are those of Mayer, Wagner, Dragendorff, and Bertrand. The color reagents mostly consist of dehydrating or oxidizing agents or a combination of the two.

B. Method of Degradation

The most widely used process in degradative studies of alkaloids is exhaustive methylation, known as Hofmann degradation, which involves the pyrolysis of a quaternary ammonium hydroxide to form an olefin and a tertiary base.

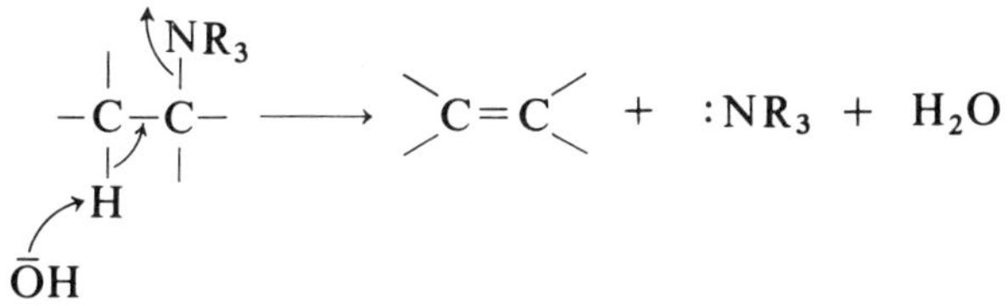

$$-\underset{\underset{OH}{H}}{C}-\underset{}{C}- \longrightarrow \quad C=C \quad + \quad :NR_3 \quad + \quad H_2O$$

When the nitrogen atom constitutes part of a ring, two degradations must be carried out in order to ensure its complete removal. It should also be borne in mind that when the exhaustive methylation of cyclic compounds might be expected to give 1, 4-dienes, the alkaline conditions of the reaction may result in the migration of one of the double bonds to give a 1, 3-diene, e.g., the exhaustive methylation of N-methylpiperidine gives 1, 3-pentadiene (piperylene) and not 1, 4-pentadiene.

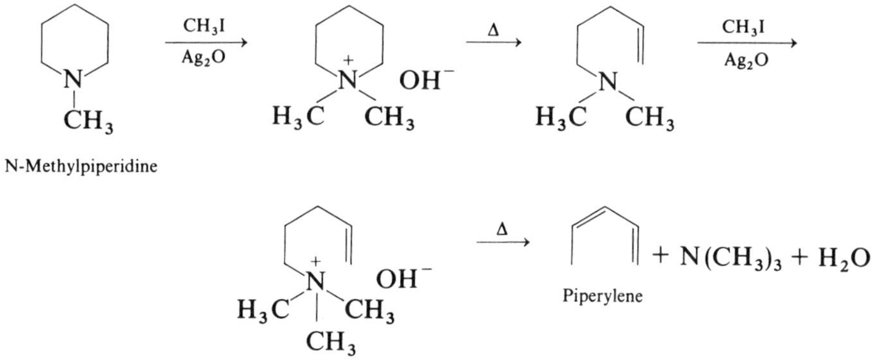

228 CHAPTER 4 Nitrogenous Compounds

The diene can then be easily hydrogenated to form a saturated hydrocarbon.

In cases where the normal Hofmann degradation fails to bring about ring fission of cyclic amines, the Emde degradation, involving catalytic reduction of a quaternary salt by sodium amalgam or sodium in liquid ammonia, may be applied. For example, attempted Hofmann degradation of N-methyltetrahydroquinoline methohydroxide results in regeneration of the parent base, while Emde reduction with sodium amalgam affords the ring-opened amine.

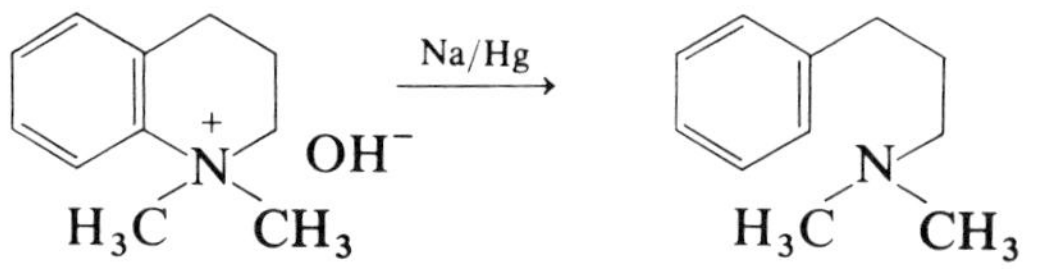

A particular method of degradation is used in the case of alkaloids having diphenyl ether linkages, e.g., the *bis*-benzylisoquinoline alkaloids. They can be cleaved into two fragments by reduction with sodium in liquid ammonia. Thus, the structure of the alkaloid dauricine was established by the reductive cleavage of *O*-methyl-dauricine.

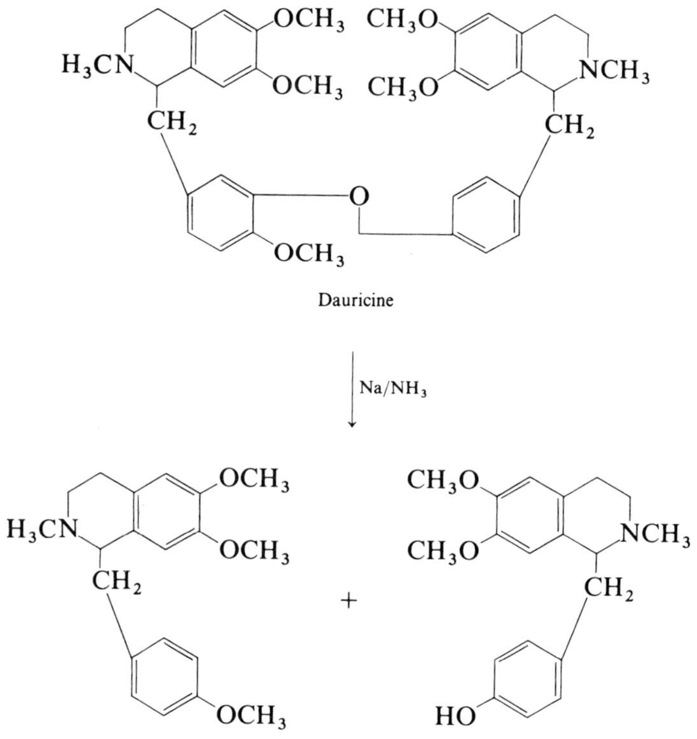

Dauricine

l-1-(p-methoxybenzyl)-6,7-dimethoxy-2-methyl-1,2,3,4-tetrahydroisoquinoline

l-1-(p-hydroxybenzyl)-6,7-dimethoxy-2-methyl-1,2,3,4-tetrahydroisoquinoline

I. Alkaloids   229

It should be mentioned that physical methods, in particular spectroscopic (UV, IR, NMR, ESR, MS) methods and X-ray diffraction, are now widely used in tackling structural and stereochemical problems relating to such natural products.

The existing classification of alkaloids is based on a mixture of chemical and botanical criteria. Thus, the pyrrolidine, pyridine, and indole alkaloids are classified on the basis of their chemical structure. The Senecio alkaloids on the other hand are so called because the genus *Senecio* provides the greatest number of species containing alkaloids derived from a pyrrolizidine moiety.

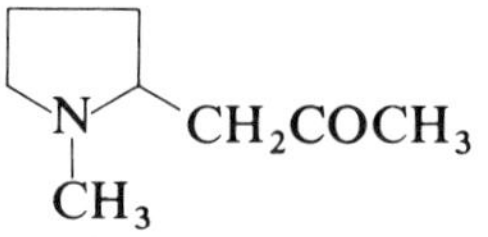

Hygrine
(Pyrrolidine group)

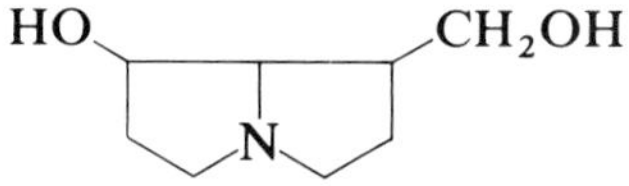

Platynecine
(Pyrrolizidine group)

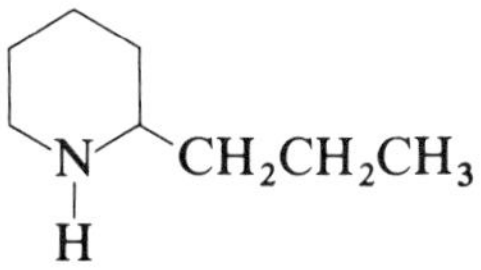

Coniine
(Pyridine group)

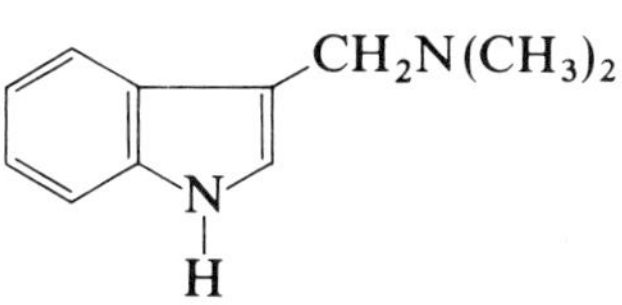

Gramine
(Indole group)

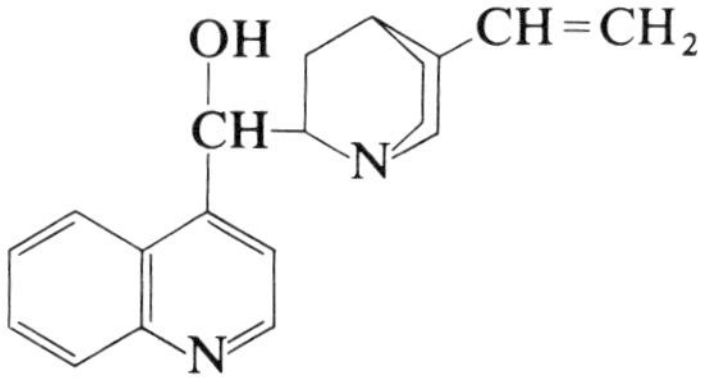

Cinchonine
(Quinoline group)

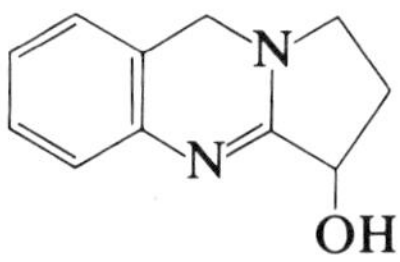

Vasicine
(Quinazoline group)

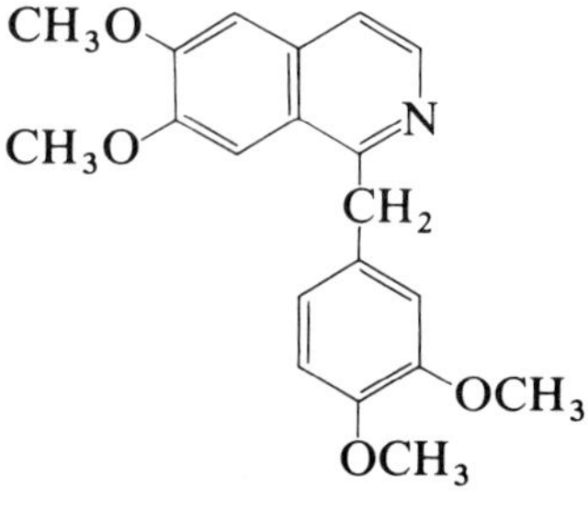

Papaverine
(Isoquinoline group)

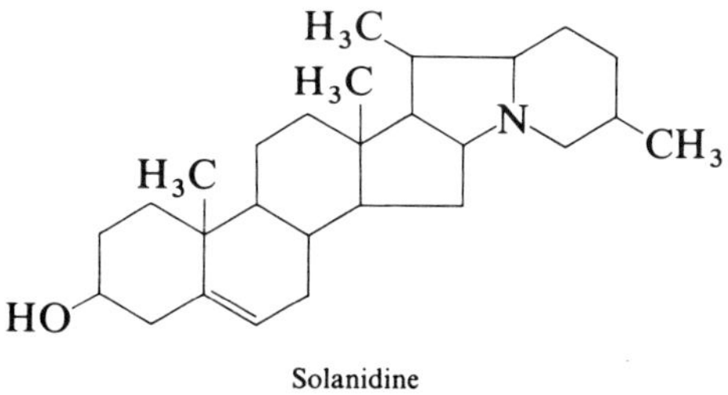

Solanidine
(Steroid group)

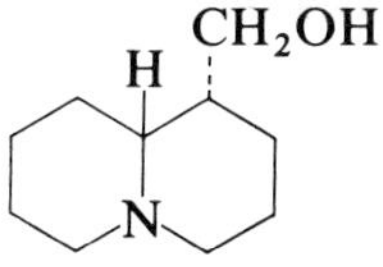

Lupinine
(Quinolizidine group)

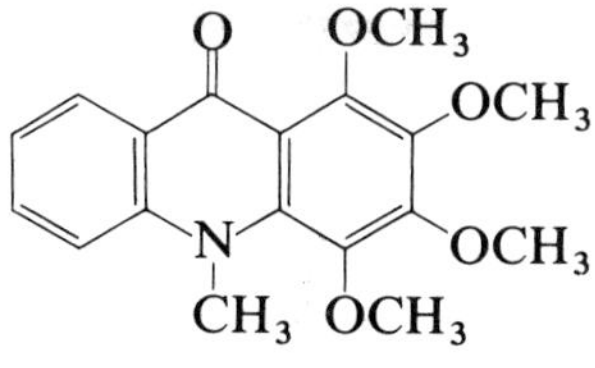

Melicopicine
(Acridine group)

The structural types found in different classes of alkaloids are so diverse that it is impossible to develop a single biogenetic hypothesis to include all alkaloids.

Using tracer techniques, it has been shown that the heterocyclic nuclei of many alkaloids are derived from only a limited number of amino acids, e.g., ornithine, lysine, phenylalanine, and tryptophan, or products of their breakdown.

$$R-CH(NH_2)-COOH \xrightarrow{-CO_2} R-CH_2-NH_2$$
$$\xrightarrow{[O], -CO_2} R-CH_2-CHO$$

The structural diversity of alkaloids in nature can be attributed mainly to the involvement of C_3, C_4, and C_5 compounds such as acetone, acetoacetic acid, and acetone dicarboxylic acid, the methylation of oxygen or nitrogen atoms by the S-methyl group of methionine, and the oxidative coupling of phenols in the ortho and para positions of the phenyl.

C. Isolation of Caffeine from Tea

1. Introduction

Alkaloids having a purine nucleus form a small but important group of natural products. They are often not classified as alkaloids because of their almost universal distribution in living matter and their mode of biosynthesis, which shows no relationship to amino acids from which most alkaloids arise.

Cafferine is one of the most important naturally occurring methyl derivatives of xanthine. Its concentration in a variety of teas, including black and green teas, depends both on climatic and topographic conditions of growth and on processing methods, and was found to vary from 2.0 to 4.6%. Thus, Chinese black tea contains 2.6–3.6%, Brazilian, 2.2–2.9%, and Turkish, 2.1–4.6% caffeine.

In addition to caffeine, other purines have also been reported to be present in small quantities. Michl and Haberler (1) separated the following compounds from black tea: caffeine, 2.5%, theobromine, 0.17%, theophylline, 0.013%, adenine, 0.014%, and traces of guanine, xanthine, and hypoxanthine.

Caffeine was also isolated by Friese (2) from seeds of *Genipa americana* (2.25%) and by Stenhouse (3) from coffee beans.

Cola nitida is the principal source of kola nuts, which are important because of their caffeine (3–5%) and theobromine contents.

Caffeine is used as a stimulant of the central nervous system. It has myocardial and diuretic effects, and relaxes the smooth muscle of the bronchi. Caffeine is a less potent diuretic than theobromine.

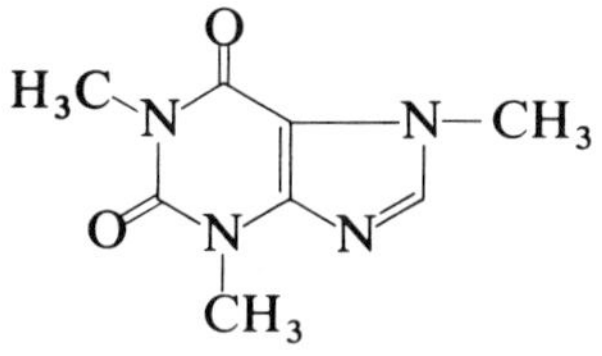

Caffeine

2. Principle

The experiment below involves the isolation of caffeine from tea leaves by extraction with hot sodium carbonate solution, and neutralization and extraction with dichloromethane.

3. Apparatus

Sublimation apparatus

4. Materials

Sulfuric acid   Dichloromethane
Tea   Sodium carbonate
Celite

5. Time

4–5 hours

6. Procedure

Twenty g finely powdered tea leaves are placed in a 400 ml beaker, 5 g sodium carbonate and 100 ml water is added, and the mixture is heated by a Bunsen burner for 20 min. Water is occasionally added to keep the volume of the solution constant. The hot solution is then filtered, and the filtrate is neutralized cautiously, with stirring, with sulfuric acid 10% solution.

It is then filtered through a thin layer of a filter aid (Celite), placed on a Buchner funnel padded with a wet filter paper, and washed with 20 ml dichloromethane. The two-phase filtrate is then placed in a separating funnel. The organic lower layer is separated, and the aqueous layer is extracted twice with two 40-ml portions of dichloromethane. The two organic (CH_2Cl_2) layers are then combined, and the solvent is evaporated. Crude caffeine is crystallized from a very small quantity of hot acetone or water. The pure crystalline silky needles of caffeine (0.5 g) melt at 235°C.

a. Murexide test
A few crystals of caffeine and 3 drops of nitric acid are placed in a small porcelain dish and evaporated to dryness. Addition of 2 drops of ammonium hydroxide imparts a purple coloration.

b. Thin layer chromatography of xanthine derivatives
The crude tea extract is spotted on a thin-layer plate covered with Silica Gel GF_{254}. The developing solvent is chloroform–ethanol, 9.5:0.5. Detection: (1) $UV_{254\ nm}$—dark spots; (2) Iodine—HCl spray.
   Solution A (prepared by dissolving 1 g potassium iodide and 2 g iodine in 100 ml ethanol (96%), is sprayed on the plate, followed (after 1 min) by solution B (prepared by mixing 5 ml (25% HCl solution) with 5 ml ethanol (96%). Theophylline yields a pink spot, R_f 0.1; theobromine, a violet spot, R_f 0.2; and caffeine, yellow-brown spot, R_f 0.6.

c. IR spectrum of caffeine (solid film)
   3134 cm^{-1}: C—H stretching
   2850: N—CH_3 groups
   1705: C=O stretching
   1660: C=C stretching of tetrasubstituted double bond
   1604, 1548, 1440: pyrimidine system
   1470, 1358: asym. and sym. CH_3 bending
   1230, 1197, 1020: C—N stretching
   740: C—H deformation

d. UV spectrum of caffeine

   λ_{max}^{water} 278 mμ (logε 4.03)

   Purines generally exhibit high-intensity selective absorption at 260 mμ. Substitution by alkyl groups causes a displacement to longer wavelengths.

e. NMR spectrum of caffeine

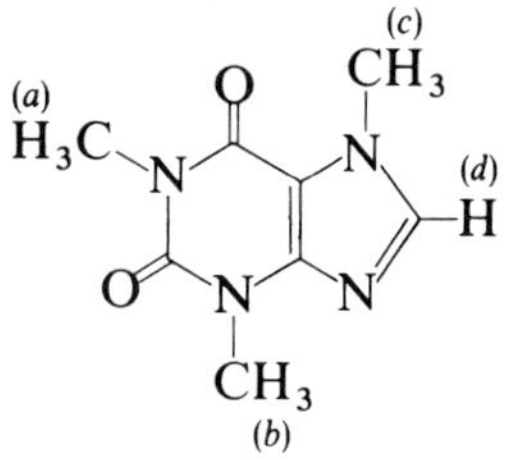

Position	δ (ppm)
a	3.35
b	3.55
c	3.95
d	7.60

f. Caffeine

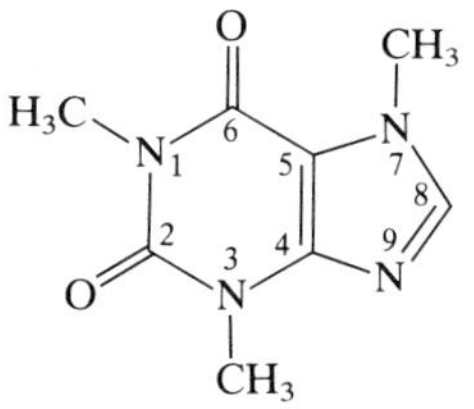

[Structure of caffeine with numbered positions: N1 (H3C), C2 (=O), N3 (CH3), C4, C5, C6 (=O), N7 (CH3), C8, N9]

g. ¹³C-NMR
C-1 30.4
C-3 32.3
C-7 35.8

h. Mass Spectrum
$C_8H_{10}N_4O_2$ Mol. wt. 194
m/z 194   m/z 137 (M-CO)
m/z 165 (M-CH$_3$)   m/z 109 (tropylium ion)

References

1. Michl, H., and Haberler, F. (1954). *Monatsh. Chem.* **35**, 770.
2. Freise, F. W. (1935). *Pharm. Zentr.* **76**, 704.
3. Stenhouse, J. (1854). *Ann.* **89**, 244.

Recommended Reviews

1. Atta-ur-Rahman and Choudhary, M. I. (1990). Purine alkaloids. *In* "The Alkaloids," Vol. 38, (A. Brossi, ed.), p. 225. Academic Press. New York.
2. Bokuchava, M. A., and Kobeleva, N. I. S. (1982). Tea aroma. *In* "Food Flavours, Part B." (I. D. Morton and A. J. Macleod eds.), p. 49, Elsevier, Amsterdam, The Netherlands.
3. Graham, H. (1983). Tea. *In* "Kirk-Othmer Encyclopedia of Chemical Technology," Vol. 22, p. 628.
4. Stahl, W. H. (1962). The chemistry of tea and tea manufacturing. *In* "Advances in Food Research" (C. O. Chichester, E. M. Mrak, and G. F. Stewart, eds.), Vol. 11, p. 202. Academic Press, New York.

D. Isolation of Piperine from Black Pepper

1. Introduction

Piperine occurs in the unripe fruit (black pepper) and the kernel of the ripe fruit (white pepper) of *Piper nigrum* and in the fruit of aschanti (*Piper clusii*). It is also found in long pepper (*Piper longum*) (1) and in the seeds of *Cubeba censii* (2). Piperine has also been isolated from *Piper famechoni* (3) and *Piper chaba* (4).

The piperine content of black pepper varies from 6 to 9%. Piperine is tasteless, so some investigators have suggested that the stereoisomeric chavicine, which accompanies piperine in the pepper, is possibly responsible for the particular taste of pepper.

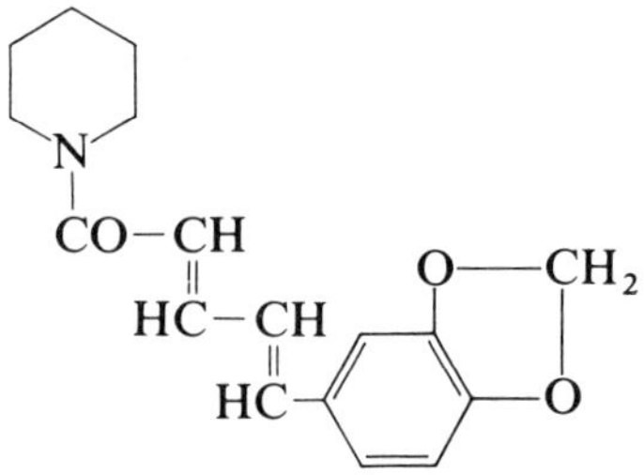

Piperine

2. Principle

In the following procedure, the piperine is extracted from black pepper with ethanol and may be converted to a red TNB (1,3,5-trinitrobenzene) complex.

Although piperine is tasteless at first, it does ultimately produce a burning sensation and sharp aftertaste. The initial tastelessness of piperine may be a consequence of its extremely low solubility in water. The crude ethanol extract contains, in addition to piperine and chavicine (its geometric isomer), some acidic, resinous material. In order to prevent co-precipitation of piperine and the resin acids, dilute ethanol KOH solution is added to the concentrated extract to keep acidic material in solution and/or as solid gummy material that is precipitated in the vessel.

3. Apparatus

Soxhlet extractor
Vacuum distillation assembly

4. Materials

Black pepper
Ethanol, 95%
Potassium hydroxide
1,3,5-Trinitrobenzene

5. Time

3 hours

6. Procedure

Ten g black pepper is ground to a fine powder and extracted with 150 ml 95% ethanol in a Soxhlet extractor for 2 hr. The solution is filtered and concentrated *in vacuo* on a water bath at 60°C. Alcoholic potassium hydroxide (ten ml 10%) is added to the filtrate residue and after a while decanted from the insoluble residue. The alcoholic solution is left overnight, whereupon yellow needles of mp 125–126°C are deposited; yield, 0.3 g.

Piperine forms a solid complex with 1,3,5-trinitrobenzene in a ratio 1:1, in the form of red needles, mp 130°C.

a. Thin layer chromatography of black pepper extract

The crude extract is spotted on a thin-layer plate (Silica Gel GF_{254}) and developed with benzene–ethyl acetate, 2:1. Detection: (1) UV_{365} shows blue fluorescence of piperine; (2) Spraying with anisaldehyde–sulfuric acid reagent, prepared by mixing anisaldehyde (0.5 ml) with glacial acetic acid (10 ml), methanol (85 ml), and concentrated sulfuric acid (5 ml). This solution is sprayed on the plate, which is then heated at 110°C for 10 min. Piperine appears as a yellow spot, R_f 0.25.

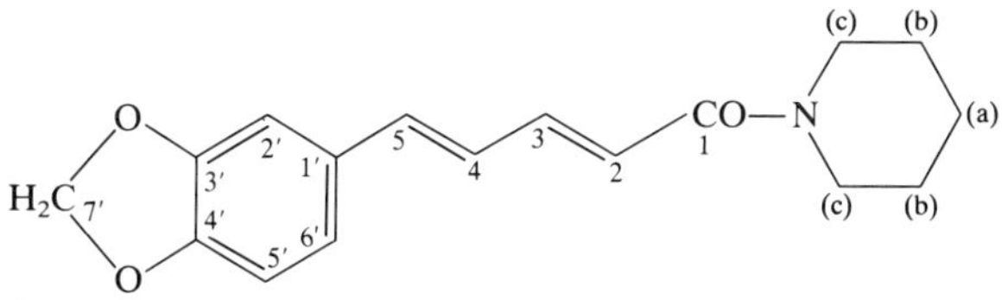

b. ^{1}H-NMR

δ_H 5.93 (2H, 7′), 7.40 (1H, 3), 6.43 (1H, 2), 3.57 (4H, c), 1.62 (4H, b), 1.62 (4H, a).

c. ^{13}C NMR

C-1 164.5	C-2′ 104.9
C-2 119.4	C-3′ 147.4
C-3 141.6	C-4′ 147.4
C-4 124.6	C-5′ 107.6
C-5 137.3	C-6′ 121.7
C-1′ 132.2	C-7′ 100.6

d. Mass spectrum

$C_{17}H_{19}NO_3$, Mol. wt. 285.

m/z 285(65%) 201(100%) 173(42%) 143(32%)
     202(29%) 174(25%)   171(26%) 115(92%)

e. IR spectrum of piperine (in KBr disk)
  3000 cm^{-1}: aromatic C—H stretching
  1635, 1608: sym. and asym. stretching of C=C (diene)
  1608, 1580, 1495: aromatic stretching of C=C (phenyl ring)
  1635: stretching of —CO—N$<$
  Methylenedioxy group:
  2925, 2840: CH$_2$ asym. and sym. stretching, aliphatic C—H stretching
  1450: CH$_2$ bending
  1250, 1190: asym. stretching of =C—O—C
  1030: sym. stretching of =C—O—C
  930: most characteristic, probably related to C—O stretching
  1132: in-plane bending of phenyl CH
  995: C—H bending for *trans* —CH=CH—
  850, 830, 805: out-of-plane C—H bending, 1,2,4-trisubstituted phenyl, two adjacent hydrogen atoms.

f. UV spectrum of piperine

$\lambda_{\max}^{\text{EtOH}}$ 245 mμ (logε 4.47)

1-Phenylbutadiene in conjugation with —CO—R and methylenedioxy chromophores.

References

1. Peinemann, (1896). *Arch. Pharm.* **234**, 204.
2. Herlant, M. (1895). *Chem. Zentr.* **66**, 319.
3. Barille, A. (1902). *Compt. Rend.* **134**, 1512.
4. Bose, P. K. (1936). *Chem. Zentr.* **107**, 4315.

Recommended Review

1. Marion, L. (1950). The pyridine alkaloids. *In* "*The Alkaloids,*" (R. H. F. Manske, and H. L. Holmes, eds.), Vol. 1, p. 167. Academic Press, New York.

E. Degradation of Piperine

1. Introduction

As early as 1849, piperine was hydrolyzed with aqueous sodium hydroxide (1), and later with nitric acid (2), to yield a volatile base, piperidine, and piperic acid (3). Spring and Stark [4] used ethanolic potassium hydroxide for hydrolysis of piperine. The presence of two double bonds in piperic acid was proved by its catalytic reduction to tetrahydropiperic acid and by the preparation of tetrabromopiperic acid by bromination with bromine in carbon

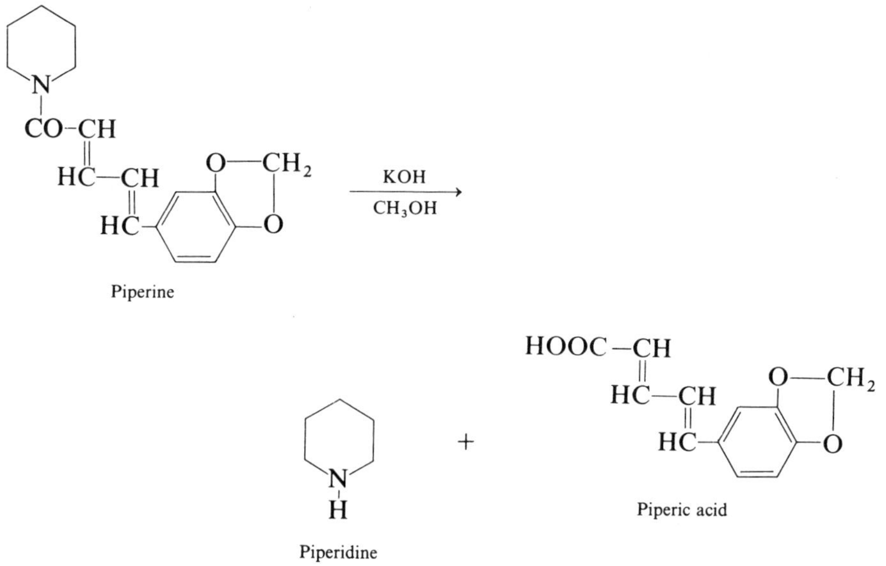

I. Alkaloids

[Piperine structure]

Piperine

$\xrightarrow{\text{KOH} \atop \text{CH}_3\text{OH}}$

Piperidine + Piperic acid

disulfide. The structure of piperic acid was deduced from its oxidation by potassium permanganate to yield oxalic acid, piperonal, and piperonylic acid.

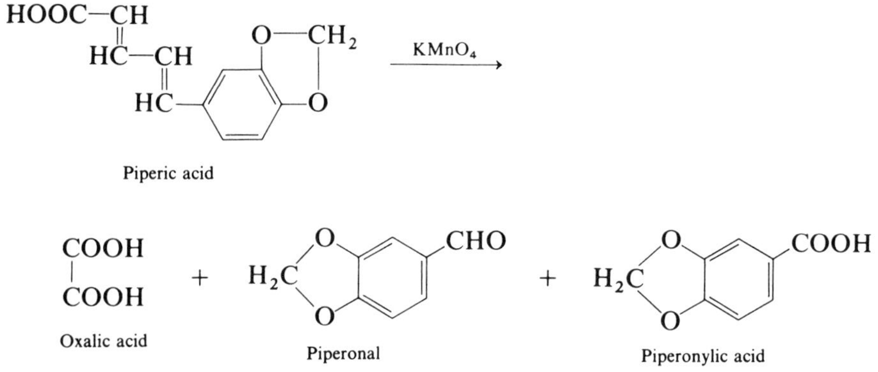

Piperic acid $\xrightarrow{\text{KMnO}_4}$

Oxalic acid + Piperonal + Piperonylic acid

Mild oxidation (KMnO$_4$ in basic solution) yields piperonal, piperonylic acid, and tartaric acid.

Piperine can be synthesized from the chloride of piperic acid and piperidine.

2. Principle

In the following degradation, piperine is treated with ethanolic potassium hydroxide, yielding piperic acid and piperidine (isolated as the hydrochloride).

3. Apparatus

Vacuum distillation assembly

4. Materials

Ethanol, 95%
Hydrochloric acid
Piperine
Potassium hydroxide

5. Time

3 hours

6. Procedure

Piperine (1 g) and 10 ml 10% alcoholic potassium hydroxide are refluxed for 90 min. The ethanolic solution is evaporated to dryness under reduced pressure, the receiver being cooled in an ice–salt bath. The solid potassium piperinate is suspended in hot water and acidified with hydrochloric acid. The voluminous yellow precipitate is collected on a Buchner funnel, washed with cold water, and recrystallized from ethanol to yield piperic acid as yellow needles of mp 216–217°C. The strongly basic ethanolic distillate in the receiver is saturated with hydrochloric acid and evaporated to dryness to give piperidine hydrochloride, which melts at 244°C after recrystallization from ethanol.

References

1. Wortheim, T. (1849). *Ann.* **70**, 58.
2. Anderson, T. (1850). *Ann.* **75**, 80.
3. Strecker, A. (1858). *Ann.* **105**, 317.
4. Spring, F. S., and Stark, J. (1950). *J. Chem. Soc.* 1177.

Recommended Review

1. Marion, L. (1950). *The pyridine alkaloids. In "The Alkaloids,"* (R. H. F. Manske, and H. L. Holmes eds.), Vol. 1, p. 167. Academic Press, New York.

F. Isolation of Strychnine and Brucine from the Seeds of *Strychnos nux vomica*

1. Introduction

Strychnine, $C_{21}H_{22}O_2N_2$, and brucine, $C_{23}H_{26}O_4N_2$, were first isolated by Pelletier and Caventou in 1819 from the seeds and bark of *Strychnos nux vomica*. *Nux vomica* seeds contain about 3% alkaloids, of which a little more than half is made up of strychnine, the other alkaloids being brucine, colubrine, vomicine, and pseudostrychnine. In recent years it was found that Australian (1) and Congolese (2) *Strychnos* species also contain strychnine, brucine, and other alkaloids.

Strychnine is intensely bitter; one part of strychnine is capable of imparting a bitter taste to 500,000 parts of water. It is a virulent poison and is used medicinally as a tonic and stimulant.

Brucine is even bitterer than strychnine. Its uses are similar to those of strychnine but its action is so much milder that it is used on a very large scale for denaturation of ethanol.

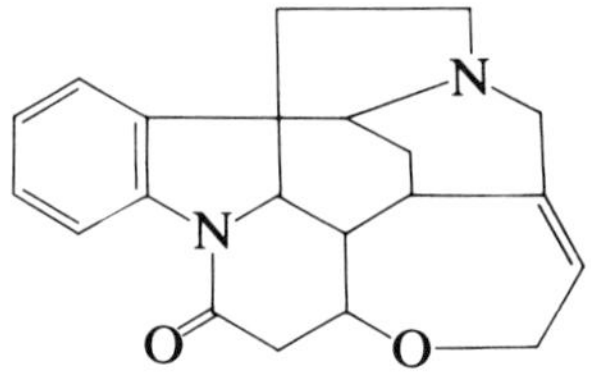

Strychnine 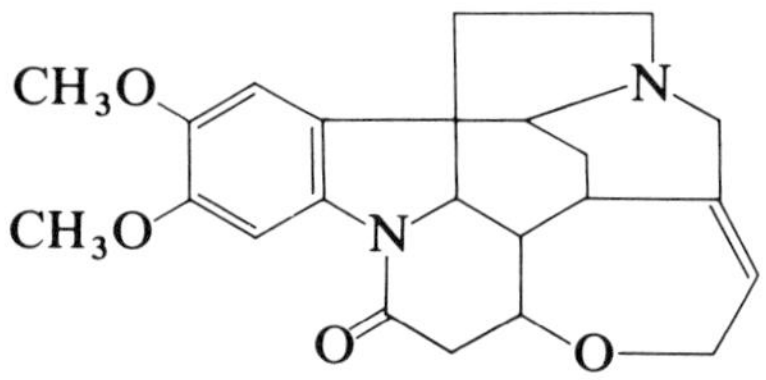   Brucine

2. Principle

In the following procedure, a mixture of strychnine and brucine is extracted with chloroform from the crushed nuts, which have been treated overnight with calcium hydroxide. Crystallization of the crude mixture from 50% ethanol yields crystalline strychnine, which is converted to its sulfate. Brucine (which is more soluble in alcohol) is recovered from the mother liquor, also as the sulfate.

3. Apparatus

Soxhlet extractor
Vacuum distillation assembly

4. Materials

Acetone	Ethanol, 95%	Sodium carbonate
Acetic acid, glacial	Nitric acid	Sodium hydroxide
Calcium hydroxide	*Nux vomica* seeds	Sulfuric acid
Chloroform	Potassium dichromate	

5. Time

Treatment of seeds with calcium hydroxide: 12 hours
Isolation of strychnine and brucine: 4–5 hours

6. Procedure

Groundnuts (200 g) are mixed thoroughly with 200 ml suspension of 10% calcium hydroxide in water and left overnight at room temperature. After air drying, the slurry is extracted with chloroform in a Soxhlet extractor for 3 hr. The chloroform solution is then extracted several times with 5% sulfuric acid solution, and subsequently basified with 10% aqueous sodium hydroxide. After cooling, the crystals are filtered, 1.5 volumes of 50% ethanol are added, and the mixture is refluxed until most of the solid has dissolved. After addition of a little activated charcoal, the solution is filtered hot and left overnight. The crystals of strychnine are filtered and washed with a little 50% ethanol. The mother liquor and washings are kept for the isolation of brucine.

a. Preparation of strychnine sulfate

The crude strychnine is dissolved in 9 volumes of boiling water, and 15% sulfuric acid solution is added slowly, with stirring, until the reaction is slightly acid to Congo red. Activated charcoal is added and the solution is refluxed for 1 hr and filtered hot. The sulfate, which crystallizes on cooling, is filtered and washed with cold water.

b. Pure strychnine

Strychnine sulfate is dissolved in 15 volumes of water at 80°C and neutralized with 10% aqueous sodium carbonate; after addition of charcoal, the solution is filtered hot. Strychnine precipitates on addition of aqueous sodium carbonate and cooling. The precipitate is filtered on a Buchner funnel and washed with cold water. Recrystallization from ethanol yields a product of mp 286–288°C.

c. Strychnine

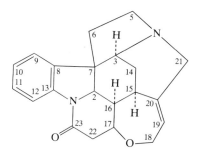

d. ^{13}C NMR

C-2 60.1   C-5 50.2   C-7 51.9
C-3 60.1   C-6 42.8^a C-8 132.6

C-9 122.3	C-14 26.8	C-19 127.7
C-10 124.3	C-15 31.5	C-20 140.3
C-11 128.6	C-16 48.2	C-21 52.7
C-12 116.2	C-17 77.5	C-22 42.3^a
C-13 142.2	C-18 64.6	C-23 169.4

^{a} Shifts may be interchanged.

e. **Mass Spectrum**

$C_{21}H_{22}N_2O_2$, Mol. wt. 334.

m/z 334(100%) 120(37%) 77(35%) 44(44%)
    144(32%)   107(33%) 55(33%) 41(34%)

f. **UV Spectrum of strychnine**

λ_{max}^{EtOH} 240 mμ (logε 4.15); 270(3.92); 294(3.72).

The UV absorption spectrum of strychnine closely resembles that of hexahydrocarbazole, as the formula might have led us to expect.

g. **Brucine sulfate**

The mother liquor remaining after strychnine separation is concentrated *in vacuo* on a water bath until most of the alcohol has been removed. The residue is acidified to pH 6 with dilute sulfuric acid and then concentrated to a volume of 3 to 4 ml. After standing in a refrigerator overnight, the product is filtered and washed with cold water. Brucine sulfate is purified by dissolution in 4.5 volumes of hot distilled water and boiled with a little charcoal for 1 hr. It is filtered hot and left in a refrigerator for several days. Brucine is recovered from the sulfate in a manner analogous to that outlined for strychnine; recrystallization from aqueous acetone yields crystals of mp 178°C.

Warning! Extreme caution should be exercised in carrying out this experiment. The hands must be washed thoroughly because of the high toxicity of these alkaloids.

h. **Brucine**

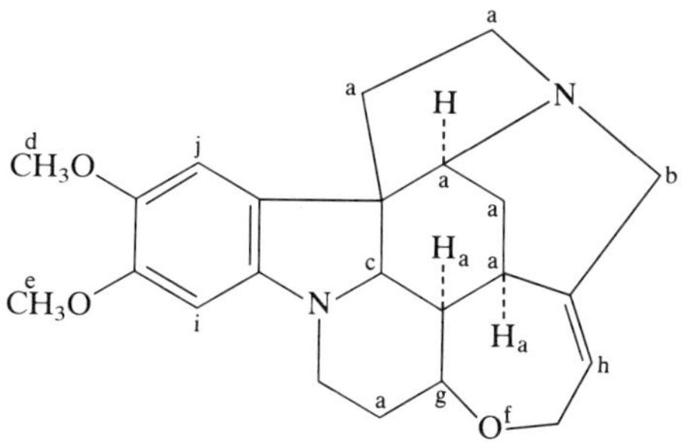

i. ¹H NMR
   a. 1.15–3.70 (11H)
   b. 2.04 2H
   c. 3.82 1H
   d. 3.88 3H
   e. 3.92 3H
   f. 4.12 2H
   g. 4.30 1H
   h. 5.91 1H
   i. 6.69 1H
   j. 7.83 1H

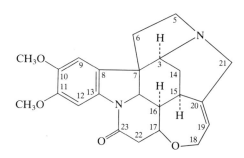

j. ¹³C NMR

C-2 60.3	C-11 149.3	C-19 127.2
C-3 59.9	C-12 101.1	C-20 140.6
C-5 50.1	C-13 136.0	C-21 52.7
C-6 42.3	C-14 26.8	C-22 42.3
C-7 51.9	C-15 31.5	C-23 168.9
C-8 123.6	C-16 48.3	OMe 56.2
C-9 105.7	C-17 77.8	OMe 56.4
C-10 146.2	C-18 64.6	

References

1. Anet, F. A. L., Hughes, G. K., and Ritchie, E. (1953). *Australian J. Chem.* **6,** 58.
2. Jaminet, F. (1953). *J. Pharm. Belg.* **8,** 339.

Recommended Reviews

1. Cromwell, B. T. (1955). Alkaloids of Strychnos spp. In *"Modern Methods of Plant Analysis,"* (K. Paech, and M. V. Tracey, eds.), Vol. 4, p. 480. Springer-Verlag, New York.
2. Robinson, R. (1952). Molecular structure of strychnine, brucine, and vomicine. In *"Progress in Organic Chemistry,"* (J. W. Cook, ed.), Vol. 1, p. 1. Butterworth, London, England.

G. Synthesis of Mescaline

1. Introduction

Mescaline, an active hallucinogen, was first synthesized in 1919 by Spath, who used 3,4,5-trimethoxybenzoyl chloride as his starting material (1). Subsequently, a number of improved syntheses have been reported. Slotta and Heller (2) prepared mescaline by Hofmann degradation of trimethoxyphenylpropionamide, and Kinder and Peschke (3) by condensation of 3,4,5-trimethoxybenzaldehyde with potassium cyanide, acetylation, and catalytic reduction.

Hahn and Wassmuth (4) used elemicine as their starting material, and Hahn and Rumpf (5) later described the preparation of mescaline by catalytic reduction of ω-nitrotrimethoxystyrene. Benington and Morin (6) reduced 3,4,5-trimethoxy-β-nitrostyrene with lithium aluminum hydride and obtained an 86% yield of mescaline.

The most striking physiological affect of mescaline is the production of visual color hallucinations, which can be induced by doses of 0.3 to 0.4 g mescaline sulfate. Auditory, gustatory, and other sensory hallucinations have also been recorded.

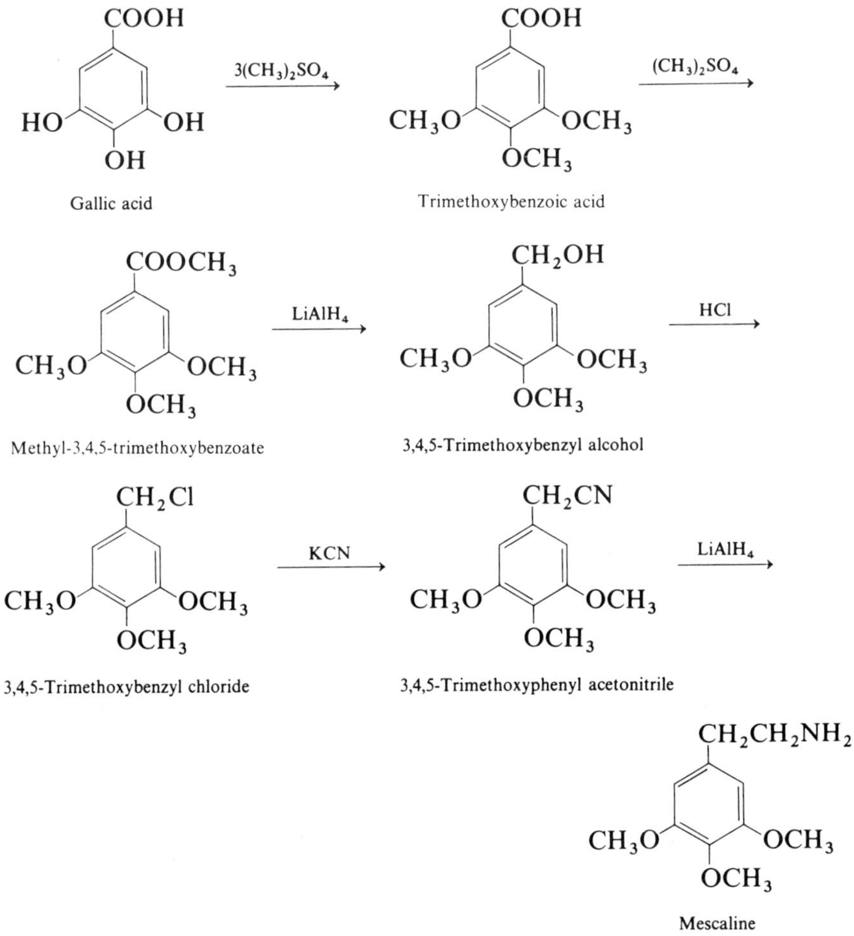

Gallic acid $\xrightarrow{3(CH_3)_2SO_4}$ Trimethoxybenzoic acid $\xrightarrow{(CH_3)_2SO_4}$

Methyl-3,4,5-trimethoxybenzoate $\xrightarrow{LiAlH_4}$ 3,4,5-Trimethoxybenzyl alcohol $\xrightarrow{HCl}$

3,4,5-Trimethoxybenzyl chloride $\xrightarrow{KCN}$ 3,4,5-Trimethoxyphenyl acetonitrile $\xrightarrow{LiAlH_4}$

Mescaline

244 CHAPTER 4 Nitrogenous Compounds

2. Principle

The synthesis (7, 8) of mescaline involves the following stages: gallic acid→ 3,4,5-trimethoxybenzoic acid→ methyl ester of trimethoxybenzoic acid→ 3,4,5-trimethoxybenzyl alcohol→ 3,4,5-trimethoxybenzyl chloride→ 3,4,5-trimethoxyphenyl acetonitrile→ mescaline.

3. Apparatus

Reflux assembly
Vacuum distillation assembly

4. Materials

Dimethyl sulfate
Dry ether
Ether
Gallic acid
Hydrochloric acid
Lithium aluminum hydride

Potassium cyanide
Picric acid
Potassium hydroxide
Sodium carbonate
Sodium sulfate, anhydrous
Sulfuric acid

5. Time

9–10 hours

6. Procedure

a. 3,4,5,-Trimethoxybenzoic acid

To a cold solution of 80 g sodium hydroxide in 500 ml water in a 1-liter flask is added 50 g gallic acid. The flask is immediately tightly stoppered, and the mixture is shaken occasionally until all the acid has dissolved. Dimethyl sulfate[1] 89 g (67 ml) is then added and the flask is shaken for 20 min while being cooled by means of cold water in order to ensure that the temperature does not rise above 30 to 35°C. Occasionally the stopper is raised to release any built-up pressure. A second portion of 89 g dimethyl sulfate is then added, and shaking is continued for an additional 10 min. During this second addition, the temperature may rise to 40 to 45°C. The flask is then fitted with a reflux condenser, and the contents are boiled for 2 hr. In order to saponify the small amount of ester produced, a solution of 20 g sodium hydroxide in 30 ml water is then added, and boiling is continued for 2 additional hr. The

---

[1] The toxic nature of methyl sulfate must always be borne in mind! Ammonia is a specific antidote for methyl sulfate and should be kept at hand to destroy any of the ester accidentally spilled.

reaction mixture is then cooled and acidified with dilute hydrochloric acid. The precipitated trimethoxybenzoic acid is filtered with suction and washed well with cold water. The product melts at 157 to 160°C and weighs 50–52 g. It may be purified by recrystallization from 2 liters boiling water with the use of decolorizing carbon, the filtration being carried out in a steam-jacketed funnel. In this way, 41–43 g colorless needles melting at 167°C are obtained.

b. Methyl ester of 3, 4, 5-trimethoxybenzoic acid

To a solution prepared from 50 g 3, 4, 5-trimethoxybenzoic acid, 10 g sodium hydroxide, 27 g sodium carbonate, and 300 ml water are added, with stirring, 47 ml methyl sulfate during the course of 20 min. The reaction mixture is refluxed for half an hour. The crude ester (33 g) precipitates from the cold mixture. From the filtrate 19 g of starting material are recovered upon acidification with dilute hydrochloric acid. The ester is further purified by dissolving it in the minimal amount of methanol and treatment with Norit. Usually it is necessary to repeat this treatment to obtain a colorless crystalline product that melts at 80 to 82°C.

c. 3, 4, 5-Trimethoxybenzyl alcohol

To a suspension of 6.9 g lithium aluminum hydride in 300 ml anhydrous ether is added, in the course of 30 min, a solution of 34 g methyl ester of 3, 4, 5-trimethoxy-benzoic acid in 450 ml ether. The solid that forms is carefully decomposed with 75 ml ice water. After decantation of the ether, 375 ml ice-cold 10% sulfuric acid is added. The product is extracted with 200 ml ether. After drying over sodium sulfate, the combined extracts are freed of ether, and the residue is distilled. Bp 135–137°C (0.25 mm); yield 22 g (73%).

d. 3, 4, 5-Trimethoxybenzyl chloride

A mixture of 20 g 3, 4, 5-trimethoxybenzyl alcohol and 100 ml ice-cold concentrated hydrochloric acid is shaken vigorously until a homogeneous solution is obtained. Within a few minutes a turbidity develops, followed by heavy precipitation of a gummy product. After 3 hr and dilution with 100 ml ice water, the aqueous layer is decanted and extracted with three 60-ml portions of benzene. The gummy organic residue is then dissolved in the combined benzene extracts. The benzene solution is washed with water and dried over sodium sulfate. It is then transferred to a distilling flask, and the benzene is removed under reduced pressure. The red, semisolid residue is suspended in a small amount of ice-cold ether and filtered through a chilled funnel. The crystalline product, after washing with small portions of cold ether, weighs 7.8 g. After standing in a refrigerator, the combined filtrates yield more crystals. The total yield is 10.5 g (48%). After four recrystallizations from benzene, colorless needles are obtained, mp 60–62°C.

e. 3,4,5-Trimethoxyphenyl acetonitrile

A mixture of 9 g potassium cyanide in 35 ml water, 60 ml methanol, and 9.7 g 3,4,5-trimethoxybenzyl chloride is heated for 10 min at 90°C. The solvents are partially removed under reduced pressure. The residue is then extracted with 90 ml ether in three portions. The combined extracts are washed and dried over sodium sulfate. After removal of the drying agent, the ether solution is warmed on a steam bath, and the ether is removed by a stream of air. On chilling, the residue yields scale-like crystals. Recrystallization from ether gives rectangular prisms: yield 2.5 g (27%); mp 76–77°C.

f. Mescaline

In 150 ml anhydrous ether is suspended 0.85 g lithium aluminum hydride powder. Two grams of 3,4,5-trimethoxyphenyl acetonitrile in 150 ml anhydrous ether are added, with stirring, during the course of 15 min. After stirring for an additional 15 min, 10 ml ice water is introduced carefully. A mixture of 10 g sulfuric acid in 40 ml water is then added at a moderate rate. The aqueous layer is separated and treated with concentrated sodium hydroxide. The brown oil is extracted with three 30-ml portions of ether. The combined extracts are washed once with water and dried over potassium hydroxide pellets. To the decanted ether solution is added a mixture of 1 g sulfuric acid and 25 ml ether. The white precipitate is washed several times with ether; yield, 1.2 g (40%). After two recrystallizations from 95% ethanol, the colorless, long, thin plates soften at 172°C and melt at 183°C. The picrate prepared from the acid sulfate melts at 217°C (dec.) after triple recrystallization from ethanol. The chloroplatinate prepared from the free base melts at 184 to 185°C.

g. UV spectrum of mescaline sulfate

λ_{max}^{EtOH} 264, 266, 269 mμ.

The spectrum is similar to that of phenylethylamine sulfate, but shows a bathochromic shift obviously due to the presence of the methoxyl group in the benzene ring, especially that in the *para* position.

h. Mass Spectrum

Mescaline $C_{11}H_{17}NO_3$, Mol. wt. 211.

m/z 211(17%)	182(59%)	167(25%)	44(100%)
183(7%)	181(34%)	151(6%)	30(70%)

References

1. Spath, E. (1919). *Monatsh. Chem.* **40,** 129.
2. Slotta, K. H., and Heller, H. (1930). *Ber.* **63,** 3029.

3. Kindler, K., and Peschke, W. (1932). *Arch. Pharm.* **270,** 410.
4. Hahn, G., and Wassmuth, H. (1934). *Ber,* **67,** 696.
5. Hahn, G., and Rumpf, F. (1938). *Ber.* **71,** 2141.
6. Benington, F., and Morin, R. D. (1951). *J. Am. Chem. Soc.* **73,** 1353.
7. Tsao, M. U. (1951). *J. Am. Chem. Soc.* **73,** 5495.
8. Donrow, A., and Petsch, (1952). *Arch. Pharm.* **285,** 323.

Recommended Review

1. Reti, L., (1953). β-Phenylethylamines. *In* "The Alkaloids," (R. H. F. Manske, and H. L. Holmes eds.), Vol. 3, p. 313. Academic Press, New York.

H. Gas Chromatography of Alkaloids

1. Introduction

In 1958, Quin (1) used gas-liquid chromatography to study the alkaloidal composition of tobacco smoke. More recently, Lloyd and coworkers (2) have shown that a large number of alkaloids, even those of high molecular weight, can be successfully separated by gas chromatography if a suitable stationary phase is used.

Nonpolar silicone polymers, particularly silicone rubber SE–30, possess great thermal stability and have been used successfully for gas chromatography of alkaloids (3–6). However, polar liquid phases often have distinct advantages over nonpolar phases. Because of greater solute–solvent interaction, they are more selective and may often effect separation of substances that do not separate on non-polar columns.

2. Principle

The following experiment (7) uses a nonpolar silicone polymer (SE–30) and a moderately polar cyanosilicone (XE–60). Because of its high selectivity, the latter appears to be the stationary liquid of choice for most alkaloid separations. Thus, atropine and scopolamine, morphine, codeine, and thebaine, quinine and quinidine, as well as cinchonine and cinchonidine, are readily separated on the more polar stationary phase (XE–60).

3. Apparatus

Gas-chromatograph
Microsyringe
Spiral glass column

4. Materials

Atropine	Hydrastine
Caffeine	Morphine
Chloroform	Narcotine
Cinchonidine	Papaverine
Cinchonine	Pilocarpine
Cocaine	Quinidine
Codeine	Quinine
Cyanosilicone XE–60	Scopolamine
Gas Chrom P	Silicone polymer SE–30
Hexamethyldisilazane	Strychnine
Homatropine	Thebaine

5. Time

4 hours

6. Procedure

a. Preparation of column

A spiral glass column 3 ft long and with an internal diameter of 0.07 in is used. The solid support is Gas Chrom P, 80–100 mesh, which is washed with concentrated hydrochloric acid and deactivated by treatment with 5% hexamethyldisilazane in toluene for 10 min. The silanized support is then filtered on a Buchner funnel, washed with toluene, and coated with 1% stationary liquid (SE–30 or XE–60) in toluene. The support is then refiltered and dried at 80°C for 2 hr. Columns are packed by gradual addition of the coated support accompanied by repeated tapping.

b. Preparation of samples

The alkaloids are dissolved in analytical grade acetone or chloroform to give a concentration of 0.5 to 1% and introduced with a Hamilton microsyringe.

c. Operating conditions

The columns are operated at optimal flow rates. The temperature of the injection port should be about 300°C.

The following table summarizes the retention values of the alkaloids on two stationary phases at three different temperatures.

Relative Retention Values of Alkaloids

Alkaloid	SE–30 (1%) 175°	SE–30 (1%) 225°	XE–60 (1%) 220°
Atropine	0.51	—	0.59
Caffeine	0.15	—	0.32
Cinchonidine	—	2.12	3.28
Cinchonine	—	2.06	3.04
Cocaine	0.52	—	0.48
Codeine[a]	1.00	1.00	1.00
Homatropine	0.31	—	0.39
Hydrastine	—	5.95	17.70
Morphine	—	1.26	2.71
Narcotine	—	11.60	20.70
Papaverine	—	3.59	6.63
Pilocarpine	0.40	—	2.20
Quinidine	—	3.41	6.63
Quinine	—	3.41	6.95
Scopolamine	0.84	—	1.40
Strychnine	—	6.89	14.70
Thebaine	—	1.55	1.54

[a] Codeine is used as the standard for calculation of relative retention values; it is arbitrarily given the value 1.00. Its actual retention times are 16.4 min (175°C), 1.7 min (225°C), and 3.6 min (220°C).

References

1. Quin, L. D. (1958). *Nature* **182**, 865.
2. Lloyd, H. A., Fales, H. M., Highet, P. F., Vanden Heuvel, W. J. A., and Wildman, W. C. (1960). *J. Am. Chem. Soc.* **82**, 3791.
3. Brochmann-Hanssen, E., and Svendsen, A. B. (1962). *J. Pharm. Sci.* **51**, 1095.
4. Parker, K. D., Fontan, C. R., and Kirk, P. L. (1963). *Anal. Chem.* **35**, 346.
5. Arndt, R. R., Baarschers, W. H., Douglas, B., Shoop, E. C., and Weisbach, J. A. (1963). *Chem. Ind. (London)* 1163.
6. Mule, S. J. (1964). *Anal. Chem.* **36**, 1907.
7. Brochmann-Hanssen, E., and Fontan, C. R. (1965). *J. Chromatogr.* **19**, 296.

Recommended Review

1. Burchfield H. P., and Storrs, E. E. (1962). Chromatography of alkaloids. *In* "Biochemical Applications of Gas Chromatography." p. 366. Academic Press, New York.

I. Thin-Layer Chromatography of Opium Alkaloids

1. Introduction

Many investigators failed to group the alkaloids in one analytical system because of the great variety of ring systems and their differing basicities. Opium alkaloids were first separated on thin layers by Borke and Kirch (1) in 1953. Machata (2), Baumler and Rippstein (3), Bayer (4), and Steele (5) have used silica gel G for the same purpose. Tiechert, Mutschler and Rochelmeyer (6) reported the separation of opium alkaloids using layers of alkaline silica gel G. Winkler and Awe (7) and Ikram, Niana, and Islam (8) separated opium alkaloids on thin layers of alumina G.

Most alkaloids are detected by observation of the fluorescence under ultraviolet light and by spraying with reagents such as Dragendorff, ceric sulfate, and iodo-platinic acid.

2. Principle

Thin-layer chromatography is preeminently suitable as a rapid method for screening plants for alkaloid content.

3. Apparatus

TLC applicator
Glass plates (20 × 20 cm)
Spray gun

4. Materials

Acetone	Methanol	Potassium iodide
Bismuth subnitrate	Methyl ethyl ketone	Silica gel G
Chloroform	Morphine	Tartaric acid
Codeine	Narcotine	Thebaine
Diethylamine	Papaverine	Xylene

5. Time

2–3 hours

6. Procedure

a. Preparation of plates

The suspension for five plates (20 × 20 cm) is prepared by shaking 25 g silica gel G and 50 ml water for 30 sec; it is then spread on the plates with an applicator to give a layer of 0.25 mm thickness. After 30 min at room temperature, the plates are activated in an oven at 120°C for 30 min.

b. Development

The samples of the alkaloids are dissolved in chloroform and spotted by means of micropipettes along a line 2 cm above the rim of the plate. Development is carried out in either of the following two solvent systems: (1) chloroform–acetone–diethylamine, 5:4:1; (2) xylene–methyl ethyl ketone–methanol–diethylamine, 20:20:3:1.

c. Detection

The dried plates are first examined under ultraviolet light and then sprayed with a modified Dragendorff's reagent which is a mixture of the following two solutions: (1) 17 g bismuth subnitrate and 200 g tartaric acid in 800 ml water; (2) 160 g potassium iodide in 400 ml water.

Retention Characteristics of Alkaloids

Alkaloid	R_f in Solvent System	
	1	2
Morphine	0.10	0.12
Codeine	0.38	0.26
Thebaine	0.65	0.45
Papaverine	0.67	0.59
Narcotine	0.72	0.74

References

1. Borke, N. L., and Kirch, E. R. (1953). *J. Am. Pharm. Assoc., Sci. Edit.* **42**, 627.
2. Machata, G. (1960). *Mikrochim. Acta.* 79.
3. Baumler, J., and Rippstein, S. (1961). *Pharm. Acta Helv.* **36**, 382.
4. Bayer, J. (1964). *J. Chromatogr.,* **16**, 237.
5. Steele, J. A. (1965). *J. Chromatogr.,* **19**, 300.
6. Teichert, K., Mutschler, E. and Rochelmeyer, H. (1960). *Deut. Apotheker Ztg.,* **100**, 283.
7. Winkler, W., and Awe, W. (1961). *Arch. Pharm.* **294**, 301.
8. Ikram, M., Miana, G. A., and Islam, M. (1963). *J. Chromatogr.* **11**, 263.

Recommended Review

1. Waldi, D. Alkaloids. *In* Stahl, E. "Thin-layer Chromatography," p. 279. Academic Press, New York. (1965).

J. Questions

1. What initial steps would you take to establish the structure of an unknown natural alkaloid?

2. How are alkaloids distinguished from proteins? What is the biogenetic relationship between the two groups of compounds?
3. Why is tannic acid used as an antidote in cases of poisoning by alkaloids?
4. List six heterocyclic structures that constitute the basis of important alkaloids. Indicate an alkaloid derived from each.
5. Many alkaloids, e.g., quinine, strychnine, and brucine, are used in the resolution of racemic mixtures of optically active acids. Explain the reason for this method and describe in detail how you could resolve racemic mandelic acid.
6. Write equations illustrating Hofmann's exhaustive methylation of coniine (I), lobelanine (II), and papaverine (III).

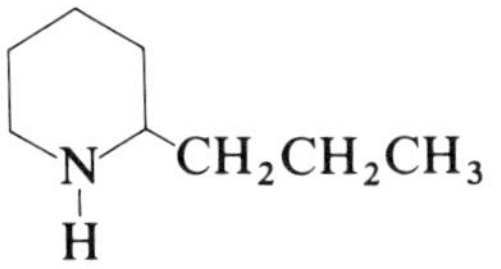

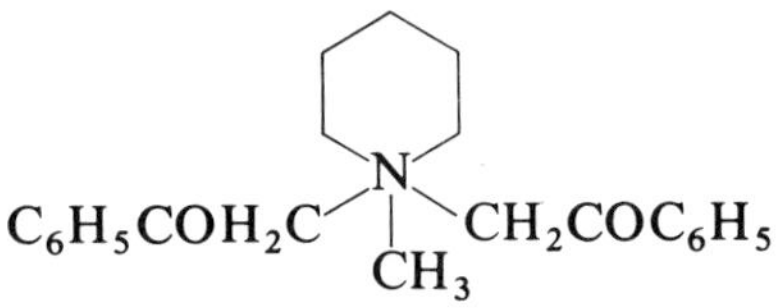

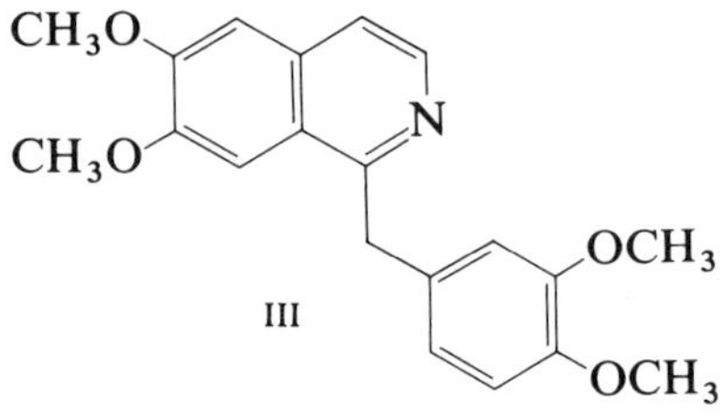

7. Which products are obtained when nicotine is (a) reduced by sodium in ethanol; (b) oxidized by nitric acid?
8. When cigarette smoke is blown through a white handkerchief, a stain is left. Is this stain due to nicotine? Explain.
9. Capsaicin, the alkaloid from red pepper, has the following structure:

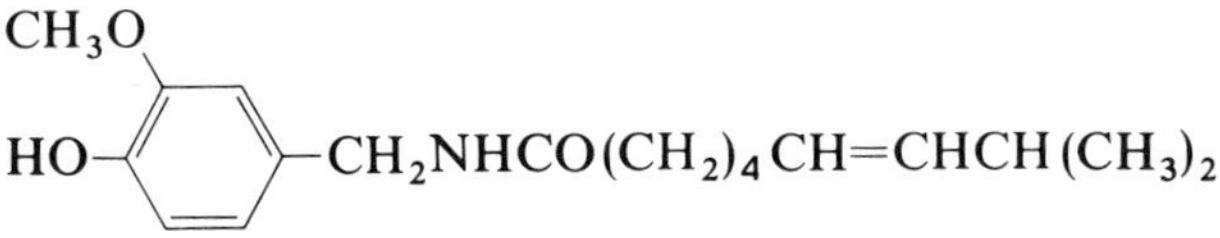

By which reactions could you prove this formula?

10. How can nicotine (I) and papaverine (II) be synthesized?

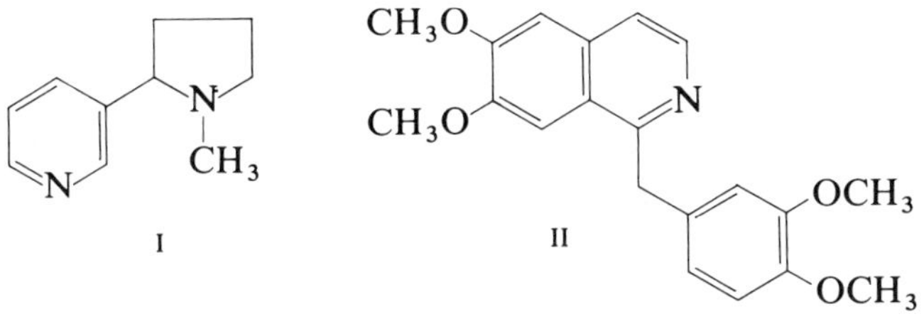

11. Cusparine, $C_{19}H_{17}NO_3$ (from Angostura bark), gives 4-methoxy-quinoline-2-carboxylic acid and piperonylic acid (3, 4-methylenedioxy-benzoic acid) on oxidation with potassium permanganate. Suggest a structure for cusparine.
12. How are the following alkaloids interrelated chemically: morphine, codeine, heroine, and apomorphine?
13. Write equations for the synthesis of tropinone by Robinson's method. Why is this synthesis (like many similar ones) termed synthesis under physiological conditions?
14. What are the sources and structures of uric acid, theophylline, theobromine, and caffeine?
15. What is the pathway of biosynthesis of the following alkaloids: papaverine, harmane, mescaline?

K. Recommended Books

1. Bentley, K. W. (1957). *"The Alkaloids, Parts 1–2."* Interscience, New York.
2. Bernfeld, P. (1963). *"Biogenesis of Natural Products."* Pergamon Press, London, England.
3. Hendrickson, J. B. (1965). *"The Molecules of Nature."* Benjamin, New York.
4. Manske, R. H. F. (1960–1977). *"The Alkaloids."* 16 Volumes. Academic Press, New York.

II. AMINO ACIDS

A. Introduction

The amino acids are colorless crystalline compounds that are generally soluble in water, but sparingly soluble in organic solvents. They are amphoteric compounds, their reactions with acids and bases being given by the following equation:

$$\overset{+}{N}H_3 \cdot CHR \cdot COOH \underset{}{\overset{H^+}{\rightleftarrows}} \overset{+}{N}H_3 \cdot CHR \cdot COO^- \overset{OH^-}{\longrightarrow} NH_2 \cdot CHR \cdot COO^- + H_2O$$

However, they are normally written in their classical form $RCH \cdot NH_2 \cdot COOH$. If R is a group other than hydrogen, the acid will have an asymmetric carbon atom, and can therefore occur in an optically active form. The isomers having the L-configuration are most widely distributed in nature.

For each amino acid in solution there is a characteristic pH value at which it exists only in the zwitterion form. This pH is known as the isoelectric point. At this point the amino acid has its lowest solubility.

The amino acids are classified as follows:

aliphatic neutral, such as glycine, $CH_2(NH_2)COOH$
aliphatic acidic, such as aspartic acid, $HOOC \cdot CH_2 \cdot CH(NH_2) \cdot COOH$
aliphatic basic, such as lysine, $NH_2 \cdot (CH_2)_4 \cdot CH(NH_2) \cdot COOH$

aromatic, such as phenylalanine, ⟨C₆H₅⟩—$CH_2 \cdot CH(NH_2) \cdot COOH$

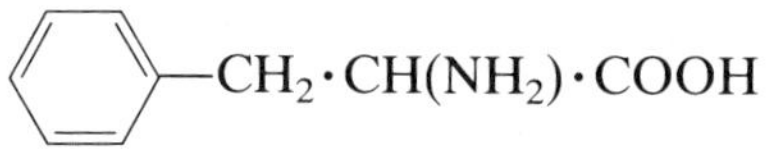

heterocyclic, such as tryptophan, [indole]—$CH_2 \cdot CH(NH_2) \cdot COOH$

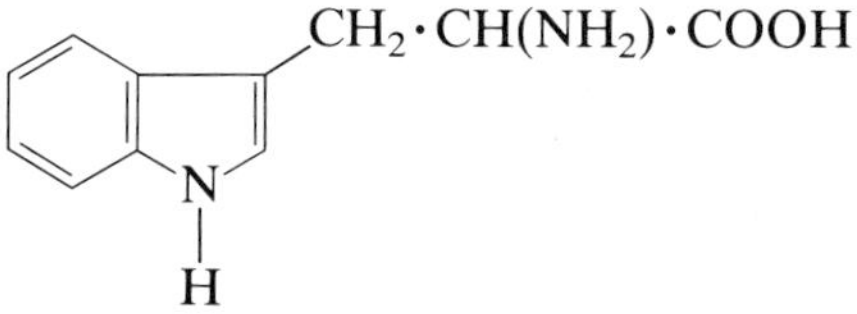

sulfur-containing (aliphatic, neutral), such as methionine,
$CH_3 \cdot SCH_2 \cdot CH_2 \cdot CH(NH_2) \cdot COOH.$

Twenty-six amino acids are obtained from proteins by hydrolysis with acids, alkalis, and enzymes. Altogether, about 170 amino acids have been found so far in nature, either in the free form or linked to other compounds (except proteins). Such amino acids have been detected in the circulatory system of animals, as intermediates in metabolic processes or as components of antibiotics, and in plants. Some of these acids have the *unnatural* D-configuration.

Several of these are listed below.

γ-Methylene-glutamic acid, $HOOC \cdot C(=CH_2) \cdot CH_2 \cdot CH(NH_2) \cdot COOH$
α-Aminoadipic acid, $HOOC \cdot CH_2 \cdot CH_2 \cdot CH_2 \cdot CH(NH_2) \cdot COOH$

α-Amino-β-phenylbutyric acid, ⟨C₆H₅⟩—$CH(CH_3) \cdot CH(NH_2) \cdot COOH$

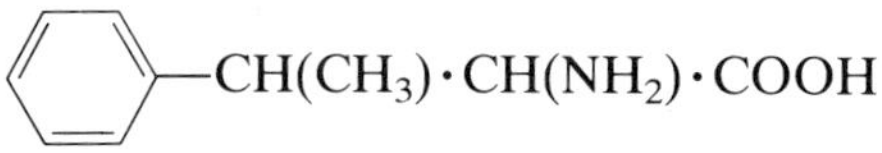

L-Homoserine, $HO \cdot CH_2 \cdot CH_2 \cdot CH(NH_2) \cdot COOH$
S-Methyl-L-cysteine, $CH_3 \cdot SCH_2 \cdot CH(NH_2) \cdot COOH$
Butyrine, $CH_3 \cdot CH_2 \cdot CH(NH_2) \cdot COOH$

II. Amino Acids

Some of these acids possess structural features uncommon in natural products, e.g.,

azetidine-2-carboxylic acid, (azetidine ring)-COOH with NH

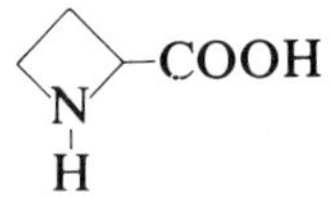

and methylenecyclopropylglycine, $CH_2=C-CH\cdot CH(NH_2)\cdot COOH$ with cyclopropane ring containing CH_2.

With only few exceptions, such as β-alanine, $NH_2\cdot CH_2\cdot CH_2\cdot COOH$ and γ-aminobutyric acid, $NH_2\cdot CH_2(CH_2)_2\cdot COOH$, the natural amino acids are α-amino acids.

Mixtures of amino acids may be separated by fractional distillation of the esters and by selective precipitation of salts. However, the most useful methods are column, paper, thin-layer, HPLC, and gas chromatography of the free or esterified amino acids, as well as ionophoresis.

1. Synthesis of α-Amino Acids

a. Amination of α-halogenated acids

$$RCH_2\cdot COOH \xrightarrow{Br_2} RCHBr\cdot COOH \xrightarrow{2NH_3} RCHNH_2\cdot COOH + NH_4Br$$

The α-bromo acids are sometimes synthesized through the corresponding malonic acid:

$$HOOC\cdot CHR\cdot COOH \xrightarrow{Br_2} HOOC\cdot CBrR\cdot COOH \xrightarrow{\Delta} RCHBr\cdot COOH + CO_2$$

Better methods are those using the esters of phthalimido-, benzamido-, formamido-, or acetamido-malonic acid, the blocking groups being removed by hydrolysis in the final step of the synthesis.

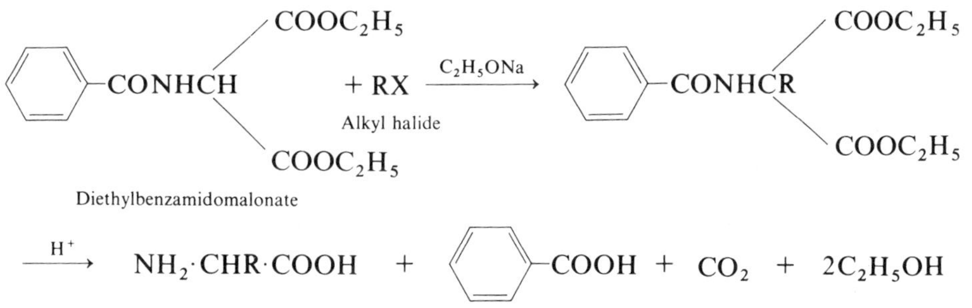

Ph-CONHCH(COOC$_2$H$_5$)$_2$ + RX $\xrightarrow{C_2H_5ONa}$ Ph-CONHCR(COOC$_2$H$_5$)$_2$

Diethylbenzamidomalonate + Alkyl halide

$\xrightarrow{H^+}$ $NH_2\cdot CHR\cdot COOH$ + Ph-COOH + CO_2 + $2C_2H_5OH$

b. Strecker synthesis

A cyanohydrin is treated with ammonia, and the resulting aminonitrile is then hydrolyzed.

$$RCHO \xrightarrow{KCN} RCH(OH)CN \xrightarrow{NH_3} RCH(NH_2)CN \xrightarrow{H^+} RCH(NH_2)COOH$$

c. Reductive amination

Reduction of an α-keto acid in the presence of NH_3 to the desired amino acid may be accomplished by chemical reducing agents or catalytically. The mechanism of the reaction probably occurs via the imino acid.

$$RCOCOOH + NH_3 \longrightarrow \begin{bmatrix} R \cdot C \cdot COOH \\ \parallel \\ NH \end{bmatrix} \longrightarrow RCH(NH_2)COOH + H_2O$$

d. Erlenmeyer azlactone synthesis

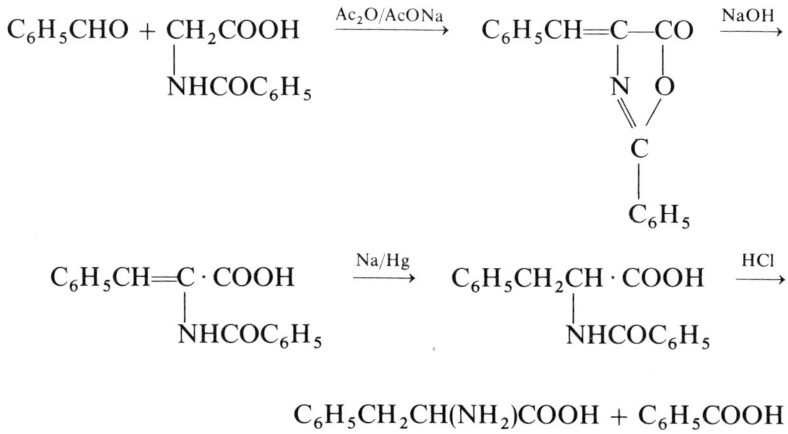

$$C_6H_5CHO + \underset{\underset{NHCOC_6H_5}{|}}{CH_2COOH} \xrightarrow{Ac_2O/AcONa} C_6H_5CH=\underset{\underset{\underset{\underset{C_6H_5}{|}}{C}}{\overset{N}{\diagdown}\overset{O}{\diagup}}}{C-CO} \xrightarrow{NaOH}$$

$$\underset{\underset{NHCOC_6H_5}{|}}{C_6H_5CH=C \cdot COOH} \xrightarrow{Na/Hg} \underset{\underset{NHCOC_6H_5}{|}}{C_6H_5CH_2CH \cdot COOH} \xrightarrow{HCl}$$

$$C_6H_5CH_2CH(NH_2)COOH + C_6H_5COOH$$

2. Resolution Procedures

Various procedures for the resolution of synthetic racemic amino acids into their active antipodes have been employed.

a. Biological procedures

Asymmetric oxidation or decarboxylation by microorganisms, such as *Penicillium glaucum*, whereby only one of the antipodes is converted into the corresponding α-keto acid or amine, while the other remains and can be isolated. More recently, enzymes have been employed for this purpose. Lately, a good separation of amino acids has been obtained by gas-liquid chromatography.

b. Diastereoisomeric salts

This method is based on the formation of diastereoisomeric salts from the racemic mixture and optically active bases (quinine, strychnine, brucine, cinchonine, chloramphenicol).

$$(+)B + (dl)A \rightarrow (+)B \cdot (+)A + (+)B \cdot (-)A$$

The salts have different physical properties, e.g., different rotations. The mixture of the salts formed is separated by a series of fractional crystallizations, followed by polarimetric measurements.

Recently, thin-layer and gas-liquid chromatography have been used successfully for the resolution and separation of racemic amino acids.

3. Biosynthesis of Amino Acids

It has long been known that there is a close interrelationship between the metabolisms of carbohydrates and amino acids. The starting materials for the biosynthesis of amino acids are certain products of carbohydrate metabolism, e.g., pyruvic acid, oxaloacetic acid, and α-ketoglutaric acid. These compounds can be converted, by transamination or reductive amination, to alanine, aspartic acid, and glutamic acid. Glutamic and aspartic acids are in turn the precursors of many amino acids. Thus, glutamic acid can be converted into ornithine. Ornithine in turn is a precursor of citrulline and arginine. Aspartic acid can give rise to β-alanine, homoserine, and threonine. The biosynthesis of tyrosine and phenylalanine by the shikimic acid pathway has been established.

4. Peptides

Peptides are polyamides of amino acids and may be represented as follows:

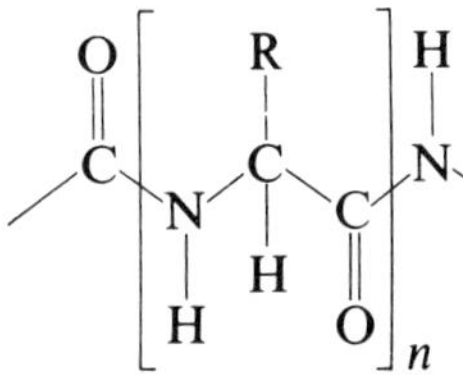

A number of biologically active peptides are found in nature. Among the simplest are the tripeptide glutathione, present in yeast, and folic acid, present in liver, spinach, and yeast.

Examples of more complex peptides are gramicidin S, an antibiotic that contains 22 amino acid residues in combination with two molecules of ethanolamine, and oxytocin—the major uterine-contracting and milk-ejecting hormone of the posterior pituitary. Vasopressin, a naturally occurring nonapeptide hormone, raises blood pressure by vasoconstriction. A number of alkaloids containing peptides have been isolated from rye-ergot. Many peptide antibiotics, such as penicillins, are now known.

$$\text{HOOC·CH(NH}_2\text{)·CH}_2\text{·CH}_2\text{·CONH·CH·CONH·CH}_2\text{·COOH}$$
$$|$$
$$\text{CH}_2\text{SH}$$

Glutathione (γ- glutamylcysteinglycine)

Penicillin G

The first stage in the determination of the structure of a peptide is the identification of its amino acid residues. This is done by hydrolysis and the use of paper, thin-layer, HPLC, and vapor-phase chromatography, and electrophoresis. Another important method is the end-group analysis, which involves the reaction of either the terminal amino or carboxyl groups.

The main objectives of peptide synthesis are to prepare the biologically active natural peptides and to clarify the correlation between their structure and biological activity.

The standard technique is based on building up long chains by repeated condensations of individual amino acids. The following difficulties are encountered in such methods: the selection of suitable blocking groups that can be effectively removed without disrupting the peptide bond; activation of the carboxyl group for coupling without side reactions; selection of conditions that will avoid racemization; isolation and purification of intermediate products before using them for the next step.

A new method of peptide synthesis was recently proposed by Merrifield. In his solid-phase peptide synthesis, a peptide chain can be synthesized in a stepwise manner from one end, while it is attached by a covalent bond at the other end to an insoluble solid support. After the synthesis is complete, the peptide chain can be cleaved from the solid phase to which it has been anchored.

Merrifield's method was recently modified by Shemyakin and his coworkers, who used a polymeric support in solution.

Katchalski and his coworkers have used insoluble polymeric reagents for peptide synthesis, and have synthesized bradykinin and glutathione. Katchalski's method calls for the addition of a solution of free peptide ester to an insoluble, N-blocked active ester of an amino acid. Purification of intermediate peptides can be effected during the synthesis, since these peptides are liberated in solution.

In the Merrifield synthesis, peptide purification can be carried out only after detachment of the final product from the polymeric carrier.

It is hoped that these accelerated techniques will make possible the synthesis of large polypeptides and perhaps even of proteins.

B. Isolation of Glutamine from Red Beet

1. Introduction

The first indication of the possible existence of glutamine in nature was reported by Scheibler (1) in 1868 during his investigation of the isolation of asparagine and aspartic acid from beet sugar molasses. In 1883, Schulze and Bosshard (2) isolated pure glutamine from an aqueous extract of beet roots. The isolation of L-glutamine from gliadin was accomplished by Damodarm (3). Glutamine can also be obtained from many other plant tissues, but beet roots are the most readily available material and are also one of the richest sources of this compound. Furthermore, most tissues additionally contain a considerable amount of asparagine, which can be separated only by troublesome fractional crystallization. The quantity of asparagine in beet root is so small that it does not raise any problems.

Glutamine is an anticonvulsant, but is applied only in cases refractory to drug therapy.

$$HOOC \cdot CH(NH_2) \cdot CH_2 \cdot CH_2 \cdot CONH_2$$
Glutamine

2. Principle

In the following procedure (4), the content of natural glutamine in the root tissue of the common red beet (*Beta vulgaris*) is increased by treating the plants with dilute ammonium sulfate solution. The roots are first frozen and then thawed in order to reduce the permeability of the cells. They are then ground and extracted with cold water, and the extract is treated with basic lead acetate to precipitate interfering substances. After filtration, the glutamine is precipitated with mercuric nitrate. The precipitate is decomposed with hydrogen sulfide and filtered, and the solution is neutralized with ammonium hydroxide and concentrated *in vacuo* at low temperatures. The glutamine is crystallized in the presence of 60% alcohol and purified by recrystallization. The quantity of glutamine obtained depends on its content in the original tissue.

It should be noted that mercuric nitrate is not a specific precipitant for glutamine. Asparagine is also precipitated quantitatively, accompanied by small amounts of other amino acids, e.g., arginine, lysine, tyrosine, and cystine.

3. Apparatus

Blender Vacuum desiccator
Meat mincer Vacuum distillation assembly

4. Materials

Ammonium hydroxide Lead acetate (basic)
Ammonium sulfate Lead oxide (litharge)
Celite Mercuric oxide (red)
Ethanol Nitric acid
Ether Red beet
Hydrogen sulfide Sodium hydroxide
Lead acetate

5. Time

Soaking of beet roots: 24 hours
Laboratory procedure: 8–9 hours

6. Procedure

a. Extraction

One kg red beet roots is soaked for 1 day in 5 liters $0.2M$ ammonium sulfate solution, and then washed thoroughly and kept in cold storage. The coarsely sliced frozen roots are ground in a meat mincer. The cold pulp is warmed at room temperature, and the juice is expressed with a hydraulic press or a blender, or convenient quantities of the tissue may even be wrapped in strong cotton cloth and the juice wrung out by hand. In both cases, a liberal volume of wash water should be used. The total effluent (about 1.5 liters) is treated with basic lead acetate reagent (prepared by dissolving 18 g lead acetate in about 70 ml hot water to which 11 g lead oxide (litharge) is then added in powdered form). The mixture is boiled with stirring for half an hour, cooled, filtered, and diluted with water to 100 ml. Alternatively, an approximately 25% solution of commercial basic lead acetate may be used in slight excess (approximately 80 ml) and immediately filtered on a Buchner funnel fitted with filter paper covered by a thin layer of Celite. The precipitate is packed down hard and washed three times with water, a total of 600 to 800 ml being used. The clear yellow filtrate is treated with mercuric nitrate reagent (prepared by slowly adding 22 g high-grade red mercuric oxide to 16 ml concentrated nitric acid with stirring). Water (16 ml) is then added, and the mixture is refluxed for 3 or 4 hr or until the oxide has completely dissolved. The solution is cooled and treated with $1N$ sodium hydroxide until a faint permanent opalescence is noted; it is then diluted to 100 ml and filtered. It is kept in a dark bottle.

The suspension is then neutralized to approximately pH 6 by addition of 10% sodium hydroxide, accompanied by vigorous stirring. The operations up to this point require about 6 hr. It is usually convenient to allow the precipitate to settle overnight. The clear supernatant liquid is siphoned off. The precipitate is transferred to a Buchner funnel fitted with a filter paper covered with a thin layer of Celite, and is washed three times with a total of 500 ml water. The precipitate is suspended in the minimal necessary amount of water, all lumps being broken up, and transferred to a 1-liter filter flask. To this suspension is added 0.25 ml $4N$ sulfuric acid. The flask is placed on a shaking machine, and hydrogen sulfide is passed through for about 2 hr with continuous agitation. Air is then passed through the suspension for a short time to remove most of the excess hydrogen sulfide, and the precipitate is filtered and, washed with 100 to 200 ml water. The clear, pale yellow filtrate is transferred to a 1-liter flask and concentrated *in vacuo* for about 20 min, at a bath temperature of 60°C, to remove the last traces of hydrogen sulfide. The solution is then neutralized to approximately pH 6 by the careful addition of 3 to 5 ml $15N$ ammonium hydroxide. Equal portions of the solution are then poured into two 500-ml flasks and concentrated *in vacuo*, at a bath temperature of 60 to 65°C, to a total volume of about 100 ml. The concentrated solutions are combined, treated with a little Norit at 60°C, and filtered until perfectly clear. Concentration is then continued in a 250-ml flask until a sludge of crystals is obtained. Once the solution has become fairly concentrated, the temperature should be carefully controlled.

After the vacuum is released, the contents of the flask are warmed to 60 to 70°C, whereupon most of the crystals dissolve. Two volumes of warm alcohol are then added, and the flask is chilled overnight. The crystals are filtered, washed with cold 50% ethanol and then with 95% ethanol and ether, and dried over sulfuric acid in a vacuum desiccator.

The quantity of glutamine obtained depends on its content in the original tissue. Approximately 3.5 g crude material can be obtained from the 1 kg beet roots which was treated with ammonium sulfate.

b. IR spectrum of glutamine (in CCl_4)

   3403 cm^{-1}: N—H stretching (anhydrous glutamine)
   1690: C=O stretching of —CONH$_2$ (amide I band)
   1639: COO$^-$ asym. stretching
   1586: NH$_2$ in-plane bending of the amide group
   1487: asym. CH$_2$ bending
   1428: sym. COO$^-$ stretching
   1411: C—N stretching
   1361: sym. CH$_2$ bending
   1336: C—H bending

c. NMR spectrum of glutamine

$$\overset{-}{O}OC-CH(\overset{+}{N}H_3)-(CH_2)_2-CONH_2$$
$$\phantom{\overset{-}{O}OC-}(b)(c)(a)(d)$$

Position	δ (ppm)
a	2.74
b	4.53
c	7.80
d	8.26

References

1. Scheibler, C. (1868). *Ber.* **2**, 296.
2. Schulze, E., and Bosshard, E. (1883). *Ber.* **16**, 312.
3. Damodarm, M. *Biochem. J.* **26**, 235 (1932).
4. Vickery, H. B., and Pucher, G. W. (1949). *Biochem. Prepns.* **1**, 44.

Recommended Review

1. Archibald, R. M. (1945). Chemical characteristics and physiological roles of glutamine. *Chem. Revs.* **37**, 161.

C. Preparation of D-Tyrosine from L-Tyrosine

1. Introduction

The resolution of tyrosine was first carried out by Fischer (1). Treatment of N-benzoyl-DL-tyrosine with brucine yielded an insoluble salt corresponding to the L-isomer, while treatment with cinchonine yielded an insoluble salt corresponding to the D-isomer. D-Tyrosine has also been obtained by the action of brucine on formyl-DL-tyrosine (2), and on N-acetyl-DL-tyrosine (3).

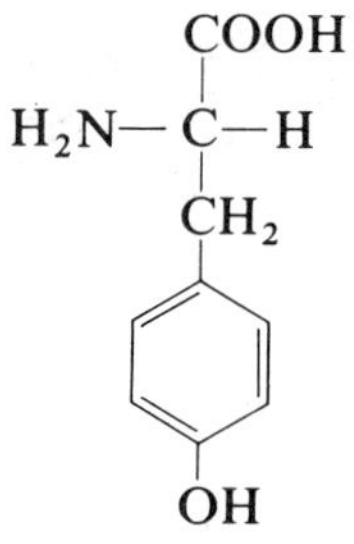

Tyrosine

2. Principle

In the following procedure (4), the natural isomer, L-tyrosine, is racemized with excess acetic anhydride. The diacetyl-DL-tyrosine produced is hydrolyzed to N-acetyl-DL-tyrosine monohydrate.

Crystallization from ethanol of the brucine salts of the monohydrate yields the salt of the D-isomer. Upon decomposition N-acetyl-D-tyrosine is obtained. Hydrolysis of this compound gives D-tyrosine; overall yield, 65%.

3. Apparatus

Reflux assembly
Vacuum distillation assembly

4. Materials

Acetic acid	Chloroform	Sodium hydroxide
Acetic anhydride	Ethanol, absolute	Sulfuric acid
Acetone	Ethanol, 95%	L-Tyrosine
Brucine	Hydrochloric acid	

5. Time

4 overnight periods for crystallization
8–9 hours, laboratory procedure

6. Procedure

a. N-Acetyl-DL-tyrosine monohydrate

L-Tyrosine (18 g) is suspended in 100 ml of water in a 500-ml beaker, to which is added 100 ml 2 N sodium hydroxide exactly standardized against 6N sulfuric acid. The solution is stirred continuously, and 66 ml redistilled acetic anhydride is added in ten 6.6-ml portions at 10-min intervals. The warm reaction mixture is placed in a water bath (60 to 70°C) for 6 hr. It is then *exactly* neutralized with 6N sulfuric acid, and the mixture is concentrated *in vacuo*. To remove additional acetic acid, the distillation is continued after the separate addition of two 20-ml portions of water. The thick syrup and sodium sulfate are extracted with one 50-ml portion and four 10-ml portions of acetone. The extracts are combined and evaporated to dryness *in vacuo*, and the residue is dissolved in 20 ml water. The O-acetyl group is hydrolyzed by the addition of 80 ml 2 N sodium hydroxide, the reaction being tested in order to ensure that it is strongly alkaline to phenolphthalein. After 30 min at room temperature, the calculated amount of 6 N sulfuric acid required to exactly neutralize the alkali is added. The solution is concentrated, giving crude N-acetyl-DL-tyrosine as a thick syrup. This is dissolved in 60 ml water, and the solution is allowed to stand in the cold until crystallization occurs (about 24 hr). Occasional stirring promotes more rapid crystallization and gives a material that is more readily filtered. The yield of crystalline *N*-acetyl-DL-tyrosine monohydrate is 21 g; mp 94–95°C.

b. Resolution of *N*-acetyl-DL-tyrosine

In a 250-ml Erlenmeyer flask are placed 12 g *N*-acetyl-DL-tyrosine monohydrate, 19.7 g anhydrous brucine, and 160 ml absolute ethanol. The mixture is heated to boiling to ensure complete solution. The flask is allowed to stand at room temperature until crystals begin to separate, whereupon the solution is stirred well, and the container is scratched to promote further crystallization. The best yields are obtained if the stoppered container is stored in the cold, with occasional stirring, for 2 or 3 days. During this time interval, a little more than half of the total salt is obtained. The crystals are filtered off quickly, pressed firmly, and washed with 5 ml cold absolute ethanol. The air-dried material should at this stage weigh 16.5–17 g. The salt is purified by 3 or 4 recrystallizations from 95% ethanol. The recrystallized, air-dried material, mp 155°C, contains from 7 to 12% water of hydration. Because the completely anhydrous salt is hygroscopic, dehydration is achieved by drying *in vacuo* at 100°C. The dried sample is weighed and dissolved in 25 ml 95% ethanol, and the optical activity is determined. A 0.5% solution of the anhydrous salt gives $[\alpha]_d - 42.3°$. An average yield of 26 g pure anhydrous salt is obtained.

c. Acetyl-D-tyrosine

A quantity of air-dried salt representing 12.3 g anhydrous compound is dissolved by warming in 20 volumes water. The solution is rapidly cooled to approximately 40°C, and 20 ml 2 *N* sodium hydroxide is added with thorough stirring. After 12 hr in the cold, the brucine tetrahydrate is filtered off and thoroughly washed with 5 portions of cold water. The combined alkaline filtrates are extracted five times with chloroform to remove the last traces of brucine. The calculated amount of 6 *N* sulfuric acid is added, and the solution is evaporated to dryness *in vacuo*. The residue is extracted with acetone, and the acetone solution is evaporated to dryness. A small portion of the residue is recrystallized from a little water, giving acetyl-D-tyrosine, mp 153–154°C, $[\alpha]_D - 48.3°$.

d. D-tyrosine

The acetone-soluble residue of acetyl-D-tyrosine is dissolved in 44 ml 5 *N* hydrochloric acid and hydrolyzed by gently refluxing the solution for 2.5 hr. The excess acid is removed by evaporating to dryness, adding three portions of water, and reevaporating to dryness. The residue is dissolved in 40 ml water and treated with Norit. The solution is made slightly alkaline with sodium hydroxide and then just acid to litmus with a few drops of acetic acid. After 24 hr in the cold, the precipitated tyrosine is filtered off and washed with cold water until free of chloride ion, and then with 2 portions of 95% ethanol and 2 portions of ether. By this method, 3.2 g D-tyrosine is obtained from 0.02 mole brucine salt.

References

1. Fischer, E., (1900). *Ber.* **33**, 3638.
2. Abderhalden, E., and Sickel, H. (1923). *Z. Physiol. Chem.* **131**, 277.
3. Sealock, R. R. (1946). *J. Biol. Chem.* **166**, 1.
4. Sealock, R. R. (1949). *Biochem. Prepns.* **1**, 71.

Recommended Review

1. Greenstein, J. P. (1954). The resolution of racemic α-amino acids. In "Advances in Protein Chemistry." (M. L. Anson, K. Bailey, and J. T. Edsall, eds.), Vol. 9, p. 121. Academic Press, New York.

D. Synthesis of DL-β-Phenylalanine

1. Introduction

DL-Phenylalanine was first prepared in 1882 by Erlenmeyer and Lipp (1) by the action of ammonia and hydrogen cyanide on phenylacetaldehyde. The classical synthesis of phenylalanine by the condensation of benzaldehyde with hippuric acid in the presence of sodium acetate–acetic anhydride was first described by Plochl (2), and later by Erlenmeyer (3), who established the constitution of the intermediate azlactone. This procedure was later improved by Harington and McCartney (4). The method described below was introduced by Gillespie and Snyder (5) and is a simplification of the procedure of Harington and McCartney. A similar method was recently elaborated by Herbst and Shemin (6). In 1911, Wheeler and Hoffman (7) reported a synthesis of phenylalanine that involves the condensation of benzaldehyde and hydantoin. Phenylpyruvic acid, after conversion to its oximino derivative of phenylhydrazone, may be reduced to yield β-phenylalanine (8). Aminomalonic esters have been used as precursors in the synthesis of phenylalanine (9). Albertson and Tullar (10) used substituted cyanoacetic esters. In several syntheses the α-bromo intermediate is aminated at some stage of the reaction sequence (11, 12). A method involving the Curtius rearrangement was developed by Gagnon, Gaudry, and King (13). An interesting synthesis of phenylalanine from cinnamic acid was recently reported by Yukawa and Kimura (14).

β-Phenylalanine is an essential amino acid. The recommended daily uptake for a normal adult male is 2.2 g.

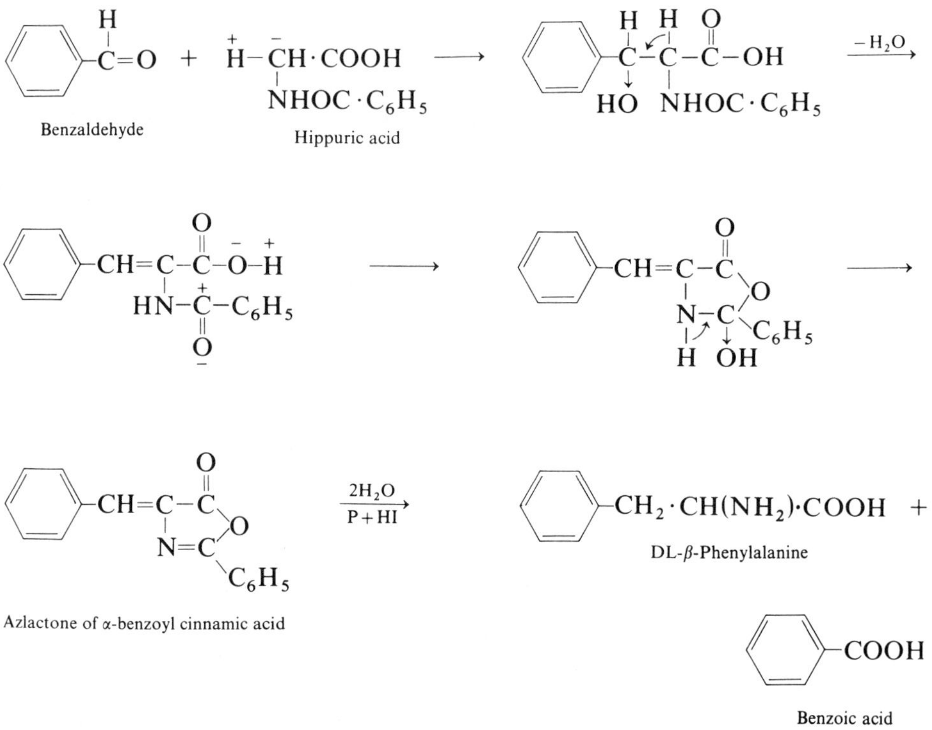

266 CHAPTER 4 Nitrogenous Compounds

$$C_6H_5-CHO + H-CH(NHOC\cdot C_6H_5)-COOH \longrightarrow C_6H_5-CH(OH)-CH(NHOC\cdot C_6H_5)-COOH \xrightarrow{-H_2O}$$

Benzaldehyde + Hippuric acid

$$C_6H_5-CH=C(COO^-)-N^+H=C(C_6H_5)-O^- \longrightarrow C_6H_5-CH=C-C(=O)-O-N(H)-C(C_6H_5)(OH) \longrightarrow$$

$$\text{Azlactone of } \alpha\text{-benzoyl cinnamic acid} \xrightarrow[P+HI]{2H_2O} C_6H_5-CH_2\cdot CH(NH_2)\cdot COOH +$$

DL-β-Phenylalanine

$$C_6H_5-COOH$$

Benzoic acid

2. Principle

In the following method (5), benzaldehyde is condensed with hippuric acid to form an azlactone. The latter, on reduction and hydrolysis, by means of red phosphorus–hydriodic acid, gives DL-β-phenylalanine and benzoic acid. Since the isoelectric pH of phenylalanine is 5.9, it is precipitated from the solution by neutralization to Congo red with dilute ammonia.

3. Apparatus

Electric hot plate
Stirring assembly
Vacuum distillation assembly

4. Materials

Acetic acid, glacial
Acetic anhydride
Ammonium hydroxide
Benzaldehyde
Ethanol, 95%
Ether
Hippuric acid
Hydriodic acid
Phosphorus, red
Sodium acetate
Sodium sulfite

5. Time

7–8 hours

6. Procedure

a. Azlactone of α-benzoylaminocinnamic acid

A mixture of 26.5 g benzaldehyde, 45 g powdered dry hippuric acid, 20.5 g freshly fused sodium acetate, and 150 g (140 ml) high-grade acetic anhydride in a 500-ml Erlenmeyer flask is heated on an electric hot plate, with constant shaking. The mixture becomes almost solid and then, as the temperature rises, gradually liquefies and turns deep yellow in color. As soon as the contents have liquefied completely the flask is transferred to a steam bath and heated for 1 hr. During this period, a part of the product separates as deep yellow crystals. When heating is completed, 200 ml alcohol is added slowly to the contents of the flask. During this addition, the flask is cooled slightly to lessen the vigor of the reaction. The reaction mixture is then cooled, and the yellow crystalline product is fltered with suction and washed on the filter with two 25-ml portions of ice-cold alcohol, and finally with two 25-ml portions of boiling water. After drying, the almost pure product weighs 39–40 g (62–64% yield); mp 165–166°C. This material is sufficiently pure for use in the preparation of phenylalanine. By crystallization from 150 ml benzene a product melting at 167 to 168°C is obtained.

b. DL-β-Phenylalanine

In a 1-liter three-necked, round-bottomed flask fitted with a reflux condenser, a mechanical stirrer, and a dropping funnel, are placed 25 g azlactone of α-benzoyl-aminocinnamic acid, 20 g red phosphorus, and 135 g (125 ml) acetic anhydride. Over a period of about 1 hr 195 g (125 ml) 50% hydriodic acid are added with stirring. During this addition the reaction mixture may solidify. If this occurs, the stirrer is stopped, and about 5 ml hydriodic acid solution is stirred into the cake with a glass rod. The mass then becomes sufficiently fluid to permit use of the mechanical stirrer. The mixture is refluxed for 3 to 4 hr and, after cooling, is filtered with suction. The unreacted phosphorus is washed on the filter with two 5-ml portions of glacial acetic acid, and discarded. The filtrate and washings are evaporated to dryness, under reduced pressure, in a 500-ml Claisen flask heated in a water bath. 100 ml is added to the residue in the Claisen flask, and evaporation to dryness is repeated. To the residue in the flask are added 150 ml water and 150 ml ether, and the mixture is shaken until solution is complete. The aqueous layer is separated and extracted three times with 100-ml portions of ether. The ether extracts are discarded. The aqueous solution is heated on a steam bath with 2 to 3 g Norit and a trace of sodium sulfite until all dissolved ether has been removed. The solution is filtered, and the filtrate is heated to boiling and neutralized to Congo red with 15% ammonia. About 25 ml ammonia is usually required. The phenylalanine separates as colorless plates which, when cold, are filtered off and washed thoroughly on the filter with two 30-ml portions of cold water.

The yield is 10.5–11 g (63–67%) of a product that decomposes at 284 to 288° C.

c. Phenylalanine

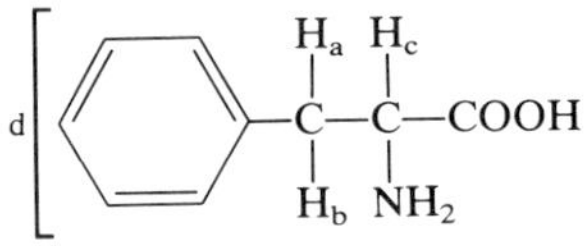

$$\left[\underset{d}{\bigcirc}-\underset{H_b}{\overset{H_a}{\underset{|}{C}}}-\underset{NH_2}{\overset{H_c}{\underset{|}{C}}}-COOH\right]$$

d. ^{1}H NMR
   a = 3.11   1H ⎤ interchange.
   b = 3.31   1H ⎦
   c = 4.01   1H
   d = 7.40   5H

e. Mass Spectrum
   $C_9H_{11}NO_2$, Mol. wt. 165.
   m/z 165(7.43%)   92(18%)   74(100%)
       120(74%)   91(55%)   65(10%)
       103(9%)   77(7%)   28(15%)

f. IR Spectrum of DL-β-phenylalanine (in nujol)

   303, 2898, 2710, 2154 cm^{-1}: $\overset{+}{N}H_3$
   1623: $\overset{+}{N}H_3$ asym. bending
   1589: COO^- asym. bending
   1503: $\overset{+}{N}H_3$ sym. bending
   1401: COO^- sym. bending
   1310: CH bending
   1212: $\overset{+}{N}H_3$ rocking
   1150: C—CN asym. stretching.

References

1. Erlenmeyer, E., and Lipp, A. (1882). *Ber.* **15**, 1006.
2. Plochl, J. (1883). *Ber* **16**, 2815.
3. Erlenmeyer, E. (1893). *Ann.* **275**, 1.
4. Harington, C. R., and McCartney, W. (1927). *Biochem. J.* **21**, 852.
5. Gillespie, H. B., and Snyder, H. R. (1943). *Org. Synth. Coll.* **2**, 489.

6. Herbst, R. M., and Shemin, D. (1943). *Org. Synth. Coll.* **2,** 421.
7. Wheeler, H. L., and Hoffman, C. (1911). *Am. Chem. J.* **45,** 368.
8. Granacher, C. (1922). *Helv. Chim. Acta* **5,** 610.
9. Galat, A. (1947). *J. Am. Chem. Soc.* **69,** 965.
10. Albertson, N. F., and Tullar, B. F. (1945). *J. Am. Chem. Soc.* **67,** 502.
11. Marvell, C. S. (1955). *Org. Synth. Coll.* **3,** 705.
12. Gaudry, R. (1945). *Can. J. Research* **23B,** 234.
13. Gagnon, P. E., Gaudry, R., and King, F. E. (1944). *J. Chem. Soc.* 13.
14. Yukawa, Y., and Kimura, S. (1957). *J. Chem. Soc. Japan* **78,** 454.

Recommended Review

1. Dunn, M. S., and Rockland, L. B. (1947). *The preparation and criteria of purity of the amino acids.* In *"Advances in Protein Chemistry"*, M. L. Anson, and J. T. Edsall, (eds.), Vol. 3, p. 296. Academic Press, New York.

E. Preparation of Glycylglycine by Mixed Carboxylic-Carbonic Acid Anhydride Method

1. Introduction

Treatment of a solution of an acylamino acid or acylated peptide with one equivalent each of an alkyl chlorocarbonate and a tertiary amine, e.g., triethylamine, under anhydrous conditions gives rise to the formation of the mixed carboxylic-carbonic acid anhydride. Condensation of the anhydride *in situ* with the alkali-metal salt of the amino acid or peptide leads to the formation of the desired peptide derivative, in addition to an alcohol and carbon dioxide. Conversion of this product to the corresponding free acid makes possible further condensation with another equivalent of amino acid by similar condensation. This method calls for adequate protection of reactive groups; this is given by the carbobenzoxy (Cbzo) group, which was introduced into peptide chemistry by Bergmann and Zervas in 1932 (1). This group is to this day the most widely used amino-protective group. Carbohenzoxy amino acids are easily prepared without racemization and are converted without difficulty into peptides by the mixed anhydride, dicyclohexylcarbodiimide, and *p*-nitrophenyl ester methods. In addition to its protective effect against racemization of the amino acid, this group possesses a further advantage in that it can be removed by a great variety of methods: catalytic hydrogenation at NTP; hydrobromic acid (gas) in glacial acetic acid (2) (the hydrobromide of the free peptide usually precipitates upon addition of ether, which also dissolves the benzyl bromide formed during the reaction); sodium in liquid ammonia. The reductive action of sodium in liquid ammonia removes not only the carbobenzoxy group but also *S*-benzyl, *N*-(imino)-benzyl, and other groups. It is preferentially employed at the end of a synthesis for the simultaneous removal of several of these groups.

270 CHAPTER 4 Nitrogenous Compounds

1. $\underset{\text{Benzyl alcohol}}{C_6H_5CH_2OH} + \underset{\text{Phosgene}}{COCl_2} \longrightarrow \underset{\text{Benzyl chloroformate}}{C_6H_5CH_2OCOCl} + HCl$

2. $C_6H_5CH_2OCOCl + \underset{\text{Glycine}}{H_2NCH_2COOH} \xrightarrow{OH^-}$

   $\underset{\text{Carbobenzoxyglycine}}{C_6H_5CH_2OCONHCH_2COOH} + NaCl$

3. $C_6H_5CH_2OCONHCH_2COOH \xrightarrow[\text{c. HCl}]{\text{a. } C_2H_5OCOCl}{\text{b. } NH_2CH_2COONa}$

   $\underset{\text{Carbobenzoxyglycylglycine}}{C_6H_5CH_2OCONHCH_2CONHCH_2COOH} + CO_2 + C_2H_5OH$

4. $C_6H_5CH_2OCONHCH_2CONHCH_2COOH \xrightarrow[\text{b. pyridine}]{\text{a. } HBr/CH_3COOH}$

   $\underset{\text{Glycylglycine}}{NH_2CH_2CONHCH_2COOH} + \underset{\text{Benzyl bromide}}{C_6H_5CH_2Br} + CO_2$

2. Principle

Benzyl chloroformate is prepared by bubbling phosgene through cold benzyl alcohol (3). Treatment of glycine with benzyl chloroformate in alkaline solution forms carbobenzoxyglycine, which is transformed by ethyl chloroformate (in the presence of triethylamine) to acid chloride. This chloride in turn condenses with sodium glycinate to yield sodium carbobenzoxyglycyl glycinate, this stage being accompanied by carbon dioxide evolution. Acidification of the product yields carbobenzoxyglycylglycine. Addition of gaseous hydrogen bromide in glacial acetic acid causes smooth nonhydrolytic cleavage of the carbobenzoxy group, yielding glycylglycine hydrobromide. The free peptide is obtained by adding excess of pyridine.

3. Apparatus

   Three-necked flask with stirrer and dropping funnels
   Vacuum distillation assembly

4. Materials

Acetic acid, glacial	Ethanol, absolute	Phosgene
Acetone	Ethyl chloroformate	Pyridine
Benzyl alcohol	Hydrobromic acid	Sodium hydroxide
Calcium chloride	Hydrochloric acid	Toluene
Chloroform	Methanol	Triethylamine

5. Time

10–12 hours

6. Procedure

a. Benzyl chloroformate

A 500-ml round-bottomed flask is fitted with a rubber stopper carrying an outlet tube and a delivery tube extending to the bottom of the flask. In the flask is placed 50 g dry toluene, and the apparatus is weighed. The flask is then cooled in an ice bath, and phosgene is bubbled into the toluene until 10.9 g has been absorbed (about 1 hr is required for this step). The outlet gases are passed through a flask containing toluene in order to remove any phosgene, and then through a calcium chloride tube to a gas trap. After the absorption of phosgene is completed, the connection to the phosgene tank is replaced by a separatory funnel. The reaction flask is gently shaken while 10.8 g (10.4 ml) redistilled benzyl alcohol is added rapidly via the separatory funnel. The flask is allowed to stand in the ice bath for half an hour and at room temperature for 2 hr.

The solution is then concentrated under reduced pressure, at a temperature not exceeding 60°C, in order to remove hydrogen chloride, excess phosgene, and the major portion of the toluene. (In order to prevent the escape of phosgene, a toluene trap is inserted between the apparatus and the water pump).

The residue weighs 20–22 g and contains 15–16 g benzyl chloroformate. The amount of benzyl chloroformate present in this solution may be estimated by preparing the amide from a small aliquot portion, or it may be safely calculated by assuming a minimum yield of 90% of the theoretical amount based on the benzyl alcohol used. **Note: All operations must be performed in a well-ventilated hood!**

b. Carbobenzoxyglycine

A solution of 7.5 g glycine in 25 ml 4 N sodium hydroxide is placed in a 200-ml three-necked flask fitted with a mechanical stirrer and two dropping funnels. The flask is cooled in an ice bath, and 17 g benzyl chloroformate and 25 ml 4 N sodium hydroxide are added simultaneously to the vigorously stirred solution over a period of 20 to 25 min. The mixture is stirred for an additional 10 min. The toluene layer is separated, and the aqueous layer is extracted once with ether.

The aqueous solution is cooled in an ice bath and acidified to Congo red with concentrated hydrochloric acid. The precipitate is filtered off, washed with small portions of cold water, and dried in air. It consists of almost pure carbobenzoxy-glycine and weighs 18–19 g (86–91%); mp 119–120°C. The material may be recrystallized from chloroform; mp 120°C.

c. Carbobenzoxyglycylglycine

A solution of 16.8 g carbobenzoxyglycine in 200 ml dry toluene is placed in a 500-ml three-necked flask fitted with a mechanical stirrer and a dropping funnel equipped with a calcium chloride tube. 8.15 g triethylamine is added to the stirred solution, whereupon the acid dissolves, leaving the salt in solution. The flask is cooled in an ice–salt bath and 8.6 g ethyl chloroformate is added to the vigorously stirred solution over a period of 15 min. The mixture is stirred for an additional 45 min, and a solution of 6 g glycine in 80 ml 1 N sodium hydroxide is added. The mixture is stirred vigorously for 2 to 3 hr, during which time it is permitted to warm to room temperature. Carbon dioxide is evolved during this interval. The toluene layer is now separated and extracted with a 5% sodium hydroxide solution. The aqueous solutions are combined and extracted three times with 50-ml portions of ether. The aqueous solution is then heated on a water bath at 50°C and the dissolved ether is removed under reduced pressure. The solution is then cooled in an ice bath and acidified to Congo red with concentrated hydrochloric acid. After about 45 min, the product precipitates and is filtered on a Buchner funnel and washed with small portions of cold water. It is dried in air and recrystallized from methanol. The product melts at 178°C; yield 70–80%.

d. Glycylglycine

Eight g carbobenzoxyglycylglycine is suspended in 25 ml glacial acetic acid in a 250-ml round-bottomed flask fitted with a gas inlet tube and a calcium chloride drying tube. The flask is placed in an oil bath heated to 60°C, and a stream of dry hydrogen bromide is bubbled into the solution for 1 hr. Carbon dioxide is evolved during the bubbling. Upon cessation of gas evolution, 150 ml dry ether is added to precipitate the glycylglycine hydrobromide. The flask is left in an ice bath for 1 hr, its contents are filtered, and the residue is washed with several portions of anhydrous ether.

The free peptide is obtained by dissolving the salt in a little water, adding excess pyridine, and precipitating with absolute ethanol. The precipitate is then filtered, washed with ether and acetone, and dried in air.

e. IR spectrum of glycylglycine (in nujol)

3300 cm^{-1}: N—H stretching (of —CONH—)
3080: bonded N—H
2920: asym. C—H stretching of CH_2
2840: sym. C—H stretching of CH_2
1680: C=O stretching of non-ionized carboxyl
1655: C=O stretching of —CONH— (amide I band)
1608: C—O asym. stretching of COO^-
1550: mainly N—H deformation mixed with C—N stretching (amide II band)

1460: CH$_2$ deformation
1395: ionized carboxyl group from zwitterion form
1330, 1275: mainly C—N stretching mixed with N—H bending (amide III band)
1015: skeletal vibration

f. NMR spectrum of glycylglycine

$$\overset{+}{\underset{(b)}{H_3N}}\underset{(a)}{CH_2CON}\underset{(c)}{HCH_2COO^-}$$

Position	δ (ppm)
a	4.29
b	7.54
c	7.95 (doublet)

References

1. Bergmann, M., and Zervas, L. (1932). *Ber.* **65**, 1192.
2. Ben-Ishay, D. (1954). *J. Org. Chem.* **19**, 62.
3. Carter, H. E., Frank, R. L., and Johnston, H. W. (1955). *Org. Synth. Coll.* **3**, 167.

Recommended Reviews

1. Albertson, N. F. (1962). *Synthesis of peptides with mixed anhydrides.* In "Organic Reactions" Vol. 12, p. 157. John Wiley, New York.
2. Boissonnas, R. A. (1965). *Selectively removable amino-protective groups used in the synthesis of peptides.* In "Advances in Organic Chemistry" Vol. 3, p. 159. Interscience, New York.

F. Thin-Layer Chromatographic Enantiomeric Resolution of Racemic Amino Acids

1. Introduction

Natural and synthetic chiral compounds (stereoisomers) show different biological and pharmacological activities. Usually the optical rotation is used as an indication of the enantiomeric purity. However, estimation of the enantiomeric (ee value) from optical rotation is not possible in cases where the α_D value of the pure enantiomer is unknown. Furthermore, a calculation of the ee value may give erroneous results in the presence of impurities.

Gas chromatographic (1, 2) and high-performance liquid chromatographic (3–5) separations of enantiomers have been extensively studied. These methods, however, are time consuming and require costly equipment, and sometimes sample derivatization is necessary.

Chiral amino acids are required for the synthesis of biologically active peptides, in asymmetric syntheses, and as pharmacological ingredients. The enantiomeric separation of racemic DL-amino acids can be performed efficiently and quickly on Chiral plates (Macherey-Nagel).

2. Principle

The following procedure (6) is based on using Chiral thin-layer plates for the separation of racemic amino acids, both natural and synthetic. The respective enantiomers can be determined at trace levels.

3. Apparatus

TLC developing chambers

4. Materials

Acetonitrile
Chiral TLC plates
  (such as Macherey-Nagel)
Glutamine
Isoleucine
Methanol
Ninhydrin (1%) spray
Phenylalanine
Proline
Tryptophan
Tyrosine
Valine

5. Time

Less than 2 hours

6. Procedure

The Chiral plates (10 × 20 cm; 0.25 thick) should be activated for 15 min at 100°C before use. Sample volume: 2 µl of the racemic mixture as a 1% solution of amino acid (in methanol or methanol–water). Developing solvent is methanol–water–acetonitrile (1:1:4) with chamber saturation. The developing time is 25 min. for 13 cm. Detection is accomplished by using 0.1% ninhydrin spray reagent followed by heating in an oven at 100°C.

The following racemic amino acids may be used: glutamine, isoleucine, phenylalanine, proline, tryptophan, tyrosine, valine.

References

1. Frank, H., Woiwoode, W., Nicholson, G., and Bayer, E. (1981). *Liebigs Ann. Chem.* 354.
2. Konig, W. A., Steinbach, E., and Ernst, K. (1984). *Angew Chem.* **96,** 516.
3. Lam, S., and Karmen, A. J. (1984). *Chromatogr.* **289,** 339.
4. Roumeliotis, P., Unger, K. K., Kurganov, A. A., and Davankov, V. A. (1983). *J. Chromatogr.* **255,** 51.
5. Allenmark, S. (1984). *J. Biochem. Biophys. Methods* **9,** 1.
6. Gunther, K., and Schickedanz, M. (1985). *Naturwissen* **72,** 149.

Recommended References

1. Stevenson, D., and Wilson, I. D., eds. (1987). "Chiral Separation." Plenum Press, New York.
2. Poole, C. F., and Poole, S. K. (1989). Modern thin-layer chromatography. *Anal. Chem.* **61**, 1257A.
3. Jost, W., and Houck, H. E. (1987). The use of modified silica gels in TLC and HPLC. *Adv. Chromatogr.* **27**, 129.

G. Reverse-Phase High-Performance Liquid Chromatography of Amino Acids

1. Introduction

The separation of amino acids by liquid chromatography has been pursued by many investigators in various fields of science, mainly in the area of amino acids and protein chemistry (1–4). The analysis of amino acids presents the chromatographer with a challenging task owing to two main difficulties: (1) Most of the amino acids are not amenable to easy detection with conventional detectors. Thus frequently they undergo some sort of either pre- or postcolumn derivatization procedure; (2) As a group, the amino acids cover a very wide range of polarities, making isocratic analysis very difficult.

2. Principle

In the following procedure (5), the charge transfer complexes of copper of the amino acids are analyzed by reverse-phase HPLC. Thus the mobile phase is made up of a sodium acetate buffer (pH 5.6) containing cupric ions and alkylsulfonate ion-pair reagent. The complexes absorb UV radiation with a maximum around 235 nm and a molar absorptivity of about 6000 (for the 1:1 complex).

a. Instrumentation

The column is a Merck Lichrosorb RP_{18} cartridge, packed with 7 μm particles. The cartridge dimensions are 250 × 4 mm. A precolumn is connected between the pump and the injector. It is refilled once a month with a mixture of Partisil 10 silica gel and Partisil 10 ODS-3.

b. Solutes

To demonstrate the influence of each of the chromatographic variables, the following amino acids are used as sample solutes: aspartic acid, glycine, histidine, lysine, arginine, valine, methionine, threonine, hydroxyproline, asparagine, and tyrosine. These amino acids cover a wide range of polarities and hydrophobicities, and they represent all types of side chains. The amount of injected amino acids is 2.4 nmol for all but the most retained solutes, for which 10 nmol are introduced to the column.

c. Choice of buffer

It has been found that the complexation of Cu(II) by amino acids, and consequently chromatographic retention, is more favorable at high pH. However, at pH greater than 6, cupric hydroxide will precipitate out unless some chelating agent, which is capable of strongly complexing copper ions, is present in the mobile phase. On the other hand, the complexation should not be too strong, for then amino acids chelation would greatly diminish, and the benefits of cupric ions would be lost. In order to overcome these problems, an acetate buffer at pH 5.6 is used.

d. Problems associated with use of Cu(II) ions

Besides the buffer issue, several other chemical and technical difficulties are attributed to the presence of cupric ions. Exposed metal parts of the system, such as tubings, pump pistons, etc., can be subjected to chemical attack by the copper ions in the mobile phase. The extent of corrosion is a function of the mobile phase pH. The corrosion problem can be greatly minimized by the following action: the mobile phase flow should not be stopped for long periods. The system, including the column, should be rinsed thoroughly with water and methanol whenever it will not be used for a while, e.g., overnight. At least once a week, the system should be washed with $0.01\ M$ EDTA in a $0.1\ M$ acetate buffer. This treatment extends the lifetime of the column to equal those of conventional reverse-phase systems.

e. Effects of alkylsulfonates

The copper complex of the amino acids, which is assumed to be in a 1:1 ratio in the range of the Cu(II) concentration of $5 \times 10^{-4}\ M$ is probably positively charged. Thus introducing an anionic species such as an alkylsulfonate should enhance the retention via an ion-pair mechanism. It was found that when hexane-, heptane-, or octanesulfonates were added to the copper-containing mobile phases, the retention times of the test amino acids were verified substantially. For the separation of all amino acids mentioned above, the Cu(II) concentration was $5 \times 10^{-4}\ M$, and the buffer acetate concentration was $0.1\ M$. The concentration of the sulfonates varied from zero to $0.005\ M$ or to $0.01\ M$. It was found that the extent of retention increase is a function not only of alkylsulfonate concentration but also of the alkyl length.

3. Apparatus

   High pressure liquid chromatograph

4. Materials

Acetate buffer	Asparagine
Arginine	Heptanesulfonate

Histidine   Partisil 10 ODS-3
Hydroxyproline   Partisil 10 silica gel
Lichrosorb RP_{18} cartridge   Threonine
Lysine   Tyrosine
Methionine   Valine

5. Time

   45 minutes

6. Procedure

   The mobile phase: 0.01 M acetate buffer (pH 5.6).
   4×10^{-4} M $CuAc_2$
   8×10^{-4} M Heptanesulfonate
   Column: RP_{18}, 250 × 4 mm
   Flow rate: 2 ml/min
   Temperature: 45°C

The elution order of the amino acids is aspartic acid, glutamine, glycine, serine, aspartic acid, hydroxyproline, threonine, histidine, proline, valine, methionine, tyrosine, leucine, arginine. Thus the acidic amino acids elute fast, while basic amino acids remain in the column for a longer period.

References

1. Grushka, E., Levin, S., and Gilon, C. (1982). *J. Chromatogr.* **235,** 401.
2. Gimpel, M., and Unger, K. (1983). *Chromatographia* **17,** 200.
3. Chinnick, C. C. T. (1981). *Analyst (London)* **106,** 1203.
4. Foucault, A., and Rosset, R. (1984). *J. Chromatogr.* **317,** 41.
5. Levin, S., and Grushka, E. (1985). *Anal. Chem.* **57,** 830.

Recommended Reviews

1. Williams, A. P. (1982). HPLC in Food Analysis. *In* "Food Science and Technology." (R. Macrae, ed.), p. 285. Academic Press, New York.
2. Pfeifer, R. F., and Hill, D. W. (1983). "Advances in Chromatography." (J. C. Giddings, E. Grushka, E. Cazes, and P. R. Brown, eds.), Vol. 22, p. 37. Marcel Dekker, New York.

H. Gas-Liquid Chromatography of Amino Acids

1. Introduction

Free amino acids possess a zwitterionic structure, so they are not volatile and must be transformed into a suitable volatile derivative before the chromatography. Many attempts at employing gas-liquid chromatography for amino acids have been recorded. The first technique was the oxidation of α-amino acids with ninhydrin to the corresponding aldehydes, which were subjected to GLC (1). Zlatkis, Oro, and Kimball (2) extended this approach to four additional aliphatic amino acids. Bier and Teitelbaum (3) explored the utility of several amines obtained from amino acids by decarboxylation. Bayer, Reuther, and Born (4) separated the methyl esters of several amino acids. More promising results have been obtained with derivatives in which both amino and carboxyl groups have been masked. Youngs (5) chromatographed the *N*-acetyl-*n*-butyl esters of some amino acids and developed a quantitative gas-liquid chromatographic method for their determination. Johnson, Scott, and Meister (6) concluded that *N*-acetyl-*n*-amyl esters give better results than other esters. Saroff and Karmen (7) have used N-trifluoroacetyl methyl esters, and these derivatives were also utilized by Cruickshank and Sheehan (8). Zomzely, Marco, and Emery (9) prepared *n*-butyl-N-trifluoroacetyl derivatives. Landowne and Lipsky (10) have used 2,4-dinitrophenyl derivatives, and Ruhlmann and Giesecke (11) *N*-trimethylsilyl esters. It is noteworthy that Pollock and his coworkers (12) recently succeeded in resolving 21 racemic amino acids by gas chromatography.

2. Principle

In the following method (6), the amino acids are converted into *N*-acetyl-*n*-amyl esters by treatment first with amyl alcohol and anhydrous hydrobromic acid, and then with acetic anhydride. The derivatives are detected by gas chromatography on columns containing a very small percentage of the liquid phase (1%) and at relatively low temperature: 125–150°C.

3. Apparatus

   Gas chromatograph
   U-shaped glass columns

4. Materials

   Acetic anhydride
   Amino acids (see the table)
   n-Amyl alcohol
   Benzene
   Chromosorb W (60–80 mesh)
   Hydrobromic acid (gas)
   Polyethylene glycol (Carbowax 1540 or 6000)

5. Time

   Conditioning of column: 24 hours
   Analysis: 3–4 hours

6. Procedure

a. Preparation of samples

1–10 mg amino acid is suspended in 20 ml amyl alcohol, and the mixture is saturated with anhydrous hydrobromic acid. This procedure takes 2–4 min. The flask containing the mixture is placed in an oil bath at 165°C, and approximately one half the alcohol is removed by distillation at atmospheric pressure. The remaining alcohol is evaporated *in vacuo* at 60°C. The residual oil or crystalline material consisting of the amino acid ester hydrobromide is mixed with 8 ml acetic anhydride and allowed to stand at 26°C for 5 min. The solution is then evaporated *in vacuo* at 60° to a syrup, which is taken up in n-amyl alcohol or benzene and used directly for chromatography. The preparation of samples by this procedure takes approximately 1 hr.

b. Preparation of columns

U-shaped glass columns, 2–8 feet in length and $\frac{1}{4}$ inch inner diameter, are packed with polyethylene glycol coated on a solid support consisting of acid-washed Chromosorb W. The support is prepared by adding a weighed amount of the solid support to a solution of Carbowax in benzene. The mixture is then stirred vigorously at its boiling point until most of the solvent has evaporated. After removal of benzene *in vacuo*, the column is packed and heated for 24 hr at 150°C, with a flow rate of argon of 150 ml/min.

Retention Times of Amino Acids

Amino Acid	Retention Time (min)
α-Alanine	14
Valine	16
α-Aminobutyric acid	17
Isoleucine	19
Norvaline	20
Leucine	22
Norleucine	23
Glycine	24
β-Alanine	26
β-Aminobutyric acid	28
Proline	29
γ-Aminobutyric acid	32
Ornithine	33
Threonine	35
Serine	39
Phenylalanine	60
Aspartic acid	84
Glutamic acid	102
Tyrosine	132

c. Operating conditions

The columns are maintained at 125 to 155°C. The flash heaters are set at 300°C, and the temperature of the detectors is maintained at 225°C. The flow rate of the carrier gas is adjusted to constant values between 60 and 240 ml/min as measured at the outlet. The samples are introduced into the column in *n*-amyl alcohol or in a benzene solution.

Derivatives of alanine, valine, isoleucine, the leucine emerge from the 8-ft column packed with Chromosorb W coated with 1% Carbowax 1540 at 125°C. Flow rate 60 ml/min. Following the appearance of leucine, the column temperature is sharply increased to 148°C, and the other amino acids emerge at this temperature. The retention times of amino acids are presented in the previous table.

References

1. Hunter, I. R., Dimick, K. P., and Corse, J. W. (1956). *Chem. Ind. (London)*, 294.
2. Zlatkis, A., Oro, J. F., and Kimball, A. P. (1960). *Anal. Chem.* **32**, 162.
3. Bier, M., and Teitelbaum, C. (1959). *Ann. N. Y. Acad. Sci.* **72**, 641.
4. Bayer, E., Reuther, K. H., and Born, F. (1957). *Angew. Chem.* **69**, 640.
5. Youngs, C. G. (1959). *Anal. Chem.* **31**, 1019.
6. Johnson, D. E., Scott, S. J., and Meister, A. (1961). *Anal. Chem.* **33**, 669.
7. Saroff, H. A., Karmen, A., and Healy, J. W. (1960). *Anal. Biochem.* **1**, 344.
8. Cruickshank, P. A., and Sheehan, J. C. (1964). *Anal. Chem.* **36**, 1191.
9. Zomzely, C., Marco, G., and Emery, E. (1962). *Anal. Chem.* **34**, 1414.
10. Landowne, R. A., and Lipsky, S. R. (1963). *Federation Proceedings* **22**, 235.
11. Ruhlmann, K., and Giesecke, W. (1961). *Angew. Chem.* **73**, 113.
12. Pollock, G. E., and Oyama, V. I. (1966). *J. Gas Chromatogr.* 126.

Recommended Reviews

1. Karmen, A., and Saroff, H. A. (1964). *Gas-liquid chromatography of the amino acids.* In "New Biochemical Separations," (A. T. James, and L. J. Morris, ed.), p. 81. Van Nostrand, London.
2. Weinstein, B. (1966). *Separation and determination of amino acids and peptides by gas-liquid chromatography.* In "Methods of Biochemical Analysis," (D. Glick, ed.), Vol. 14, p. 203. Interscience, New York.
3. Blackburn, S. (1983). "CRC Handbook of Chromatography of Amino Acids and Amines," Vol. 1. CRC Press, New York.

I. Test Tube and Glass Rod Thin-Layer Chromatography of Amino Acids

1. Introduction

Several investigators (1–5) have recently described the rapid preparation of microchromatoplates from microscope slides and from test tubes coated on their inner surface (6–7).

2. Principle

In the following method (8), test tubes coated with silica gel G on their *outer* surface and glass rods are utilized for thin-layer chromatography. This eliminates the need for special applicators and glass plates of specific dimensions, and the prolonged development periods. It also permits quantitative recovery of the components and their spectroscopic determination.

3. Apparatus

Micropipettes
Test tubes

4. Materials

α-Alanine	Isoleucine	Silica gel G
Aspartic acid	Ninhydrin	Tyrosine
Glycine		

5. Time

2 hours

6. Procedure

a. Coating

The most convenient test-tube size for routine use is 150 mm (length) × 14–15 mm (external diameter); glass rods should have a diameter of 5 to 10 mm.

The test tubes and glass rods are thoroughly cleaned with a dichromate–sulfuric acid solution, then with a dilute detergent solution, and are finally rinsed with water. A homogeneous suspension consisting of one part silica gel G and 2 parts water is prepared in a test tube or a cylinder. The test tubes (or rods) to be coated are immersed in the slurry to a depth of 12 cm, removed, and held for a few seconds above the slurry until draining has almost stopped. They are then inverted and dried at room temperature for 15 min. The coated test tubes or rods are activated in an oven at 120 to 130°C for 45 min, cooled, and kept in a desiccator.

b. Spotting of samples and chromatography

The acids (1–2 mg/ml water) are applied with a micropipette to the thin layer 1 cm above the closed end of the test tube or rod. The solvent is allowed to evaporate, and the test tube or rod is lowered carefully into a larger test tube containing a few ml developing solvent (water–ethanol–acetic acid, 1:5:0.1, containing 10 mg ninhydrin). After about 30 to 45 mins, the solvent front is marked with a pencil. The test tube is dried at 110 to 120°C for 10 min, whereupon amino acids appear as violet and pink spots.

Detection of Amino Acids

Amino Acid	Color	R_f
Aspartic	violet	0.11
Glycine	pink	0.31
α-Alanine	pink	0.45
Isoleucine	pink	0.70
Tyrosine	pink	0.73

References

1. Anwar, M. H. (1963). *J. Chem. Educ.* **40**, 29.
2. Hansbury, E., Ott, D. G., and Perrings, J. D. (1963). *J. Chem. Educ.* **40**, 31.
3. Rollins, C. (1963). *J. Chem. Educ.* **40**, 32.
4. Peifer, J. J. (1962). *Mikrochim. Acta* 529.
5. Samuels, S. (1966). *J. Chem. Educ.* **43**, 145.
6. Lie, K. B., and Nyc, J. F. (1962). *J. Chromatogr.* **8**, 75.
7. BeMiller, J. N. (1964). *J. Chem. Educ.* **41**, 608.
8. Ikan, R., and Rapaport, E. (1967). *J. Chem. Educ.* **44**, 297.

Recommended Review

1. Brenner, M., Niederwieser, A., and Pataki, G, (1965). Amino acids and derivatives. *In* Stahl, E. "Thin-Layer Chromatography," p. 391. Academic Press, New York.

J. Thin-Layer Chromatography of Carbobenzoxy-Peptides

1. Introduction

Thin-layer chromatography of carbobenzoxy (Cbzo) amino acid esters or diols was reported by Zachau and Karau (1). Erhardt and Cramer (2) and Pataki (3) have shown that Cbzo-amino acids, Cbzo-peptides, and Cbzo-peptide esters can be separated from each other and from unprotected amino acids, amino acid ester hydrochlorides, and peptides on silica gel G layers.

In the case of the mentioned solvents, the order of migration is as follows: Cbzo-peptide esters > Cbzo-peptides > amino acids and peptides. The course of peptide synthesis using Cbzo compounds can be simply and rapidly followed by means of this method.

Thin-layer chromatography of Cbzo compounds has been used in the synthesis of derivatives of oxytocin (4) and other peptides (5). Compounds of higher molecular weight, e.g., polymyxin, were also chromatographed by this method (6). Recently, carbobenzoxy amino acids and peptides have been chromatographed as hydrobromides (7).

2. Principle

The procedure outlined below is according to Erhardt and Cramer (2).

3. Apparatus

Thin-layer chromatography equipment

4. Materials

Acetic acid
Ammonium hydroxide
n-Butanol
Cbzo acids (see table)
DL-Alanine
DL-Phenylalanine
Glycine
Glycylglycine
Ninhydrin
Potassium dichromate
Pyridine
Silica gel G
Sulfuric acid

5. Time

2–3 hours

6. Procedure

a. Preparation of plates

For the preparation of five plates (20 × 20 cm) covered with a coating of silica gel G 0.25 mm in thickness, 30 g silica gel G and 60 ml water are required.

b. Development

The following solvents can be used for development:

1. n-Butanol–acetone–acetic acid–ammonia (concentrated ammonia–water, 1:4)–water, 4.5:1.5:1:1:2, and
2. n-Butanol–acetic acid–ammonia–water, 6:1:1:2.

The solvents are allowed to rise about 8 to 10 cm in the course of 2 hr.

c. Detection

The plates are dried at 120 to 150°C for 10 to 15 min, and while still hot are sprayed with a solution of 0.2% ninhydrin in 95% n-butanol and 5% 2N acetic acid. They are then heated for a few minutes at 120 to 150°C and sprayed with a saturated solution of potassium dichromate in concentrated sulfuric acid, whereupon carbobenzoxy derivatives, amino acids, and peptides appear as dark gray spots.

Detection of Carbobenzoxy-Peptides

	R_f in Solvent	
Substance	1	2
Cbzo-Gly	0.81	0.74
Cbzo-Digly	0.66	0.56
Cbzo-Trigly	0.57	0.41
Cbzo-Tetragly	0.51	0.26
Cbzo-Digly-OEt	0.82	0.75
Cbzo-Trigly-OEt	0.76	0.68
Cbzo-DL-Ala	0.77	0.74
Cbzo-DL-Diala	0.72	0.68
Cbzo-DL-Phe	0.74	0.77
Gly-gly	0.17	0.04
DL-Ala	0.27	0.14
DL-Phe	0.45	0.31
Gly	0.22	0.10

References

1. Zachau, H. G., and Karau, W. (1960). *Ber.* **93**, 1830.
2. Erhardt, E., and Cramer, F. (1962). *J. Chromatogr.* **7**, 405.
3. Pataki, G. (1964). *J. Chromatogr.* **16**, 553.
4. Huguenin, R. L., and Boissonnas, R. A., (1961). *Helv. Chim. Acta* **44**, 213.
5. Schwyzer, R., and Dietrich, H. (1961). *Helv. Chim. Acta* **44**, 2003.
6. Volger, K., Studer, R. O., Legier, W., and Lanz, P. (1960). *Helv. Chim. Acta* **43**, 1751.
7. Maskaleris, M. L., Sevendal, E. S., and Kibrick, A. C. (1966). *J. Chromatogr.* **23**, 403.

K. Questions

1. What are the three general classes into which the naturally occuring amino acids may be divided? Illustrate by examples.
2. What is a zwitterion? What is meant by an isoelectric point? How is it related to the solubility of an amino acid?
3. Write equations for the appropriate preparation of the following compounds: a. alanine, from propionic acid; b. valine, using the Strecker synthesis; c. leucine, using the Gabriel synthesis.
4. Review the methods for the preparation of α-amino acids.
5. Outline the steps in the resolution of a racemic amino acid.
6. Show how L-leucine could be converted into D-leucine.

7. What are essential amino acids?
8. Explain and illustrate by examples the processes of deamination and transamination of amino acids.
9. Describe the biosynthetic pathways of glycine and serine, and of phenylalanine and tyrosine.
10. What are peptide antibiotics and hormones? Give examples.
11. Describe the methods generally used for the fractionation of peptides.

L. Recommended Books

1. Bernfeld, P. (1963). *"Biogenesis of Natural Compounds."* Pergamon Press, London, England.
2. Beyerman, H. C., van de Linde, A., and van den Brink, Maassen, W., eds. *"Peptides, Proceedings of the Eighth European Peptide Symposium."* (1967) Noordwijk, The Netherlands, Sept. 1966. North-Holland Publ. Co., Amsterdam. The Netherlands.
3. Greenstein, D. M., and Winitz, M. (1961). *"Chemistry of Amino Acids."* John Wiley, New York.
4. Meister, A. (1965). *"Biochemistry of the Amino Acid"* 2 vols. Academic Press, New York.
5. Williams, R. M. (1989). *"Synthesis of Optically Active α-Amino Acids."* Pergamon Press.

III. NUCLEIC ACIDS

A. Introduction

Nucleic acids, originally isolated from cell nuclei, are divided into the following two classes according to the nature of the sugar present: the pentose nucleic acids or ribonucleic acids (RNA), and the deoxypentose nucleic acids or deoxyribonucleic acids (DNA). Ribonucleic acid is found mainly in the cytoplasm, while deoxyribonucleic acid is confined to the cell nucleus. Much information on the structural acids has been obtained through their hydrolysis by alkalis, acids, and specific enzymes, and the isolation of the hydrolysis products by the application of paper, column, and thin-layer chromatography and counter-current distribution extraction.

The pyrimidine bases found in nucleic acids are uracil, thymine, cytosine, 5-methylcytosine, and 5-hydroxymethylcytosine.

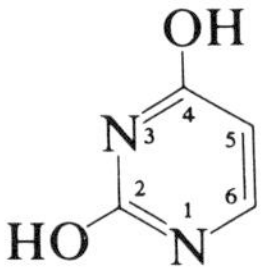

286 CHAPTER 4 Nitrogenous Compounds

Uracil

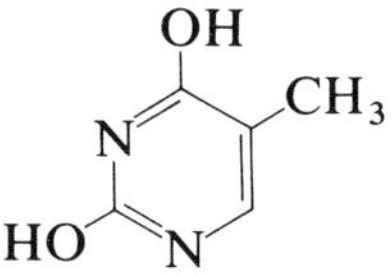

Thymine

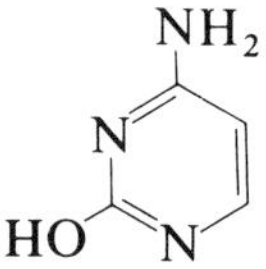

Cytosine

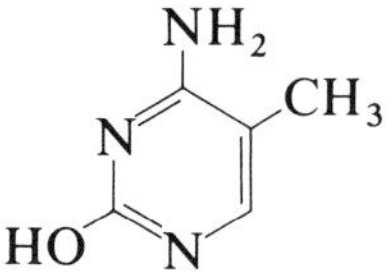

5-Methylcytosine

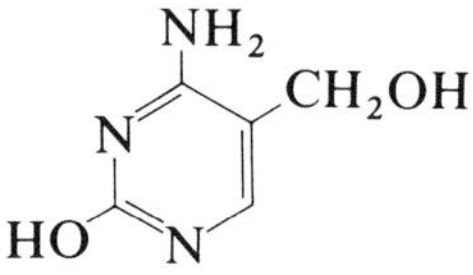

5-Hydroxymethylcytosine

The purine bases present are adenine and guanine.

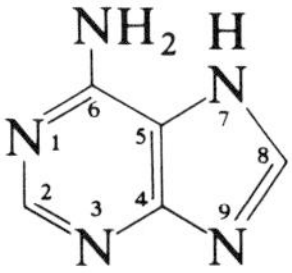

Adenine

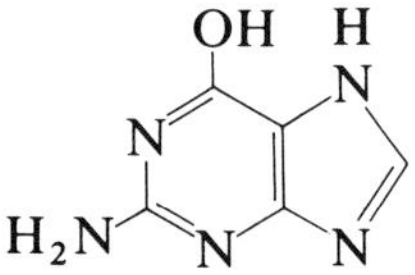

Guanine

Combination of a base with a sugar via a β-N-glycoside linkage gives rise to a nucleoside, e.g., uridine and adenosine.

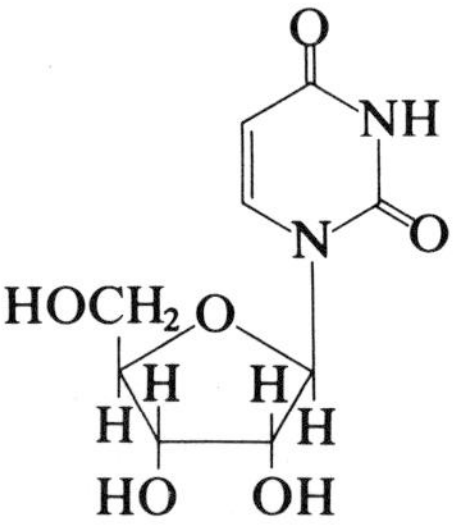

Uridine

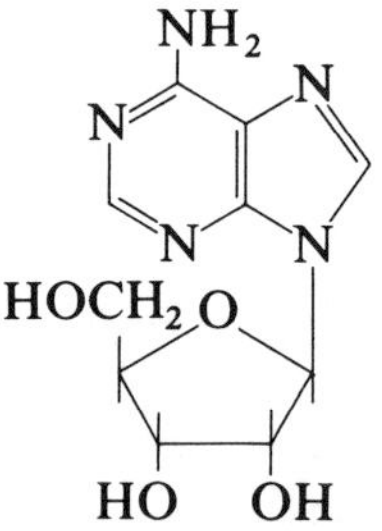

Adenosine

The pyrimidine bases are always attached to the sugar via a nitrogen at position 1 of the base, while the purine bases are linked via a nitrogen at position 9 of the base. The pyrimidine nucleosides are much more resistant to acid than are the purine nucleosides; both nucleosides are stable toward alkali. Combination of a nucleoside with phosphoric acid produces a nucleotide, e.g., cytidine 3'-phosphate and adenosine 5'-phosphate.

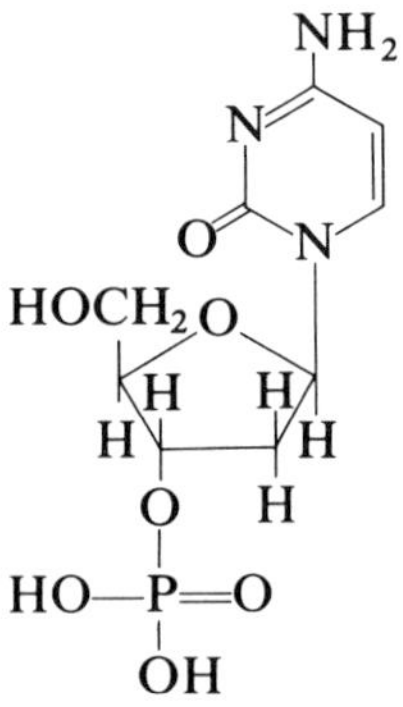

Cytidine 3'-phosphate

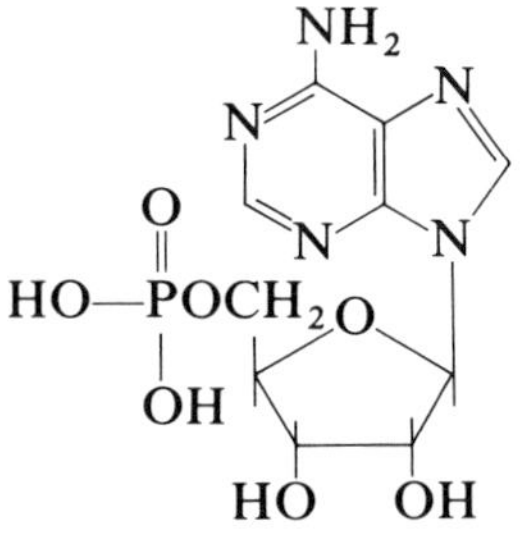

Adenosine 5'-phosphate

Since the ribose nucleosides have three free hydroxyl groups on the sugar ring (2', 3' and 5'), three monophosphates can be formed. In the case of deoxyribose nucleosides, the only free hydroxyl groups for phosphate formation are 3' and 5'.

Many biological compounds have nucleotide structures, e.g., coenzymes such as nicotinamide nucleotides, flavine-adenine dinucleotide, and coenzyme A. Cytidine 5'-triphosphate (CTP) is involved in the biosynthesis of phospholipids, and guanosine 5'-triphosphate (GTP) in the biosynthesis of proteins.

Acid hydrolysis of RNA forms bases, ribose, and phosphoric acid, while alkaline hydrolysis forms ribonucleoside 2'- and 3'-phosphates. Hydrolysis in presence of snake venom diesterase yields nucleoside 5'-phosphate.

The following are the three principal types of RNA found in living cells: ribosomal (rRNA), found in ribosomes, molecular weight $0.5–2.0 \times 10^6$; soluble or transfer RNA (sRNA), molecular weight about 25,000; and messenger RNA (mRNA), molecular weight $\geq 0.5 \times 10^6$.

The main internucleotide linkage in RNA is the phosphoester group connecting C-5' of one nucleotide with C-3' of the next.

Todd and his coworkers have shown that alkaline hydrolysis of nucleotides yields nucleoside 2'- and 3'-phosphates. Acid treatment of these phosphates forms the cyclic nucleoside 2',3'-phosphate; on hydrolysis of the latter a mixture of 2'- and 3'-phosphates is obtained.

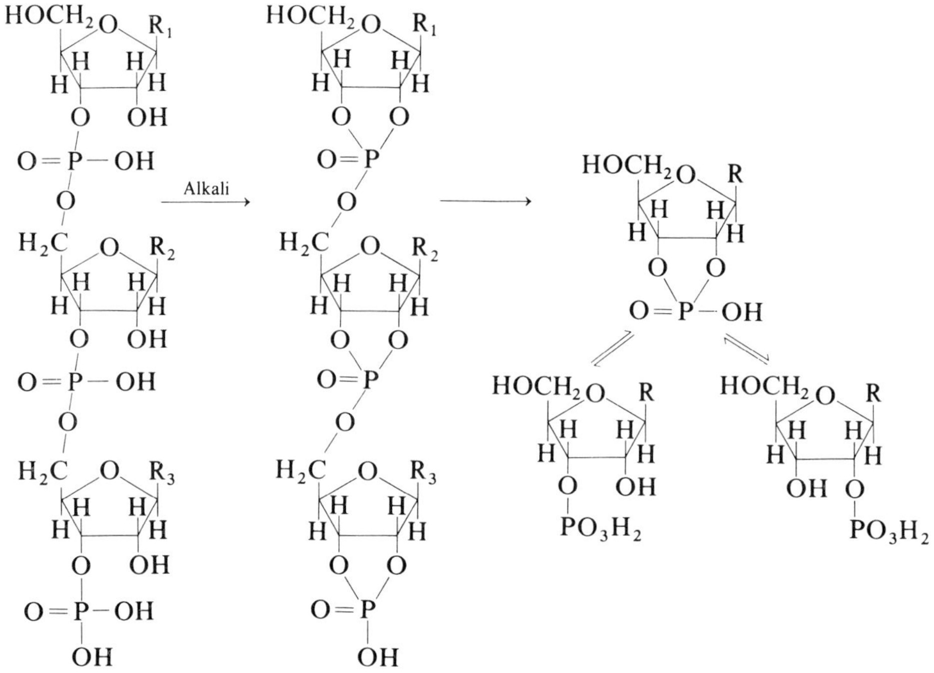

288 CHAPTER 4 Nitrogenous Compounds

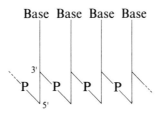

Hydrolysis of a trinucleotide by alkali. R represents a purine or a pyrimidine base

The polynucleotide chains are now represented in a schematic way as follows:

where the vertical line denotes the carbon chain of the sugar with the bases attached at C_1; each phosphate group bridges the 3'-hydroxyl group of one pentose ring and the 5'-hydroxyl of the next.

DNA exists in the cell nucleus as deoxyribonucleoprotein; the molecular weight of DNA ranges from 10^6 to 10^9. The internucleotide bond in DNA, like that in RNA, is the 3',5'-phosphodiester linkage. The absence of a hydroxyl group at C-2' makes cyclic phosphate formation impossible, and DNA thus differs from RNA in that it is not hydrolysed by alkali.

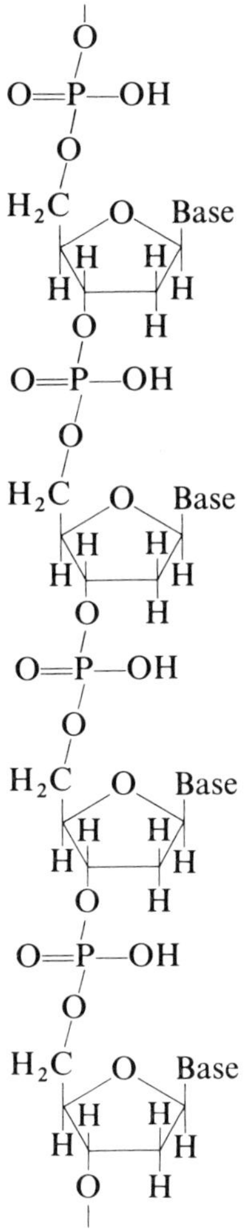

III. Nucleic Acids

$$
\begin{array}{c}
| \\
O \\
O{=}P{-}OH \\
| \\
O \\
| \\
H_2C \quad O \quad \text{Base} \\
\text{(sugar ring with H, H, H, H)} \\
O \quad H \\
| \\
O{=}P{-}OH \\
| \\
O \\
| \\
H_2C \quad O \quad \text{Base} \\
\text{(sugar ring)} \\
O \quad H \\
| \\
O{=}P{-}OH \\
| \\
O \\
| \\
H_2C \quad O \quad \text{Base} \\
\text{(sugar ring)} \\
O \quad H \\
| \\
O{=}P{-}OH \\
| \\
O \\
| \\
H_2C \quad O \quad \text{Base} \\
\text{(sugar ring)} \\
O \quad H \\
|
\end{array}
$$

Part of the polynucleotide chain in DNA

Watson and Crick have suggested that the DNA molecule is a double, dextral helix of two polynucleotide chains winding round the same axis and held together by hydrogen bonds of the purine and pyrimidine bases (Fig. 4.1).

The two phosphate-sugar chains are represented by ribbons and the pairs of bases holding the chains together are shown as horizontal rods which form, as it were, the treads of a spiral staircase.

X-ray and enzymatic studies have recently shown that sRNA consists of a chain of some 70 nucleotides, which is folded back on itself to form a single loop coiled into a double helix (Fig. 4.2).

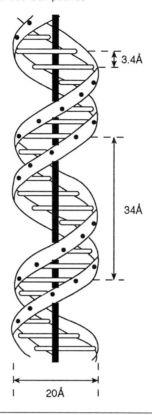

Figure 4.1 Diagrammatic representaton of the DNA molecule as proposed by Watson and Crick.

B. Isolation of Soluble Ribonucleic Acid from Baker's Yeast

1. Introduction

Soluble ribonucleic acid (sRNA) is a mixture of species ranging in molecular weight from 25,000 to 30,000 (75 to 100 nucleotides). Three methods have been used to isolate sRNA. In the first method the tissue is disintegrated and nuclei and microsomal particles are removed by high-speed centrifugation, leaving sRNA in the supernatant liquid which is extracted with phenol. Isolation of sRNA from *Escherichia coli* using this method was described by Tissieres (1), and from rabbit liver by Cantoni and colleagues (2). The second method involves the direct extraction of nucleic acids from the tissues, and subsequent separation of sRNA from rRNA. Extraction of whole cells with phenol yields a mixture of nucleic acids, from which sRNA is extracted with cold sodium chloride. Dirheimer (3) separated sRNA from rRNA by chromatography on Sephadex G-200. The third method calls for the direct extraction

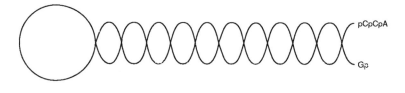

Figure 4.2 Schematic representation of the helical structure of the sRNA molecule. C, cytidine; A, adenosine; G, guanosine; p, phosphate.

of yeast cells with phenol (4, 5). The cells are not broken, and their walls permit the passage of RNA of low molecular weight into the extracting liquid. This method is particularly suitable for large-scale preparations of sRNA (6, 7).

The sRNA isolated by the above methods is usually contaminated by polysaccharides, which can be removed by extracting the nucleic acid from 1.25 M potassium phosphate (pH 7.5) with 2-methoxyethanol, followed by precipitation with ethanol and dialysis (8). The removal of the contaminating rRNA and polysaccharides has been effected with DEAE (diethylaminoethyl)-cellulose columns (9).

2. Principle

In the following procedure (7), baker's yeast is treated with aqueous phenol, which precipitates protein and DNA. The aqueous layer obtained on centrifugation contains RNA and polysaccharides. Both are precipitated by ethanol. sRNA is extracted with a buffer and purified by column chromatography on DEAE-cellulose, which removes the contaminating polysaccharides.

3. Apparatus

   Chromatography column
   Stirrer

4. Materials

   Baker's yeast   Ethanol   Potassium acetate
   DEAE-cellulose   Phenol   *tris*-HCl buffer

5. Time

   3–4 hours

6. Procedure

a. Isolation of sRNA

Baker's yeast (15 g) is mixed thoroughly with 14 ml 88% phenol and 33 ml water, and left to stand for 1 hr with occasional stirring. The mixture is then placed in a 50-ml tube and centrifuged at 10,000 rpm for 12 min. About 25 ml of the aqueous supernatant is collected by means of a rubber bulb attached to a suitable pipette. Phenol (1.5 ml, 88%) is added to the yeast extract, with stirring, and again centrifuged as before. The upper phase is separated as above, and 0.25 ml 20% aqueous potassium acetate (pH 5.2) and 50 ml 95% ethanol are added to precipitate crude sRNA. Equal portions of the preparation are then introduced into two 50-ml centrifuge tubes, allowed to stand for 5 min, and then centrifuged as above. The supernatant liquid is decanted, and the residue is transferred to a 15-ml tube by washing it from the larger tubes with small amounts of 95% ethanol. After centrifugation, ethanol is decanted, and the sRNA is dissolved in 3 ml 0.1 M *tris*-HCl buffer (pH 7.5) for column chromatography.

b. Preparation of DEAE-cellulose column

One g DEAE-cellulose is stirred with 40 ml 0.1 M *tris*-HCl buffer, and the slurry is poured into a 1 × 10 cm tube. The liquid is poured into the top of the column, 40 ml buffer is added, and the liquid is allowed to percolate at a flow rate that should not exceed 1 ml per min.

c. Chromatography of sRNA

The solution of sRNA is added to the column and allowed to pass through the cellulose. The column is then washed with 40 ml buffer. The sRNA is precipitated from the eluate by the addition of 3 volumes of ethanol (95%) and separated by centrifugation. This operation is repeated, and the sRNA is allowed to dry at room temperature.

References

1. Tissieres, A. (1959). *J. Mol. Biol.* **1**, 365.
2. Cantoni, G. L., Gelboin, H. V., Luborsky, S. W., Richards, H. H., and Singer, M. F. (1962). *Biochim. Biophys. Acta* **61**, 354.
3. Dirheimer, G., Weil, J. H., and Ebel, J. P. (1962). *Compt. Rend. Acad. Sci.* **255**, 2312.
4. Monier, R., Stephensen, M. L., and Zamecnik, P. (1960). *Biochim. Biophys. Acta* **43**, 1.
5. Osawa, S. (1960). *Biochim. Biophys. Acta* **43**, 110.
6. Monier, R. (1962). *Bull. Soc. Chim. Biol.* **44**, 109.
7. Holley, R. W. (1963). *Biochim. Biophys. Res. Commun.* **10**, 186.
8. Kirby, K. S. (1956). *Biochem. J.* **64**, 405.
9. Holley, R. W., Apgar, J., Doctor, B. P., Farrow, J., Marini, W. A., and Merrill, S. H. (1961). *J. Biol. Chem.* **236**, 200.

Recommended Reviews

1. Brown, L. *Preparation, fractionation and properties of sRNA*. In "*Progress in Nucleic Acid Research*" (J. N. Davidson, and W. E. Cohn, eds.), Vol. 2, p. 259. Academic Press, New York.
2. Kirby, K. S. (1964). *Isolation and fractionation of nucleic acids*. In "*Progress in Nucleic Acid Research*" (J. N. Davidson, and W. E. Cohn, eds.), Vol. 3, p. 1. Academic Press, New York.

C. Preparation of Thymidine 3'-Phosphate

1. Introduction

Many reagents have been employed for the synthesis of phosphate esters (1). Multifunctional reagents, e.g., phosphorus oxychloride (2, 3) and polyphosphoric acid (4), are of limited value because of the complex mixture of products they produce; their use is practical only when the reaction products

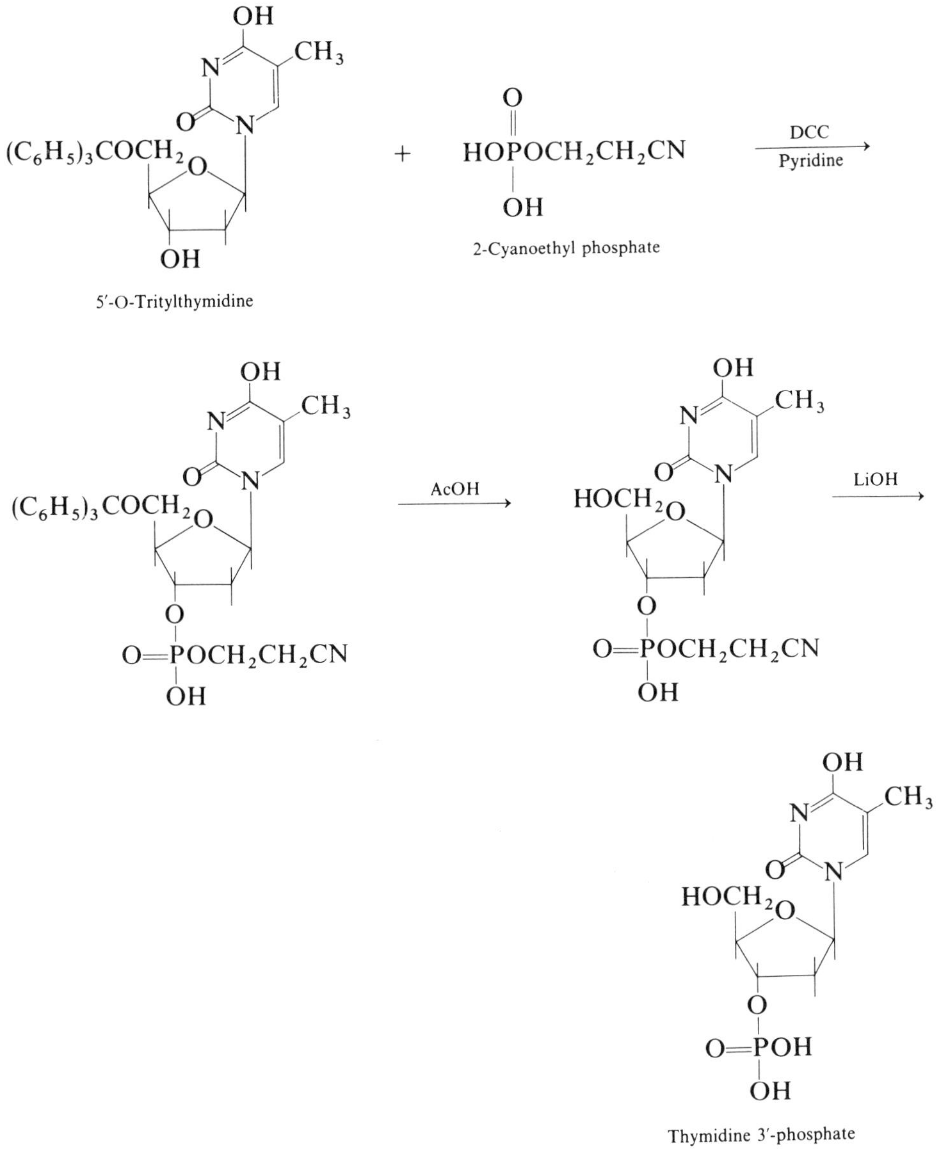

5'-O-Tritylthymidine + 2-Cyanoethyl phosphate $\xrightarrow[\text{Pyridine}]{\text{DCC}}$

$\xrightarrow{\text{AcOH}}$

$\xrightarrow{\text{LiOH}}$

Thymidine 3'-phosphate

294   CHAPTER 4   Nitrogenous Compounds

can withstand hydrolysis. Dibenzyl phosphochloridate (5) readily phosphorylates most primary alcoholic functions. The benzyl groups can be removed from the intermediate phosphotriesters by mild catalytic hydrogenolysis using a palladium catalyst. 2-Cyanoethyl phosphate is converted into an active phosphorylating agent on reacting with DCC (dicyclohexylcarbodiimide). The cyanoethyl group is readily eliminated under mild alkaline conditions (6).

Thymidine 3′-phosphate has been prepared by Michelson and Todd (7) by phosphorylation of 5′-O-tritylthymidine with dibenzyl phosphochloridate, and by Tener, Khorana, Markham, and Pol (8) with mono-p-nitrophenyl phosphodichloridate.

2. Principle

2-Cyanoethyl phosphate in the presence of DCC is an active phosphorylating agent. It breaks down on mild alkaline treatment, liberating orthophosphate,

$$\text{HO}-\underset{\underset{\text{OH}}{|}}{\overset{\overset{\text{O}}{\|}}{\text{P}}}-\text{CH}_2\text{CH}_2\text{CN} \xrightarrow{\text{OH}^-} \text{H}_3\text{PO}_4 + \text{CH}_2\!\!=\!\!\text{CHCN}$$

while the nucleotides are unaffected under these conditions.

In the following experiment (6), thymidine 3′-phosphate is prepared using 2-cyano-ethylphosphate, DCC and 5′-O-tritylthymidine. The crude product is purified by chromatography and through its barium salt.

3. Apparatus

Chromatography column
Stirrer

4. Materials

Acetic acid	Hydracrylonitrile
Acetone	Lithium hydroxide
Barium acetate	Phosphorus oxychloride
Barium hydroxide	Phosphorus pentoxide
Benzene	Pyridine
DCC	Silica gel
Dowex 50 [H$^+$]	Thymidine
Ethanol	Triphenylmethylchloride
Ether, anhydrous	

5. Time

9 days at room temperature
4 hours of laboratory work

III. Nucleic Acids 295

6. Procedure

a. Preparation of cyanoethylphosphate

Phosphorus oxychloride (30.6 g [18.4 ml]) is mixed with 200 ml anhydrous ether in a three-necked flask containing a thermometer, a stirrer, and a dropping funnel. The solution is cooled to −10°C in an ice-salt bath, and a mixture of 15.8 g anhydrous pyridine (16.1 ml) and 14.2 g hydracrylonitrile is added dropwise, with vigorous stirring, during 1 hr. Stirring is then continued for an additional hour at −10°C. The mixture is poured slowly, with stirring, into a mixture of 750 ml water, 80 ml pyridine, and 300 g ice. Barium acetate (100 g) in 300 ml water is added to this mixture, and it is set aside for 2 hr. The precipitated barium phosphate is filtered on a Buchner funnel. The ether is evaporated, and two volumes 95% ethanol are added slowly, with stirring, to the clear filtrate. After 1 hr at 0°C, the plates of barium salt are collected and washed first with 50% ethanol and then with pure ethanol; yield, 41 g.

A standard solution containing 1 mmole/ml for use in phosphorylations is prepared as follows: 16.1 g dried barium salt is dissolved in water (Dowex 50 [H^+] is added to aid solution) and washed through a column of Dowex 50 [H^+] resin; 20 ml pyridine is added to the effluent, and the solution is concentrated *in vacuo* to about 20 ml, transferred to a 50-ml volumetric flask, and diluted to the mark with pyridine. This reagent is stable for 1 month.

b. Preparation of 5'-O-tritylthymidine

Triphenylmethylchloride (3.5 g) is added to a solution of 2.5 g anhydrous thymidine in 50 ml dry pyridine, and the mixture is left at room temperature for 1 week. It is then cooled to 0°C and poured into 500 ml ice-water with vigorous stirring. The precipitate obtained is washed with water and dried *in vacuo* over phosphorus pentoxide. The product is then dissolved in 5 ml acetone and 35 ml dry benzene. After filtration, the acetone is distilled and the resulting solution is cooled. 4 g 5'-O-tritylthymidine separate as colorless needles, mp 128°C. $[\alpha]_D$ + 19.2° (in ethanol, 95%).

c. Thymidine-3'-phosphate

A solution of 560 mg 5'-O-tritylthymidine and 2 ml 2-cyanoethyl phosphate standard solution in 20 ml pyridine is concentrated to dryness *in vacuo* at 30°C. The residue is taken up in 20 ml dry pyridine and again concentrated to drynesss. This process is repeated a second time to ensure removal of water. Finally, the residue is dissolved in 10 ml dry pyridine, and 1.67 g DCC is added. The reaction mixture is left in a well-stoppered flask for 2 days at room temperature, during which time dicyclohexylurea separates. One ml water is then added, and after an hour at room temperature the solution is concentrated to dryness *in vacuo*. The residue is treated with 30 ml 10% acetic acid at 100°C for 20 min and concentrated to dryness. The last traces of acetic acid are removed by a second evaporation after adding 20 ml water. The residue is next treated with 40 ml 0.5N lithium hydroxide at 100°C for 45 minutes. The reaction mixture is cooled and the precipitate is removed by centrifugation. The supernatant is then passed through a column (2 × 20 cm) of Dowex 50 [H^+] resin, and the effluent is adjusted to pH 7.5 with barium hydroxide. The solution is concentrated *in vacuo* to about 15 ml and filtered to remove traces of impurity, and the barium thymidine 3'-phosphate is precipitated by the addition of two volumes of ethanol. The precipitate is collected by centrifugation and washed with 50% ethanol followed by ethanol, acetone, and ether. Finally it is dried over phosphorus pentoxide; yield, 450 mg (88%).

References

1. Cramer, F. (1960). *Angew. Chem.* **72**, 236.
2. Chambers, R. W., Moffat, J. G., and Khorana, H. G. (1957). *J. Am. Chem. Soc.* **79**, 3747.
3. Michelson, A. M., and Todd, A. R. (1949). *J. Chem. Soc.* 2476.
4. Hall, R. H., and Khorana, H. G. (1956). *J. Am. Chem. Soc.* **78**, 1871.
5. Atherton, F. R., Openshaw, H. T., and Todd, A. R. (1945). *J. Chem. Soc.* 382.
6. Tener, G. M. (1961). *J. Am. Chem. Soc.* **83**, 159.
7. Michelson, A. M., and Todd, A. R. (1953). *J. Chem. Soc.* 951.
8. Tener, G. M., Khorana, H. G., Markham, R., and Pol, E. H. (1958). *J. Am. Chem. Soc.* **80**, 6223.

Recommended Reviews

1. Brown, D. M. (1963). *Phosphorylation*. In "Advances in Organic Chemistry," (R. A. Raphael, E. C. Taylor, and H. Wynberg, eds.), Vol. 3, p. 75. Interscience, New York.
2. Todd, A. R. (1957). *Chemical synthesis of nucleosides and nucleotides*. In "Advances in Enzymology," (S. P. Colowick, and N. O. Kaplan, eds.), Vol. 3, p. 822. Academic Press, New York.

D. Chemical and Enzymatic Preparation of Cyclic Nucleotides

Ribonucleoside cyclic phosphates are the 2',3'-monophosphate esters of nucleosides. Two methods are generally used for the preparation of cyclic nucleotides: degradation of substances containing phosphodiester structures, and dehydration of nucleotides.

1. Introduction

a. Synthesis of uridine 2',3'-cyclic phosphate

The ribonucleoside 2',3'-cyclic phosphates were synthesized by Brown, Magrath, and Todd (1) by treatment of the 2'(3')-nucleotides with trifluoroacetic anhydride; better yields were obtained by Crooks, Mathias, and Robin (2). Khorana and his coworkers (3, 4) have obtained high yields by the reaction of the nucleotides with DCC in aqueous pyridine.

Other methods reported to give high yields are those of Michelson (5) involving the reaction of tri-*n*-octylamine or tri-*n*-decylamine salts of the nucleotides with tetraphenylphosphate or diphenylphosphochloridate in dioxane. By treatment of the nucleotide and a base with ethyl chloroformate in an aqueous or anhydrous medium (6), Smith, Moffat, and Khorana (7) have prepared cyclic phosphates by reaction of the ammonium salts of the nucleotides with DCC in aqueous *tert*-butyl alcohol and formamide, Ukita and Irie by reaction of the pyridine salts of the nucleotides with DCC in an anhydrous medium (8).

b. Enzymatic preparation of cyclic nucleotides

Jones (9) was the first to use pancreatic ribonuclease for the digestion of yeast RNA. It was purified and crystallized by Kunitz in 1940 and named ribonuclease (10). It has no action on DNA and is strongly antigenic. The main action of ribonuclease is to split the linkage joining the phosphate residue at C-3′ in a pyrimidine nucleotide to C-5′ in the next nucleotide sequence. If the enzyme acts on RNA for a short time, the main products are the cyclic 2′,3′-phosphates. They are slowly hydrolyzed by the ribonuclease to yield pyrimidine 3′-phosphates.

Enzyme activity may be determined spectrophotometrically (11), manometrically (12), and by other methods (13).

2. Principle (synthesis)

The following method (14) is based on dehydration of the monophosphates such as uridine 2′(3′)-phosphate with DCC in dimethylformamide.

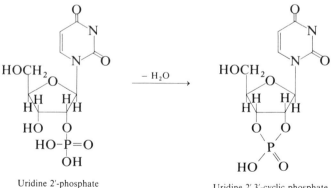

Uridine 2′-phosphate   Uridine 2′,3′-cyclic phosphate

3. Apparatus

   Centrifuge

4. Materials

   Barium acetate
   Dimethylformamide
   DCC
   Uridine 2'(3')-phosphate

5. Time

   2–3 hours

6. Procedure

To 100 mg uridine 2'(3')-phosphate and 400 mg DCC in a 100-ml flask is added 8 ml dimethylformamide. All the solid material dissolves rapidly on gentle swirling of the contents. After a few minutes the solution becomes cloudy and a precipitate appears. Approximately 40 ml water is added to the reaction mixture; the resulting pH is about 4. The mixture is passed through a fritted-glass filter, and the precipitate is washed with 5 ml water. The combined filtrates are neutralized by the careful addition of about 3 ml $0.1N$ sodium hydroxide, and evaporated under reduced pressure (water pump) until only a little dimethylformamide remains. Acetone is then added to precipitate the uridine 2',3'-phosphate as a powder and to extract the residual dimethylformamide and cyclohexylureas. This step is repeated several times, and the acetone is removed each time by decanting and pipetting. The resulting sodium salt may be left under acetone and is stable for several months in the deep freeze. Alternatively, the acetone may be removed from the sodium salt in a stream of air, or under reduced pressure, and the residue taken up in about 2 ml water. The barium salt may then be precipitated by the addition of triethylamine to pH 9 and the addition of 1 ml 1 M barium acetate, followed by 3 volumes of ethanol. The precipitate is centrifuged (or filtered), washed with ethanol followed by ether, and dried *in vacuo* over phosphorus pentoxide.

7. Principle (enzymatic degradation)

Yeast RNA is degraded by pancreatic ribonuclease (15). Cyclic nucleotides diffuse out of the dialysis bag, and the solution is concentrated and subjected to descending paper chromatography; the cyclic nucleotides form the lowest band.

8. Apparatus

   Dialysis bag
   Paper chromatography chamber

9. Materials

Dialysis cellophane bag
Isopropanol
Pancreatic ribonuclease
Yeast RNA

10. Time

4 hours

11. Procedure

Pancreatic ribonuclease (100 μg) is added to a solution of yeast RNA (500 mg in 10 ml adjusted to pH 7), and the mixture is placed in a cellophane dialysis bag in a large volume of distilled water. The dialysate is collected at intervals and concentrated *in vacuo*. It is then chromatographed, together with some reference substances, on Whatman No. 3 MM paper in a solvent consisting of isopropanol (70 parts), water (30 parts), and 0.35 ml ammonia solution (sp gr 0.88). The solvent is allowed to flow until the lowest band almost reaches the bottom of the sheet (descending technique). The compounds are detected by viewing the bands at 254 mμ.

References

1. Brown, D. M., Magrath, D. J., and Todd, A. R. (1952). *J. Chem. Soc.* 2708.
2. Crooks, E. M., Mathias, A. P., and Robin, B. R. (1960). *Biochem. J.* **74**, 230.
3. Dekker, C. A., and Khorana, H. G. (1954). *J. Am. Chem. Soc.* **76**, 3522.
4. Tener, G. M., and Khorana, H. G. (1955). *J. Am. Chem. Soc.* **77**, 5349.
5. Michelson, A. M. (1958). *Chem. Ind. (London)* 70.
6. Michelson, A. M. (1959). *J. Chem. Soc.* 3665.
7. Smith, M., Moffat, J. G., and Khorana, H. G. (1958). *J. Am. Chem. Soc.* **80**, 6204.
8. Ukita, T., and Irie, M. (1960). *Chem. Pharm. Bull. (Tokyo)* **8**, 81.
9. Jones, W. (1920). *Am. J. Physiol.* **52**, 203.
10. Kunitz, M. (1940). *J. Gen. Physiol.* **24**, 15.
11. Dickman, S. R., Aroskar, J. P., and Kropf, R. B. (1956). *Biochim. Biophys. Acta* **21**, 539.
12. Zittle, C. A., and Reading, E. H. (1945). *J. Biol. Chem.* **160**, 519.
13. Scheraga, H. A., and Rupley, J. A. (1962). *Advances in Enzymology* **24**, 161.
14. Czer, W., and Shugar, D. (1963). *Biochem. Prepns.* **10**, 139.
15. Markham, R., and Smith, J. D. (1952). *Biochem. J.* **52**, 552.

Recommended Reviews

1. Khorana, H. G. (1961). *Cyclic phosphate formation and its role in the chemistry of phosphate esters of biological interest. In* Khorana, H. G. "*Some Recent Developments in the Chemistry of Phosphate Esters of Biological Interest*," p. 44. John Wiley, New York.
2. Markham, R. (1957). *The preparation and assay of cyclic nucleotides. In* "*Advances in Enzymology*," (S. P. Colowick, and N. O. Kaplan eds.), Vol. 3, p. 805. Academic Press, New York.

E. Ion-Exchange Thin-Layer Chromatography of Nucleotides

1. Introduction

In recent years, special, finely powdered cellulose cation- and anion-exchange materials have been developed and used in thin-layer chromatography. Thus, Randerath has used ECTEOLA-cellulose (1) (obtained by treating sodium cellulose with epichlorhydrin and triethanolamine) and PEI-cellulose (poly(ethyleneimine)-impregnated cellulose) (2) for the separation of nucleic acid derivatives on thin layers. The application of DEAE (diethylaminoethyl)-cellulose (3) and DEAE-Sephadex (4) was recently reported. It is noteworthy that sharp and rapid separations of very small amounts of nucleic acid derivatives (0.05–1 μg) have been obtained on ion-exchange thin layers.

The procedure outlined below is based on that of Randerath (5).

2. Apparatus

    Electric mixer
    TLC applicator

3. Materials

    Cellulose powder
    Nucleotides
    Polyethyleneimine hydrochloride

4. Time

    Overnight drying of TLC plates
    3 hours

5. Procedure

a. Preparation of plates

A dialyzed 1% PEI hydrochloride solution is prepared from a 10% poly(ethyleneimine) hydrochloride solution, pH about 6.0, and dialyzed against distilled water. A suspension of 30 g cellulose powder for thin-layer chromatography in 200 ml dialyzed solution is homogenized in an electric mixer for about 30 sec. Glass plates (10 × 20 or 20 × 20 cm) are then covered by means of an applicator to a thickness of 0.5 mm. The plates are then dried overnight at room temperature. The resulting layers have a capacity of approximately 1.5 mEq nitrogen per gram cellulose. Layers of lower capacity can be prepared in the following manner without previous dialysis of the PEI solution. 10 g 50% PEI solution are diluted with 700 ml distilled water, brought to pH 6 with concentrated hydrochloric acid, and finally made up to

1 liter with distilled water. Cellulose powder (30 g) is suspended in 200 ml PEI solution, and the plates are coated with an applicator and allowed to dry overnight at room temperature. These layers have a capacity of 0.7 to 0.8 mEq nitrogen per gram cellulose.

b. Chromatography

The samples are applied at a starting line, drawn with a soft pencil, 3 cm from the lower edge of the plate. Sodium or lithium salts of nucleotides ($0.002M$ solutions) in distilled water are used. After spotting, the plates are exposed to a current of cold air for about 3 min. Ascending chromatography is carried out in closed tanks filled with solvent to a height of 0.7 to 1.0 cm. All chromatograms are developed up to a dividing line 10 cm above the starting line. The elution is carried out perpendicular to the coating direction. The development time ranges from 40 to 65 min, depending on the composition of the solvent. The plates are then dried in a stream of hot air. The compounds are located by examining the plates in incident shortwave ultraviolet light and marked on the plate with a pencil.

It is possible to separate interfering substances from nucleotides by a preliminary development with distilled water. Under these conditions, all nucleotides, with the sole exception of DPN, remain at the origin, while all water-soluble basic or neutral compounds migrate. The second development with electrolyte solutions, in order to separate the nucleotides, can be carried out either in the direction of the first development or, preferably, perpendicular to it.

Although ion-exchange chromatography is less sensitive than partition chromatography to the presence of salts in the samples to be analyzed, a large excess of undesirable anions does seriously interfere with the separations on PEI-cellulose layers. Interfering salts can be removed as follows: the samples are spotted in the usual manner, and the plates, after drying in a stream of air, are laid in a flat dish for about 10 min with anhydrous, reagent-grade methanol (300–600 ml). The layers are then dried, and chromatography is carried out as described above.

c. Elution under neutral conditions

At neutral pH, the rate of migration of nucleotides with the same base decreases in the following order: nucleotide sugars > monophosphates > diphosphates > triphosphates (see the following table).

d. Stepwise elution

A mixture of DPN, ADPG, AMP, ADP, and ATP cannot be resolved with one and the same solvent, since the differences in the adsorption affinities of these compounds for PEI-cellulose are too great. In such a case the

R_f Values of Nucleotides at Neutral pH

Compound[a]	Solvent: LiCl in Water	
	1.0 M	1.6 M
5'-AMP	0.52	0.65
5'-IMP	0.59	0.74
5'-GMP	0.40	0.51
5'-CMP	0.64	0.75
5'-UMP	0.74	0.80
ADP	0.26	0.54
IDP	0.30	0.63
GDP	0.17	0.45
CDP	0.33	0.64
UDP	0.41	0.71
ATP	0.06	0.34
ITP	0.09	0.39
GTP	0.05	0.25
CTP	0.11	0.41
UTP	0.14	0.49

[a] A, adenosine; G, guanosine; I, inosine; C, cytidine; U, uridine; M, mono; D, di; T, tri; P, phosphate.

chromatogram is developed with a series of solvents of increasing electrolyte concentration.

The discontinuous, stepwise procedure used for elution involves three transfers of the plate, without intermediate drying, from one tank to another containing a higher LiCl concentration. As compared with elution at constant concentration, development with increasing concentrations gives sharper and more circular spots. For the separation of adenine and uracil nucleotides on a 0.5-mm thick PEI-cellulose layer, the plate is developed for 1 min with 0.1M LiCl, for 5 min with 0.3M LiCl, for 15 min with 0.7M LiCl, and for 25 min with 1.5M LiCl. Developing distance, 10 cm; time, 45 min.

The choice of the solvent depends on the composition of the mixture to be analyzed; for example, if the mixture contains only nucleotide sugars and nucleoside monophosphates, LiCl molarities between 0.1 and 1.0 are suitable. If a mixture of nucleoside di- and triphosphates is to be analyzed, the concentration range would be between 0.6 and 0.8M for the first solvent and 1.7 and 2.0M for the last solvent. In most cases, two to five steps are sufficient.

References

1. Randerath, K. (1961). *Angew. Chem.* **73,** 436.
2. Randerath, K. (1962). *Biochim. Biophys. Acta* **61,** 852.
3. Dyer, T. A. (1963). *J. Chromatogr.* **11,** 414.
4. Weiland, T., Luben, G., and Determann, H. (1962). *Experientia* **18,** 430.
5. Randerath, K., and Randerath, E. (1964). *J. Chromatogr.* **16,** 111.

F. Questions

1. Define nucleoside and nucleotide and give an example of each.
2. What is the main structural difference between RNA and DNA?
3. Describe the Watson-Crick representation of DNA.
4. Compare RNA and DNA with respect to acidic and alkaline hydrolysis.
5. Describe the methods for the isolation of nucleic acids.
6. What is the biological function of sRNA?
7. Describe the nucleic acids of viruses.
8. What are nucleases, their sources, and functions in nucleic acids?
9. What is meant by a replication of DNA? Describe an experiment showing such a replication by multiplying bacteria.
10. Describe in detail the transfer of genetic information.

G. Recommended Books

1. Allen, F. W. (1962). *"Ribonucleoproteins and Ribonucleic Acids."* Elsevier, Amsterdam, The Netherlands.
2. Ayad, S. R. (1972). *"Techniques of Nucleic Acid Fractionation."* Wiley-Interscience, New York.
3. Chargaff, E. (1963). *"Essays on Nucleic Acids."* Elsevier, Amsterdam, The Netherlands.
4. Chargaff, E., and Davidson, J. N., (eds.) (1955). *"The Nucleic Acids."* Academic Press, New York.
5. Davidson, J. N. (1965). *"The Biochemistry of the Nucleic Acids."* Methuen, London. England.
6. Gerhke, C. W., and Kuo, K. C. T. (1990). *"Chromatography and Modification of Nucleosides."* Elsevier, Amsterdam, The Netherlands.
7. Jordan, D. O. (1960). *"The Chemistry of Nucleic Acids."* Butterworth, London. England.
8. Khorana, H. G. (1961). *"Some Recent Developments in the Chemistry of Phosphate Esters of Biological Interest."* John Wiley, New York.

304 CHAPTER 4 Nitrogenous Compounds

9. Michelson, A. M. (1963). *"The Chemistry of Nucleosides and Nucleotides."* Academic Press, London. England.
10. Pullman, B., and Jortner, J., (1983). *"Nucleic Acids: The Vectors of Life."* (eds.) Riedel Publishing Co.
11. Saenger, W. (1983). *"Principles of Nucleic Acid Structure."* Springer-Verlag, New York.
12. Schimmel, P. R., Soll, D., and Abelson, J. N., (1979). *"Transfer RNA: Structure, Properties and Recognition."* (eds.) Cold Spring Harbor Laboratory, Boston, Mass.
13. Townsend, L. B., ed. (1988). *"Chemistry of Nucleosides and Nucleotides."* Plenum, New York.

IV. PORPHYRINS

A. Introduction

Porphyrins are widely distributed throughout the animal and plant kingdoms. The three most important representatives are hemoglobin, the coloring matter present in the red corpuscles of blood; chlorophyll, the coloring matter of green plants and leaves; and the cytochromes, the redox enzymes. The functions of these substances differ. Hemoglobin participates in oxidative reactions, while chlorophyll acts as an intermediary in the photosynthetic process of reduction of atmospheric carbon dioxide to carbohydrates, and cytochromes promote cellular respiration.

Porphyrins have the property of combining very readily with a variety of metals to form metalloporphyrins. Thus, vanadium porphyrin complexes have been identified in shale oils, and a copper uroporphyrin constitutes

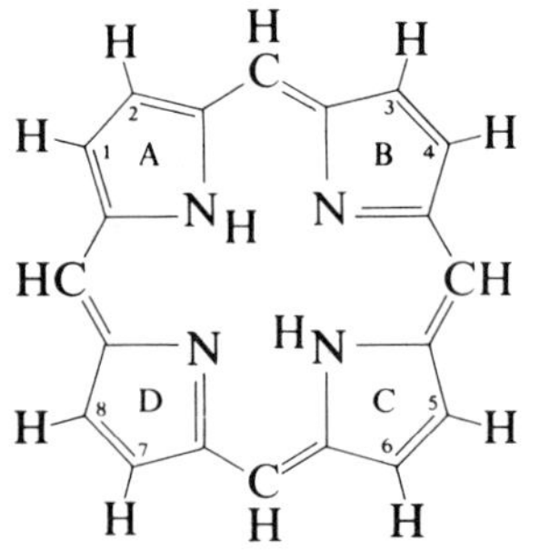

Porphin

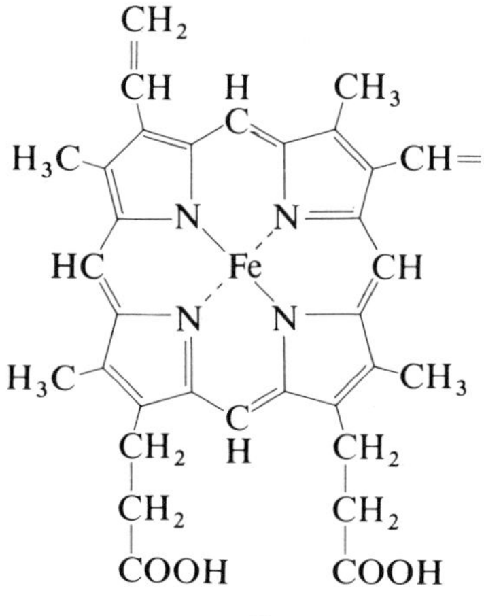

Heme

the red coloring matter of the wings of certain birds. Many complexes of porphyrins with iron, magnesium, zinc, manganese, cobalt, nickel, silver, tin, and cadmium have been prepared. Among these, the hemes are the most important.

Hemoglobin is a conjugated protein made up of a prosthetic group, the heme, and a protein part, the globin, in the proportion of 1 part heme to 26 parts globin. On treatment of hemoglobin with hot sodium chloride and acetic acid, it is decomposed to hemin, the red coloring matter. Treatment of hemin with dilute acid results in the removal of iron, thus yielding protoporphyrin. Vigorous reduction of hemin with hydriodic acid and acetic acid causes its degradation into four pyrrole derivatives.

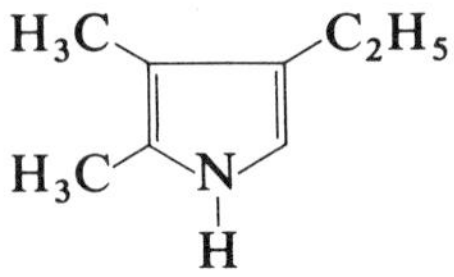

Hemopyrrole

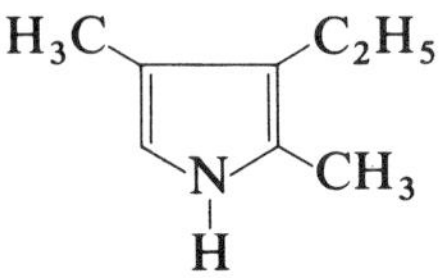

Kryptopyrrole

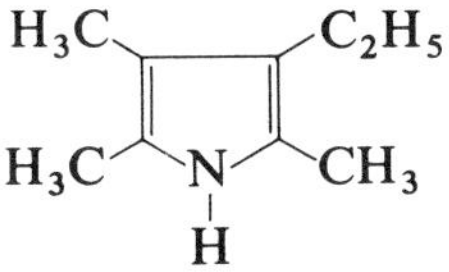

Phyllopyrrole

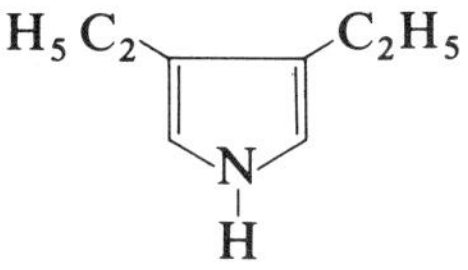

Opsopyrrole

Reduction of hemin with tin and hydrochloric acid affords four pyrrole hemocarboxylic acids.

Oxidation of hemin with chromic acid gives hematinic acid; reduction of hemin and oxidation of the resulting mesoporphyrin gives hematinic acid together with ethylmethylmaleimide.

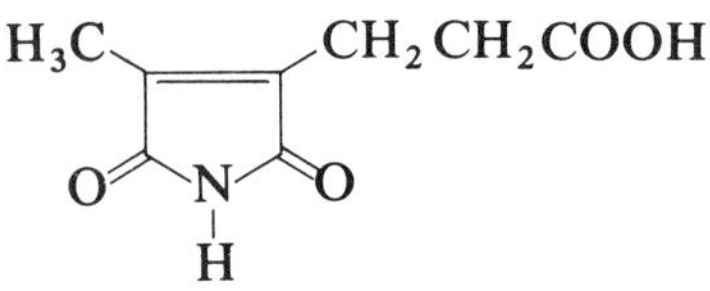

Hematinic acid

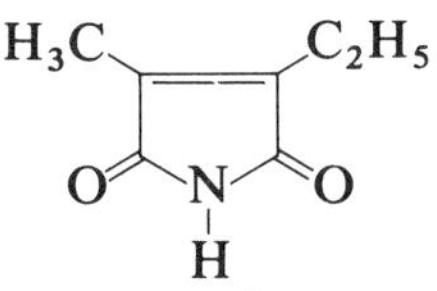

Ethylmethylmaleimide

The close chemical relationship between hemin and chlorophyll is illustrated by the fact that etioporphyrins are both decarboxylation products of protoporphyrins and degradation products of chlorophyll.

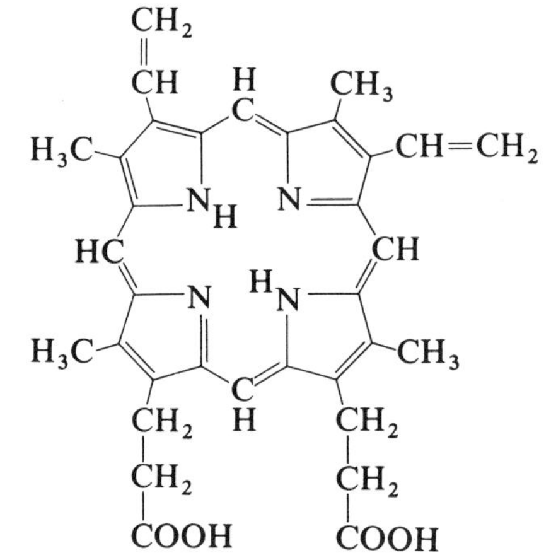

Protoporphyrin

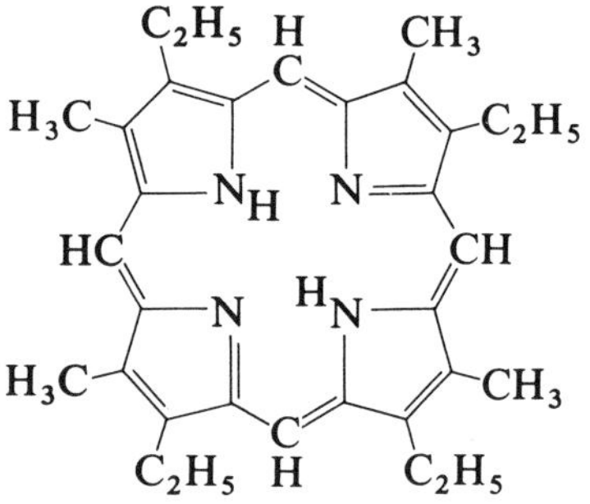

Etioporphyrin

Chlorophyll

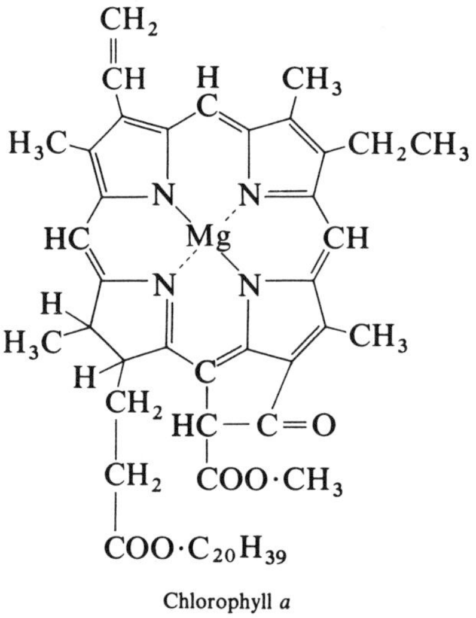

Chlorophyll *a*

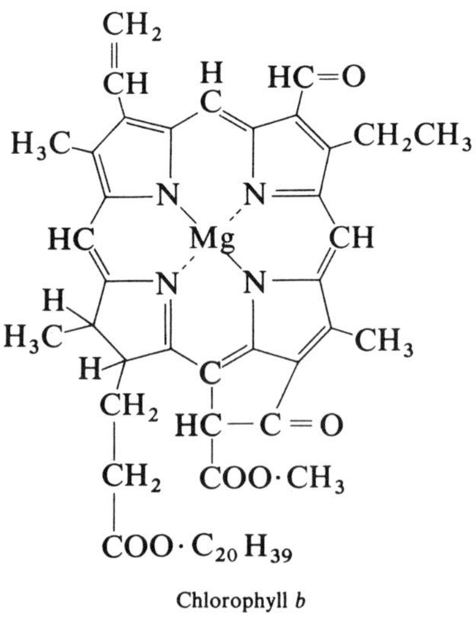

Chlorophyll *b*

Chlorophyll is the green coloring matter present in the chloroplasts of green plants, especially in their leaves. In the green leaves of higher plants the chlorophyll content constitutes about 0.1% of the fresh weight. Willstätter found that the ratio of chlorophyll *a* to chlorophyll *b* is remarkably constant, being about three to one. The molecular formula of chlorophyll *a* is $C_{55}H_{72}O_5N_4Mg$, and that of chlorophyll *b* is $C_{55}H_{70}O_6N_4Mg$. Chlorophyll *b* differs from *a* only by having an aldehyde group instead of a methyl group on carbon atom 3. The two compounds have different absorption spectra.

IV. Porphyrins   307

Alkaline hydrolysis of chlorophyll *a* affords the chlorophyllins and two alcohols—phytol and methanol. Treatment with an ethanolic solution of oxalic acid yields pheophytin.

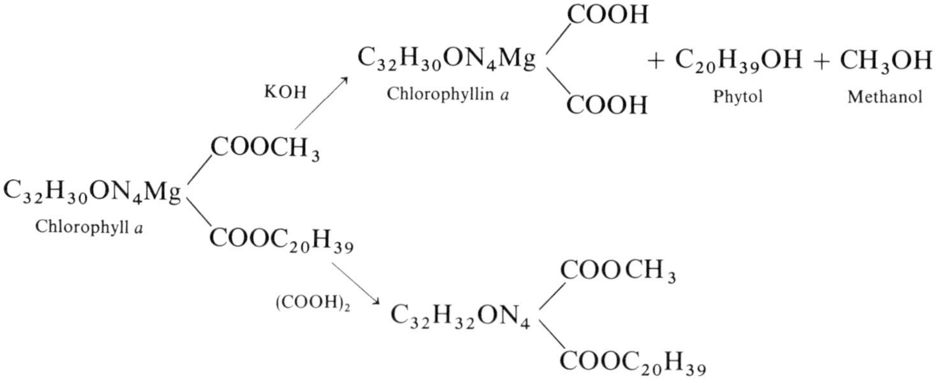

$$C_{32}H_{30}ON_4Mg \begin{array}{l} \text{COOCH}_3 \\ \text{COOC}_{20}H_{39} \end{array}$$

Chlorophyll *a*

$\xrightarrow{\text{KOH}}$ $C_{32}H_{30}ON_4Mg \begin{array}{l} \text{COOH} \\ \text{COOH} \end{array}$ + $C_{20}H_{39}OH$ + CH_3OH

Chlorophyllin *a*   Phytol   Methanol

$\xrightarrow{(\text{COOH})_2}$ $C_{32}H_{32}ON_4 \begin{array}{l} \text{COOCH}_3 \\ \text{COOC}_{20}H_{39} \end{array}$

Pheophytin *a*

Vigorous treatment of pheophytin with concentrated acids hydrolyzes the phytyl group, forming pheophorbide.

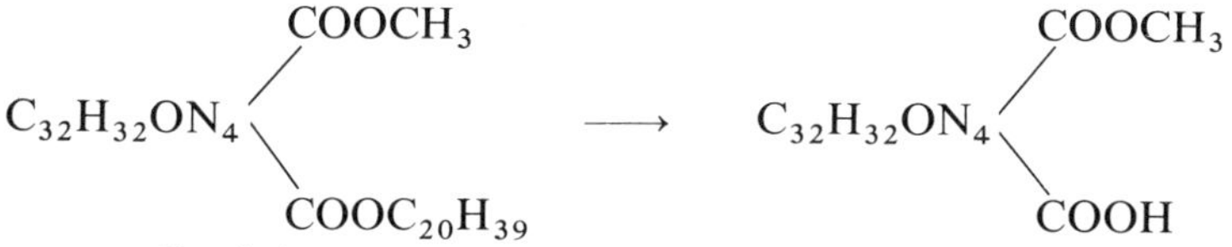

$C_{32}H_{32}ON_4 \begin{array}{l} \text{COOCH}_3 \\ \text{COOC}_{20}H_{39} \end{array}$ $\longrightarrow$ $C_{32}H_{32}ON_4 \begin{array}{l} \text{COOCH}_3 \\ \text{COOH} \end{array}$

Pheophytin *a*   Pheophorbide *a*

When pheophorbide *a* is hydrolyzed with boiling methanolic potassium hydroxide, the product is a tricarboxylic acid (chlorin e_6).

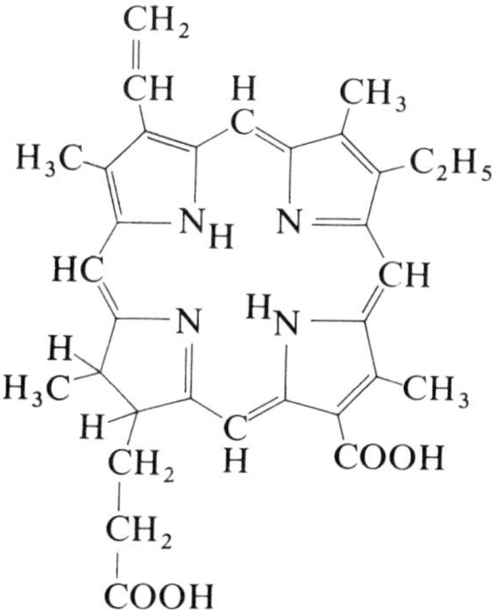

Chlorin e_6

On oxidation with chromic acid chlorin e_6 gives hematinic acid and ethylmethylmaleimide.

Several pigments similar to chlorophyll are known in nature. Bacteriochlorophyll, the pigment of purple and brown bacteria, differs from chlorophyll *a* in the 2-position, where an acetyl residue replaces the vinyl group, and also contains two additional hydrogen atoms.

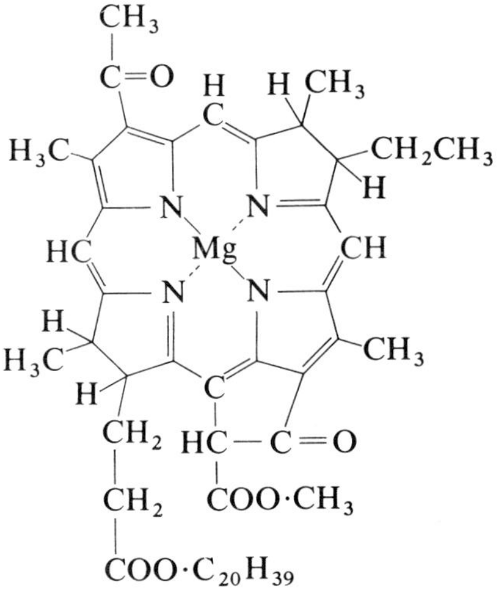

Bacteriochlorophyll

Protochlorophyll is present in pumpkin seeds and in the rinds of gourds.

Total synthesis of chlorophyll *a* was accomplished by Woodward and Strell and their coworkers in 1960.

Cytochromes promote cellular respiration. The well known cytochrome C is a hemoprotein; its molecular weight is 12.400.

2. Biosynthesis of Porphyrins

The biosynthetic pathway for porphyrins is as follows. Glycine and succinyl-coenzyme A combine to form the unstable α-amino-β-ketoadipic acid, which loses CO_2 and yields δ-aminolevulinic acid. An enzyme, δ-aminolevulinic acid dehydrase, promotes condensation of two molecules of the acid, with subsequent formation of porphobilinogen.

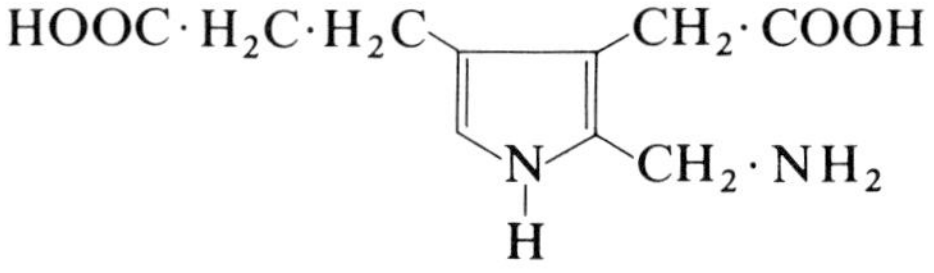

Porphobilinogen

IV. Porphyrins

This highly reactive material is readily transformed into a mixture of uroporphyrinogens. The latter are decarboxylated to coproporphyrinogen, which yields protoporphyrin. It is noteworthy that only magnesium and iron have so far been found in biologically active porphyrins. The introduction of iron to form heme and other products takes place enzymatically:

$$Fe + protoporphyrin \rightarrow protohem \rightarrow hemoglobin$$

It is also probable that an enzyme is involved in the insertion of magnesium into protoporphyrin:

$$Mg + protoporphyrin + Mg\ protoporphyrin \rightarrow chlorophylls$$

For the formation of chlorophylls, further alterations of side chains are required, including esterifications and reductions, formation of a cyclopentanone ring, and the reduction of one pyrrole ring in case of chlorophyll, or two pyrrole rings in case of bacteriochlorophyll.

B. Isolation of Hemin from Blood

1. Introduction

The classical method of Schalfejef (1), which was introduced in 1885, involves slow addition of blood to glacial acetic acid saturated with sodium chloride at 100°C. The other procedures were developed by Nencki and Zaleski (2), Piloty (3), Fischer (4), and Corwin and Erdman (5). However, the separation of hemin crystals from a rather viscous solution of proteins in acetic acid proved to be both tedious and inconvenient. Recently, Chu and Chu (6) modified the method of hemin isolation by completely removing proteins before the formation of hemin. For bulk preparations, however, the modified

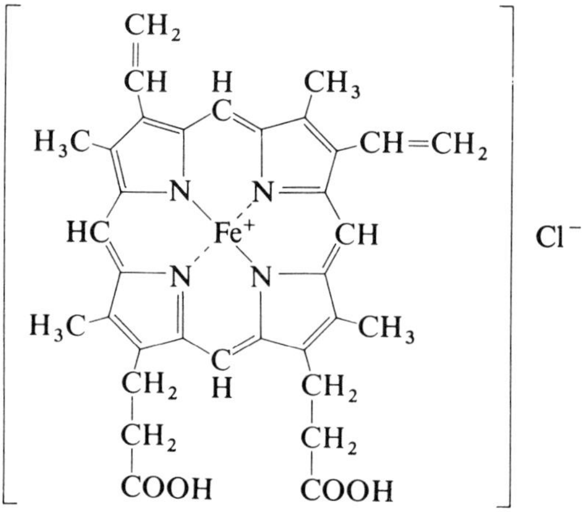

Hemin

method of Labbe and Nishida (7) can be used. A micro-scale adaptation of this method permits crystallization of hemin (50% yield) from as little as 0.1 ml blood (8).

2. Principle

In the following procedure (7), hemin is extracted with a mixture of strontium chloride in acetic acid and acetone. Heating of the extract increases protein precipitation and hence improves the purity of hemin.

3. Apparatus

Centrifuge

4. Materials

Acetic acid	Ethanol	Pyridine
Acetone	Ether	Sodium chloride
Blood	Hydrochloric acid	Strontium chloride
Chloroform		

5. Time

2–3 hours

6. Procedure

A stock solution of 2% strontium chloride hexahydrate in a glacial acetic acid is prepared. The extraction solvent, composed of one part of this solution to three parts acetone, is then prepared immediately before use to avoid crystallization of the strontium chloride. One volume blood or similar heme-containing preparation is added with stirring to 12 volumes extraction solvent. The mixture is allowed to stand for 20 to 30 min, with occasional stirring. Brief heating to the boiling point during this period appears to increase protein precipitation and improves the purity of the hemin. The mixture is then filtered with gentle suction through a medium-porosity sintered-glass filter, or by gravity through a coarse filter paper.

The residue is washed twice with one volume of extraction solvent. The combined filtrate is transferred to a beaker and heated to 100°C. The rate of concentrating the solution appears to be unimportant; however, in order to avoid bumping, porcelain chips must be added, and the temperature should not exceed 100°C. Crystallization of hemin begins as the solution concentrates and reaches completion when the latter is allowed to cool to room temperature. The hemin is centrifuged and washed twice with 50% acetic acid and water and once with ethanol and ether. The yields without further purification are in the range of 75 to 80%. The crude hemin prepared from 20 ml blood is recrystallized by stirring with 0.5 ml pyridine. After 5 to 10 min 4 ml chloroform is added, and the solution is filtered. Acetic acid containing 2% strontium chloride is added to the filtrate. The solution is then heated to 100°C and allowed to cool. The hemin is then centrifuged and treated as described above. The overall yield after recrystallization is about 60%.

References

1. Schalfejef, M. (1885). *Ber.* **18**, 232.
2. Nencki, M., and Zaleski, J. (1900). *Z. Physiol. Chem.* **30**, 390.
3. Piloty, O. (1910). *Ann.* **377**, 358.
4. Fischer, H. (1941). *Org. Synth.* **21**, 53.
5. Corwin, A. H., and Erdman, J. G. (1946). *J. Am. Chem. Soc.* **68**, 2473.
6. Chu, T. C., and Chu, E. J. (1955). *J. Biol. Chem.* **212**, 1.
7. Labbe, R. F., and Nishida, G. (1957). *Biochim. Biophys. Acta* **26**, 437.
8. Formijne, P., and Poulie, N. J. (1954). *Koninkl. Ned. Acad. Wetenschap. Proc. Sec. C.* **57**, 57.

Recommended Review

1. Falk, J. E. (1964). *"Hems. In "Porphyrins and Metalloporphyrins."* Elsevier, Amsterdam, The Netherlands.

C. Degradation of Hemin

1. Introduction

The first step in the degradation of hemin involves its conversion to the iron-free protoporphyrin. In order to remove the iron, it is first reduced to the ferrous state. If a strong acid is now added, the protons displace the coordinated ferrous ion, forming the porphyrin. The following reducing agents have been employed: iron powder (1), ferrous sulfate, palladium and hydrogen (2), stannous chloride (3), sodium amalgam (4), and hydrobromic acid in glacial acetic acid (5).

Small samples of protoporphyrin are easiest prepared by treatment of hemin with ferrous sulfate (6). For work with several grams of hemin, Grinstein's (7) method can be used. The next step in the degradation is the catalytic reduction of protoporphyrin to mesoporphyrin, which is carried out over platinum in formic acid. The mesoporphyrin is oxidized to methylethylmaleimide (derived from pyrrole rings A and B) and methyl-β-carboxyethylmaleimide or hematinic acid (derived from pyrrole rings C and D). The hematinic acid is then decarboxylated to yield methylethylmaleimide, which is ozonized to pyruvic acid and α-ketobutyric acid.

2. Principle

The procedure described below is according to Wittenberg and Shemin (8).

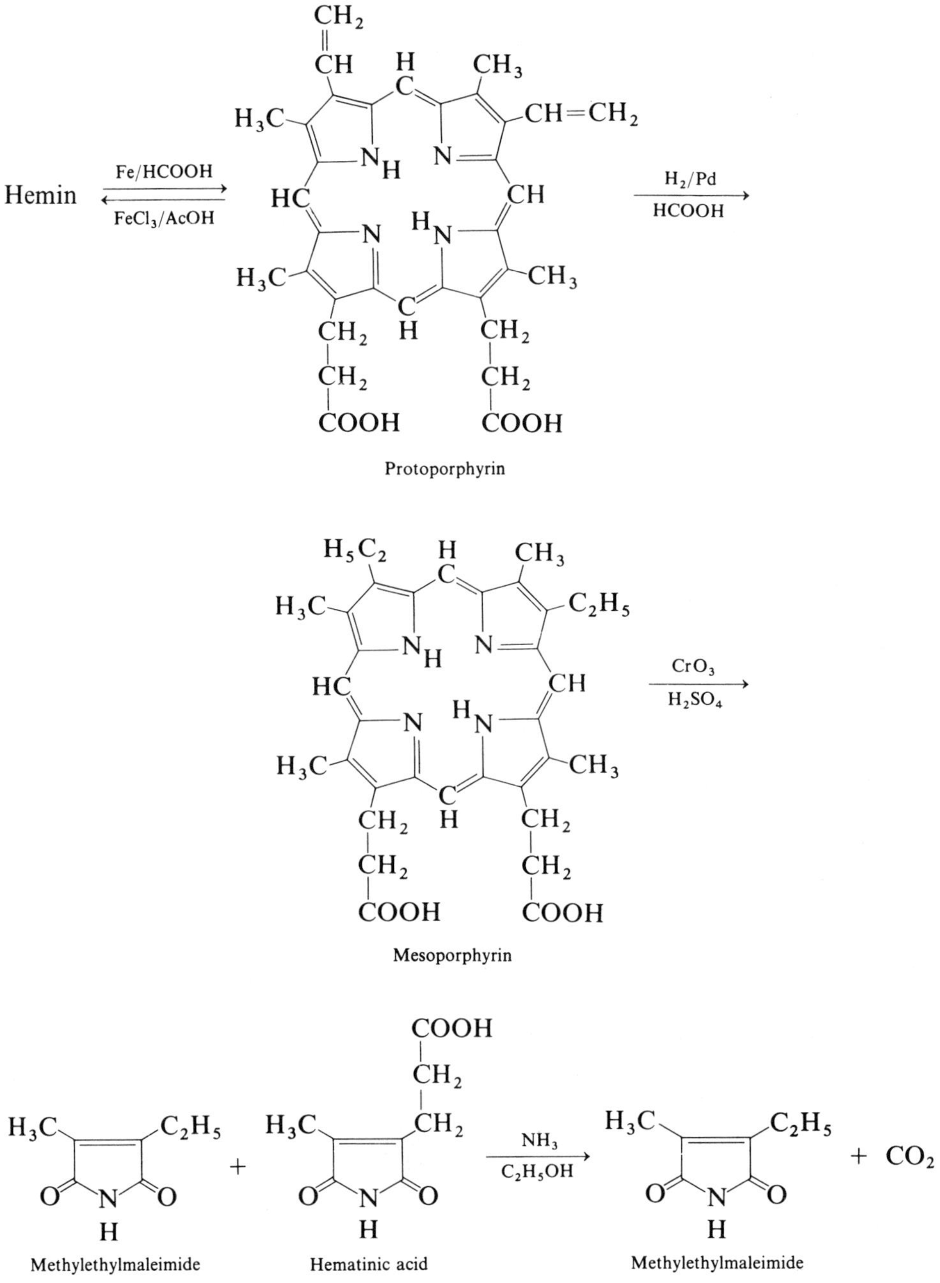

Hemin $\underset{FeCl_3/AcOH}{\overset{Fe/HCOOH}{\rightleftarrows}}$ Protoporphyrin $\xrightarrow{H_2/Pd \atop HCOOH}$ Mesoporphyrin $\xrightarrow{CrO_3 \atop H_2SO_4}$ Methylethylmaleimide + Hematinic acid $\xrightarrow{NH_3 \atop C_2H_5OH}$ Methylethylmaleimide + CO_2

Methylethylmaleimide + $O_3 \longrightarrow$ CH₃COCOOH + CH₃CH₂COCOOH
　　　　　　　　　　　　　　　　　　Pyruvic acid　　　　α-Ketobutyric acid

3. Apparatus

Ozonator
Sublimation assembly

4. Materials

Ammonium acetate	Ethyl acetate	Palladium catalyst
Ammonium hydroxide	Ferric chloride	Petroleum ether
Barium hydroxide	Formic acid (100%)	Pyridine
Chromic acid	Hemin	Sodium acetate
Drierite	Hydrochloric acid	Sodium
Ethanol, absolute	Iron, powder	bicarbonate
Ethanol, 95%	Ozone	Sulfuric acid
Ether		

5. Time

12–14 hours
Stirring overnight

6. Procedure

a. Conversion of hemin to protoporphyrin

To a boiling solution of 6 g hemin in 300 g formic acid is added 6 g iron powder in six portions over a period of 30 min. The mixture is stirred for a further 40 min and filtered. Two to three volumes water saturated with ammonium acetate are added to the filtrate. The solid material is filtered off, washed with water, and dried; yield, 4.5–5 g. It can be recrystallized from pyridine–water.

b. Conversion of protoporphyrin to hemin

One g finely powdered protoporphyrin is dissolved in 100 ml acetic acid. A few drops of concentrated hydrochloric acid and 10 ml ferric chloride (1% solution in 90% acetic acid) are added, the mixture is boiled, and solid sodium acetate is added portionwise until the solution turns brown and crystalline hemin starts to precipitate. The mixture is cooled and filtered on a Buchner funnel, and the residue is washed with acetic acid, water, ethanol, and ether; yield, 0.9 g.

c. Conversion of protoporphyrin to mesoporphyrin

Protoporphyrin (4.5 g) in 200 ml formic acid (100%) is hydrogenated at NTP in the presence of a colloidal palladium catalyst (18% Pd). After the hydrogenation is complete, the catalyst is filtered off, and the filtrate is poured into 3 volumes 30% ammonium acetate to flocculate the mesoporphyrin. The product is filtered on a Buchner funnel and washed with water, and the filtrate is extracted with ethyl acetate. The extract is added to the bulk of mesoporphyrin. The crude material is dissolved in 2% ammonium hydroxide and reprecipitated by addition of acetic acid. The collected precipitate is washed and further purified by dissolving it twice more in ammonium hydroxide and precipitating with acetic acid; yield, 3.5 g.

d. Oxidation of mesoporphyrin to methylethylmaleimide and hematinic acid

A solution of 5.1 g chromic acid in small amount of water is added, with stirring, over a period of 4 hr to a solution of 2.5 g mesoporphyrin in 170 ml 20% sulfuric acid. Stirring is continued overnight, after which period the solution turns green. The reaction mixture is diluted with water and extracted with ether for 4 to 5 hr in a continuous extractor. The ethereal solution containing methylethylmaleimide and hematinic acid is shaken once with 50 ml 5% sodium bicarbonate to remove the hematinic acid. The bicarbonate solution is then extracted six times with equal volumes of ether, and the ether extracts are combined with the original ethereal solution. The ether is removed by warming, and the residue is sublimed at 70 to 80°C/3 mm; yield, 0.5 g; mp 64–65°C.

The bicarbonate solution is immediately made acid to Congo red paper with dilute sulfuric acid. After decolorization with a small amount of charcoal, it is extracted ten times with equal volumes of ether. The ether is removed *in vacuo*. The crude hematinic acid (0.65 g) is dissolved in ether and crystallized by the gradual addition of petroleum ether; mp 114–115°C.

e. Conversion of hematinic acid to CO_2 and methylethylmaleimide

A suspension of 0.64 g hematinic acid in 10 ml almost saturated ammoniacal solution of absolute ethanol is heated in a sealed tube for 2 hr at 175°C. The reaction mixture is then diluted with water and acidified with sulfuric acid, and the evolving CO_2 is aerated into barium hydroxide solution. The yield of barium carbonate is 0.5 g.

After the aeration process, the solution is made alkaline with a dilute solution of sodium bicarbonate, and immediately extracted six times with ether. The ether is evaporated, and the residue is acidified and subjected to steam distillation. The distillate is made slightly acid with dilute sulfuric acid and extracted with ether in a continuous extractor. The ether solution is dried over Drierite and evaporated. The product is purified by sublimation; yield, 0.15 g; mp 61–63°C.

f. Ozonization of methylethylmaleimide

Ozonization is carried out by passing ozone through a chloroform solution of the imide (100 mg in 100 ml chloroform). On concentrating the solution, solid ozonide is obtained. It is crystallized from a mixture of chloroform and petroleum ether; mp 83–84°C.

Ozonide (20 mg) is taken up in 50 ml dry ethyl acetate, 10 mg Pd-CaCO$_3$ catalyst is added, and hydrogen is bubbled through the reaction mixture at NTP for 1 hr. The catalyst is filtered off and washed with ethyl acetate. The combined ethyl acetate solutions are concentrated *in vacuo*, leaving an oily residue, which is tranferred to a tube containing 50 ml saturated aqueous solution of barium hydroxide. The tube is sealed and shaken continuously at room temperature for 24 hr, after which the solution is acidified and extracted several times with ether. The solvent is evaporated *in vacuo*, and the α-keto acids are converted to the corresponding 2,4-dinitrophenylhydrazone derivatives. The resulting mixture of the DNPs of pyruvic and α-ketobutyric acids is separated by paper or thin-layer chromatography.

References

1. Fischer, H., and Putzer, B. (1926). *Z. Physiol. Chem.* **154,** 39.
2. Lemberg, R., Bloomfield, B., Caiger, P., and Lockwood, W. (1955). *Australian J. Exptl. Biol.* **33,** 435.
3. Hamsik, A. (1931). *Z. Physiol. Chem.* **196,** 195.
4. Papendiek, A. (1926). *Z. Physiol. Chem.* **152,** 215.
5. Heath, H., and Hoare, D. S. (1959). *Biochem. J.* **73,** 679.
6. Morell, D. B., Barrett, J., and Clezy, P. S. (1961). *Biochem. J.* **78,** 793.
7. Grinstein, M. (1947). *J. Biol. Chem.* **167,** 515.
8. Wittenberg, J., and Shemin, D. (1950). *J. Biol. Chem.* **185,** 103.

Recommended Reviews

1. Falk, J. E. (1964). "*Special techniques, in Porphyrins and Metalloporphyrins.*" Elsevier, Amsterdam, The Netherlands.
2. Rimington, C. R., and Kennedy, G. Y. (1962). *Porphyrins, structure, distribution, and metabolism. In "Comparative Biochemistry,"* (M. Florkin, and H. S. Mason, eds.), Vol. 4, p. 557. Academic Press, New York.

D. Thin-Layer Chromatography of Plant Pigments

1. Introduction

In view of the rapidity of thin-layer chromatography, this is a useful technique for the separation of labile compounds such as chloroplast pigments (chlorophylls *a* and *b* and carotenoids). Separations can be carried out by thin-layer adsorption or partition chromatography. The adsorption technique was used by Eichenberger and Grob (1) on silica gel G plates, and by Winterstein, Studer, and Ruegg (2) on thin layers prepared from a mixture of silica gel and calcium hydroxide, 1:4. This method can lead to the isomerization and oxidation of the pigments.

Egger (3) has used kieselgur impregnated with triglycerides for the separation of chlorophylls and their degradation products. Colman and Vishniac (4) have used thin layers of sucrose for the separation of leaf pigments.

2. Principle

The procedure outlined below is according to Anwar (5).

3. Apparatus

Thin-layer chromatography equipment

4. Materials

Acetone Glycerine Methanol
Carbon tetrachloride Isooctane Petroleum ether
Celite

5. Time

2–3 hours

6. Procedure

a. Extraction of pigments from fresh spinach leaves

The leaves are homogenized in a blender together with sufficient 50% methanol to just cover the leaves. The mixture is centrifuged, and the upper liquid layer, virtually devoid of pigment, is decanted. The green solids are mixed with an equal weight of Celite, filtered through a Buchner funnel, washed with a little acetone, and diluted with water. They are then extracted several times with petroleum ether and dried over sodium sulfate. The solution is then filtered, concentrated under reduced pressure, and kept at a low temperature.

b. Preparation of thin-layer plates

The suspension for five plates (20 × 20 cm) is prepared by shaking 25 g silica gel G and 50 ml water for 30 sec. It is then spread on the plates with an applicator to give a layer of 0.25 mm thickness. The plates are activated at 120°C for 30 min.

c. Development

The pigments are spotted by means of micropipettes along a line 2 cm above the rim of the plate. Development is accomplished with isooctane–acetone–carbon tetrachloride, 3:1:1.

The solvent is allowed to rise about 15 cm above the starting line.

d. Detection

Twelve spots are observed under ultraviolet light and their colors recorded. Carotenes travel the most rapidly, followed by chlorophyll *a* and *b* and xanthophylls. The spot colors, which are clear and bright at the end of the development period, fade rapidly and change color on exposure to light and air. These changes can be delayed for some time by covering the plate with glycerine. Each color zone is scraped off into a small test tube containing a few ml solvent and filtered. The ultraviolet spectra are measured in the range of 400–700 mμ.

References

1. Eichenberger, W., and Grob, E. C. (1962). *Helv. Chim. Acta* **45**, 974.
2. Winterstein, A., Studer, A., and Ruegg, R. (1960). *Ber.* **93**, 2951.
3. Egger, K. (1962). *Planta* **58**, 664.
4. Colman, B., and Vishniac, W. (1964). *Biochim. Biophys. Acta* **82**, 616.
5. Anwar, M. H. (1963). *J. Chem. Educ.* **40**, 29.

Recommended Review

1. Strain, H. H., and Svec, W. A. (1969). Some procedures for chromatography of the fat-soluble chloroplast pigments. *Adv. Chromatogr.* **8**, 119.

E. Isolation and Identification of Petroleum Metalloporphyrins

1. Introduction

Crude oil has its origin in the organic debris of plants, algae, bacterial fungi, and a multitude of microorganisms that have been deposited into aquatic sediments. A small amount of this organic material ultimately is converted to fossil fuels. About 99% of the organic material deposited is oxidized, either chemically or microbially, and recycled into the atmosphere as carbon dioxide. Much of the remaining 1% of the material that has been incorporated into the rocks is further altered, but only a fraction is converted into usable fossil fuels.

Organic matter that survives oxidation and microbial attack in water and the initial stages of sedimentation is incorporated into the sediments, where it undergoes additional alteration and reactions caused by increasing temperature, or thermal maturation.

Attempts to understand the transformation of organic matter in the geosphere led to the term biomarkers. Biomarkers are organic compounds, present in the geospherical record, with carbon skeletons that can be related to a precursor molecule from a specific type of organism. In essence, biomarkers are a geochemical fingerprint for an oil or other geological sample containing organic compounds. Biomarkers commonly used in petroleum exploration are hydrocarbons, such as *n*-alkanes, isoprenoids, sesquiterpanes, di-, tri-, tetra-, and pentacyclic terpanes, steranes, and various aromatic hydrocarbons. Biomarker hydrocarbons found in crude oils and extracts from source rocks generally are derived from oxygenated precursors; thus, steranes are derived from sterols.

Though it has been known from Treib's work (1936) a half century ago that metalloporphyrins occur in petroleum and shales and thus are the first biomarkers derived from chlorophyll and heme, only in the past decade has a dramatic upsurge in geoporphyrin research occurred. This was attributable mainly to the advent of mass spectroscopy.

The major metalloporphyrins of petroleum, shales, coals, and bitumens occur as complex mixtures of Ni and V=O complexes of two major skeletal types DPEP (deoxyphylloerythroetioporphyrin) and Etio (vanadyl etioporphyrin), and three others, Di-DPEP, Rhodo-Etio and Rhodo-DPEP. Porphyrins often occur in petroleum around the ppm (parts per million) level (Fig. 4.3).

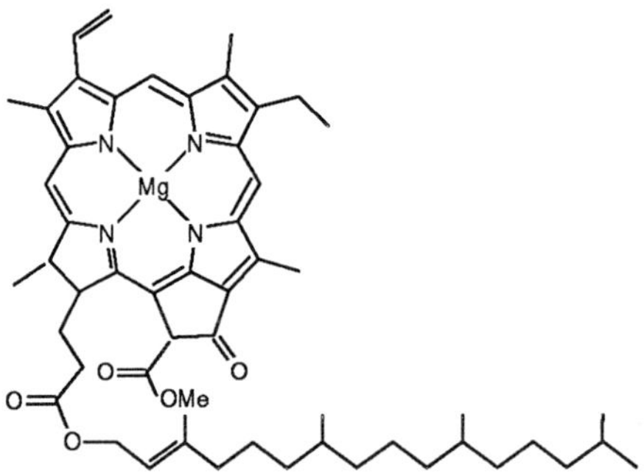

Chlorophyll-*a*

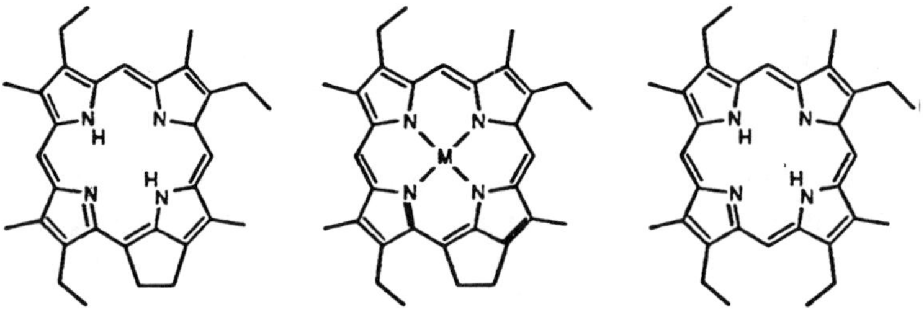

DPEP-5 Metallo-DPEP Etio-III

Figure 4.3 In the geosphere, the deoxyphylloerythroetioporphyrins with the 13,15-ethano ring (DPEP-5) and the etioporphyrins (e.g., etio-III) are major components, usually as metallo derivatives (e.g., nickel and V=O). Chlorophyll *a* is the putative ultimate biological source for most petroporphyrins.

2. Principle

A sample of crude petroleum (which can be obtained from a geological survey or a petroleum refinery) is fractionated on a silica gel column. The porphyrin fractions are detected by their characteristic absorptions in the 550–570 nm region.

The metalloporphyrins are demetallated by methanesulfonic acid, and the free porphyrins are determined by measuring their intensity at 618 nm. The demetallated porphyrins are then separated by HPLC.

Mass spectrometric characterization of the demetallated petroporphyrins is performed by direct inlet (probe-insertion) technique.

The method is based on that of Baker, Louda, and Orr (3).

3. Apparatus

Glass column
High pressure liquid chromatograph
Mass spectrometer
Stainless steel column filled with Partisil

4. Materials

Acetone	Hexane	Petroleum ether
Alumina	Methane sulfonic acid	Pyridine
Benzene	Methylene chloride	Silica gel
Chloroform	Pentane	Toluene
Ethyl acetate		

5. Time

6 hours workup followed by HPLC, GC-MS

6. Procedure

2 g petroleum crude is loaded onto a column (4 × 40 cm) filled with silica gel in benzene. Elution with benzene (fraction 1) provides a mixture of saturated hydrocarbons, aromatics, Ni-porphyrins, and nonpolar vanadyl porphyrins. Solvent polarity is increased to benzene–ethyl acetate (2:1, v/v). The majority of vanadyl pigments and N,S,O-containing compounds (i.e., resins) are collected (fraction 2). At this point the column is flushed with ethyl acetate, and the eluate (fraction 3) is examined spectrophotometrically (UV/VIS) to assure complete removal of pigments from the adsorbent.

Fraction 1 is dissolved with minimal benzene and loaded onto a column (2.5 × 30 cm) of silica gel made up in n-pentane. Elution with the following solvents yields these fractions: (1a) n-pentane (saturates and sulfur);

(1b) *n*-pentane–benzene (2:1, v/v) (Ni-porphyrins and aromatics); (1c) benzene (aromatics); (1d) benzene–ethyl acetate (3:1, v/v) (nonpolar V=O porphyrins); and (1e) ethyl acetate. All fractions are checked via UV/VIS by monitoring the bands at 550 nm ($\varepsilon = 34.280$) and 570 nm ($\varepsilon = 26.140$) for nickel and vanadyl porphyrins, respectively.

Crude Ni-porphyrins are then chromatographed over alumina (neutral grade II-III, 5–6% H_2O), eluting with 3.5 to 5% acetone in petroleum ether (30–60°C). The Ni-porphyrins are demetallated by transferring the solution background subtraction (using the scans immediately before and after the appearance of the porphyrins), as considerable amounts of nonporphyrinic material is present. **Note: All operations must be carried out in a hood!**

Molecular ions assigned to etioporphyrins are $M^+ = m/e\ 310 + 14\ n$; to rhodoporphyrins: $M^+ = m/e\ 358 + 14\ n$; to DPEP series: $M^+ = m/e\ 308 + 14\ n$.

References

1. Eglinton, G., Hajibrahim, S. K., Maxwell, J. R., and Quirke, J. M. E. (1980). *In* "Advances in Organic Geochemistry." (A. G. Douglas and J. R. Maxwell, eds.), p. 193. Pergamon Press, New York.
2. Baker, E. W., and Louda, J. W. (1986). *In* "Biological Markers." (R. B. John, ed.), p. 125. Elsevier, New York.
3. Baker, E. W., Louda, J. W., and Orr, W. L. (1987). *Org. Geochem.* **11**, 303.
4. Barwise, A. G. J., and Whitehead, E. V. (1980). *In* "Advances in Organic Geochemistry. (A. G. Douglas and J. R. Maxwell, eds.), p. 181. Pergamon Press, New York.
5. Hajibrahim, S. K., Quirke, J. M. E., and Eglinton, G. (1981). *Chem. Geol.* **32**, 173.

Recommended Reviews

1. Baker, E. W., and Palmer, S. E. (1978). Geochemistry of porphyrins. *In* "The Porphyrins." (D. Dolphin, ed.) Vol. 1, p. 486. Academic Press, New York.
2. Eglinton, G., Maxwell, J. R., Evershed, R. P., and Barwise, A. J. G. (1985). Red pigments in petroleum exploration. *Interdisciplinary Science Revs.* **10**, 221.
3. Louda, J. W., and Baker, E. W. (1986). The biogeochemistry of chlorophyll. *In* "Organic Marine Geochemistry," ACS Symp. Ser. (M. L. Sohn, ed.) p. 305.
4. Philp, R. P., and Jung-Nan, Qung. (1988). Biomarkers, *Anal. Chem.*, **60**, 887A.
5. Rullkötter, J. (1987). Organic geochemistry. *In* "Encyclopedia of Physical Science and Technology," Vol. **6**, p. 53. Academic Press, New York.

F. Questions

1. What are the chemical differences between chlorophyll *a*, chlorophyll *b*, protochlorophyll, and bacteriochlorophyll?
2. Outline the principal steps in the degradation of hemin.
3. Name some natural metalloporphyrins and their sources.

4. What are the functions of chlorophyll in the photosynthetic process?
5. Compare the biosynthesis of heme and chlorophylls.
6. How are metals inserted into, and withdrawn from, a porphyrin?

G. Recommended Books

1. Bentley, K. W. (1960). *"The Natural Pigments."* Interscience, New York.
2. Blauer, G., and Sund, H. (1985). *"Optical Properties and Structure of Tetrapyrroles."* W. de Gruyter, New York.
3. Falk, J. E. (1964). *"Porphyrins and Metalloporphyrins."* Elsevier, Amsterdam, The Netherlands.
4. Falk, H. (1989). *"The Chemistry of Linear Oligopyrroles and Bile Pigments."* Springer-Verlag, New York.
5. Lasceller, J. (1964). *"Tetrapyrrole Biosynthesis and Its Regulation."* Benjamin, New York.
6. Terrien, J., Truffant, G., and Charles, J. (1957). *"Vegetation and Chlorophyll."* Hutchinson, London, England.

V. PROTEINS

A. Introduction

The name protein is derived from the Greek word proteios, meaning primary. Proteins are high-molecular polypeptides that occur in the protoplasm of all animal and plant cells. The protein content of tissues varies widely; thus, blood plasma contains about 7% and egg yolk, about 15%. The following are the protein contents of seeds of various plants: 2% in potatoes, 1.6% in beets, and 1.8% in asparagus. In soy beans, proteins constitute 37%, in peanuts, 26%, and in almonds, 21%. All proteins yield amino acids upon hydrolysis; acid hydrolysis is preferred, since alkaline hydrolysis causes racemization. Proteolytic enzymes are often used because they cause neither racemization nor decomposition of sensitive amino acids. However, their action is slow. Enzymes are macromolecular catalysts of biological origin. It has been proved that enzymes are proteins.

They can be classified into four main groups:

Hydrolyzing enzymes (hydrolases), e.g., esterases, peptidases, and amidases, catalyze the cleavage of ester linkage.

"Adding and removing" enzymes (desmolases), e.g., dehydrases, isomerases, and aldolases, catalyze the addition or removal of some groups of a substrate, without hydrolysis, oxidation, or reduction.

Transferring enzymes (transferases), e.g., transglycosylases, transaminases, and transphosphorylases, catalyze the transfer or shift of a group from one molecule to another.

Oxidizing and reducing enzymes (oxireductases), e.g., oxidases, aerobic and anaerobic dehydrogenases, and transelectronases (cytochromes), catalyze biological oxidations.

Denaturation is one of the most characteristic properties of proteins. The most important factors governing the rate of denaturation are the temperature and pH of the solution. Chemical reagents that accelerate denaturation are urea, guanidine salts and synthetic detergents, and organic solvents such as alcohol and acetone. Physical agents are heat, X-rays, ultraviolet light, and high pressure.

Protein may be precipitated from solution by a variety of cations (Zn^{2+}, Cd^{2+}, Hg^{2+}, Fe^{3+}, Cu^{2+}, and Pb^{2+}) and anions (of tungstic, phosphotungstic, trichloroacetic, picric, and tannic acids).

The molecular weights of proteins vary over a wide range: lactalbumin, about 17,000; gliadin, 27,000; insulin, 38,000; egg albumin, 42,000; hemoglobin, 68,000; serum globulin, 170,000; edestin, 300,000; and hemocyanin, 6,600,000. The molecular weights of proteins have been determined by means of osmotic pressure, light and X-ray scattering, viscosity, diffusion, sedimentation in an ultracentrifuge, and by chemical analysis.

The proteins may be classified as follows:

Simple proteins are those that yield only amino acids upon hydrolysis.

1. Albumins are soluble in water and are coagulated by heat. Typical examples are egg albumin and lactalbumin.
2. Globulins are insoluble in water but are soluble in dilute salt solution. They are precipitated by half-saturating their solution with ammonium sulfate, and are coagulated by heat. Typical examples are ovoglobulin, myosin, arachin, amandin, and serum globulin. Immunoproteins belong to the γ-globulins and are called antibodies. Their function is to protect the mammalian body against pathogenic invaders—foreign proteins, bacteria, or viruses, which are known as antigens. When antigens penetrate into an organism, the latter responds by forming antibodies that combine specifically with the antigen. The combination of large antigen and antibody molecules can be seen under the electron microscope.
3. Glutelins are soluble in dilute acids and alkalis, and insoluble in neutral solvents. Typical examples are glutenin of wheat and oryzenin of rice.
4. Prolamines are soluble in 70 to 80% alcohol, and insoluble in water, dilute salt solution, or absolute alcohol. Typical examples are zein of corn, gliadin of wheat, and hordein of barley.

5. Scleroproteins are insoluble in water or salt solution, but are soluble in strong acids or alkalis. Typical examples are keratins of hair, horns, and hoofs, elastin of connective tissues, collagen of bones, fibroin and sericin of silk.
6. Histones are soluble in water and insoluble in dilute ammonia. They are readily soluble in dilute acids and alkalis. Typical examples are globin, thymus histone, and gadus histone of codfish sperm.
7. Protamines are the simplest of the proteins. They are basic and yield mainly basic amino acids upon hydrolysis. The protamines are soluble in water, dilute ammonia, acids, and alkalis. Typical examples are salmine of salmon sperm, clupeine of herring sperm, and cyprinine of carp sperm.

Conjugated proteins are composed of a simple protein combined with a non-protein group, known as the prosthetic group.

1. Chromoproteins are composed of proteins united with colored prosthetic groups such as hemoglobin or chlorophyll. Other examples are cytochromes and flavoproteins.
2. Lipoproteins are formed by combination of proteins with lipids such as lecithin or fatty acids. They are found in blood, milk, egg yolk, and the chloroplasts.
3. Metalloproteins are proteins that contain heavy metals such as Fe, Co, Mn, Zn, Cu, Mg, etc. Many enzymes belong to this group.
4. Mucoproteins are composed of proteins and mucopolysaccharides. They are found in serum, human urine, and albumin.
5. Nucleoproteins are composed of proteins and nucleic acids. The plant viruses and many of the animal viruses are ribonucleoproteins, since they consist chiefly of protein and RNA. The tobacco mosaic virus is the best known nucleoprotein. Its molecular weight is about 40 million, while that of the influenza virus is 350 million, and that of the smallpox virus, several billion.
6. Phosphoproteins contain phosphoric acid. They occur in casein and egg yolk.

Various methods have been developed for determining the sequence of amino acids in peptides and proteins. Thus, the N-terminal amino acid sequence can be determined by treatment with 2,4-dinitrofluorobenzene or phenylisocyanate, and the C-terminal amino acids can be determined by the formation of a thiohydantoin ring or, preferably, by enzymic cleavage with carboxypeptidase.

Proteins can be resolved into a mixture of long peptides by controlled digestion with certain enzymes, e.g., trypsin.

Pauling and Corey have suggested that proteins are arranged in a helix with 3.7 amino acid residues per turn. Each turn corresponds to a length of

5.4Å in the direction of the long axis of the helix, and each amino acid residue advances the chain in the direction of the long axis of the helix by 1.5Å. This helix has been designated the α-helix.

Keratins are the structural proteins of hair, wool, feathers, horn, nails, and hoofs. Keratin from human hair or wool contains 11–12% cystine; hence its great stability, which is due to the large number of —S—S— cross-links between the peptide chains. Reduction of keratin forms keratin which contains cysteine residues; this process can be reversed by oxidation of keratin. Temporary reduction of hair by thioglycol and subsequent oxidation by the oxygen of the air are principles underlying the cosmetic production of permanent waves. The length of keratin fibers depends on their water content; thus when α-keratin is treated with water, the stretched β-keratin is formed. A pleated-sheet structure for β-keratin has been put forward by Pauling and Corey.

Stretching of wool forms parallel peptide chains which are linked to each other by numerous hydrogen bonds. Wool shrinks on heating in water. Since the —S—S— cross-links are not cleaved by heating, this reaction is attributed to the cleavage of other interchain bonds.

Protein biosynthesis is at present a central interest of biochemists and biophysicists.

B. Isolation of Lysozyme from Albumen

1. Introduction

Lysozyme is one of the smallest proteins. It has the ability to break down (lyse) the cells of certain bacteria, whence its name.

Lysozyme has been obtained from albumen as the insoluble flavianate by procedures involving solvent fractionation and precipitation (1,2). More recently it was isolated by adsorption on bentonite and elution with 5% pyridine solution adjusted to pH 5.3 (3), by direct crystallization at pH 9.0–9.5 (4), by use of an ion-exchange resin (5), and by chromatography on hydroxyapatite (calcium phosphate adsorbent) (6).

Lysozyme has a molecular weight of approximately 17,500 and is a basic protein with an isoelectric point of 10.5 to 11.0. It is regarded as an antibiotic since it is able to kill microorganisms and even to dissolve living bacterial bodies. Lysozyme is found in various animal species and is also produced by bacteria, e.g., *Bacillus subtilis* and *Staphylococcus aureus*.

2. Principle

In the following procedure (7), lysozyme is adsorbed on bentonite together with other albumen globulins. The contaminating proteins are removed from the bentonite by washing successively with alkaline phosphate and 5% aqueous pyridine. The lysozyme is then eluted from the bentonite with 5% aqueous pyridine adjusted to pH 0.5 with sulfuric acid. The eluate is dialyzed. Amorphous lysozyme is obtained by lyophilization of the solution.

3. Apparatus

   Centrifuge
   Blender

4. Materials

   Bentonite   Pyridine
   Eggs   Sodium chloride
   Phosphate buffer (pH 7.5)   Sodium hydrogen carbonate
   Potassium chloride

5. Time

   Adsorption and elution: 2–3 hours
   Dialysis: 24 hours

6. Procedure

To 300 ml homogenized albumen (one dozen eggs of average size) are added 50 ml 10% bentonite suspension in 1% potassium chloride. The mixture is stirred vigorously for 3 to 5 minutes until a smooth suspension is obtained. The clay is separated by centrifugation and washed twice with 100-ml portions of 0.5 M phosphate buffer (pH 7.5) and three times with 150-ml portions of 5% aqueous pyridine. The clay is separated by centrifugation after each washing, and the supernatants, containing inactive proteins, are discarded. Lysozyme is eluted by washing the clay twice with 100-ml portions of 5% aqueous pyridine solution, which has been adjusted to pH 5 with sulfuric acid. The eluates are dialyzed against running tap water until no odor of pyridine remains. The volume increases considerably during the dialysis, which lasts for 24 hours. The pH of the final solution is approximately 6. Amorphous, but essentially pure lysozyme is obtained by lyophillization of the solution. The yield is 0.7–0.8 g. Crystalline lysozyme or its salts are obtained from the amorphous material by one of the following procedures:

1. To 10 ml 5% solution of amorphous lysozyme is added 0.5 g sodium hydrogen carbonate (final pH 8.0–8.5), and the solution is allowed to stand at room temperature until lysozyme carbonate crystallizes.
2. To 10 ml 5% solution of amorphous lysozyme is added 0.5 g sodium chloride, and the pH is adjusted to 9.5 to 10.0 with sodium hydroxide.

The solution is stored at 4°C. Crystalline isoelectric lysozyme separates.

3. A 5% solution of amorphous lysozyme is adjusted to pH 4.5 with hydrochloric acid, and 5% solid sodium chloride is added. The solution is stored at 4°C, and crystalline lysozyme chloride is deposited within 4 to 5 days.
4. A 2% solution of isoelectric lysozyme in hydrobromic acid is adjusted to pH 6.0, and 5% solid potassium bromide is added. On standing at room temperature, the solution deposits crystals of lysozyme bromide.

a. IR Spectrum of lysozyme

3440 cm^{-1}: hydroxylamino groups
3076: aromatic C—H vibration of the tyrosine group or overtone of the amide II band
1652: C=O stretching of —CO—NH—
1527: amide II, N—H deformation
1438, 1390: methylene and methyl bending
1290, 1242: C—N stretching in secondary amides
1098: aliphatic hydroxy amino acid

References

1. Wolff, L. K. (1927). *Z. Immunitatforsch.* **50**, 88.
2. Meyer, K., Thompson, R., Palmer, J., and Korazo, D. (1936). *J. Biol. Chem.* **113**, 303.
3. Alderton, G., Ward, W. H., and Fevold, H. L. (1945). *J. Biol. Chem.* **157**, 43.
4. Alderton, G., and Fevold, H. L. (1946). *J. Biol. Chem.* **164**, 1.
5. Tallan, H. H., and Stein, W. H. (1953). *J. Biol. Chem.* **200**, 507.
6. Tiselius, A., Hjerten, S., and Levin, O. (1956). *Arch. Biochem. Biophys.* **65**, 132.
7. Fevold, H. L., and Alderton, G. (1949). *Biochem. Prepns.* **1**, 67.

Recommended Review

1. Fevold, H. L. (1951). Egg proteins. In "Advances in Protein Chemistry," (M. L. Anson, J. T. Edsall, and K. Bailey, eds.), Vol. 6, p. 187. Academic Press, New York.

C. Isolation and Separation of Proteins of Groundnuts

1. Introduction

The proteins of groundnuts were first investigated by Ritthausen in 1880. He extracted the proteins from the oil-free groundnut meal with aqueous sodium chloride and weakly basic solutions, and precipitated them by acidification. He considered the solids he obtained to be a homogeneous compound. The investigations of Johns and Jones (1) have indicated that the total protein of groundnuts consists of globulins and a very small amount of heat-coagulable albumin. By means of ammonium sulfate fractionation they separated the globulins into two fractions, arachin and conarachin. Later, Jones and Horn (2) stated that arachin could be prepared from 10% sodium chloride extract by dilution until the extract became cloudy, followed by saturation with CO_2 or by addition of two volumes of saturated ammonium sulfate to three volumes of the extract. Conarachin could be isolated by complete saturation with ammonium sulfate.

Irving, Fontaine, and Warner (3) carried out electrophoretic analyses of groundnut meal and concluded that it contained at least three, and probably four, components. Similar observations were made by Karon, Adams, and Altschul (4).

2. Principle

The method described below is based on (5).

3. Apparatus

4. Materials

Groundnuts
Petroleum ether
Sodium chloride
Sodium hyroxide
Sulfuric acid

5. Time

2–3 hours

6. Procedure

a. Isolation of arachin and conarachin

Fifty g blanched groundnuts and 200 ml light petroleum are refluxed for 1 hr and filtered in order to expel the oil. The solid residue is dried in air and then extracted with 0.1% sodium hydroxide solution. The solution of sodium proteinate is filtered off, and the protein is precipitated by addition of acid or SO_2 until the isoelectric range (around 5) is reached. The protein is allowed to settle, filtered and washed free from salt and is then dried.

Arachin and conarachin can be prepared by two alternative methods.

1. Dried protein (4 g) is shaken with 100 ml 2% aqueous sodium chloride to give a solution of conarachin. The residue is extracted with 100 ml 10% aqueous sodium chloride to give a solution of arachin and an insoluble residue.
2. Protein (4 g) is extracted with 100 ml 10% aqueous sodium chloride to give a solution of arachin and conarachin, and an insoluble solid. The solution is diluted with four times its volume of water in order to precipitate arachin; conarachin remains in solution. Conarachin may be precipitated from the aqueous solution by heating.

b. Paper chromatography

The round-paper chromatographic technique is used for the separation and identification of arachin and conarachin: one drop of a solution of groundnut protein in 10% aqueous sodium chloride is placed in the center of the paper. Aqueous sodium chloride, 10%, serving as the developing solvent, is placed in an uncovered petri dish. Thus, the water from the paper chromatograph evaporates and the salt solution spreads upward. After about 15 min, the chromatography is complete. The paper is dried and immersed in a 0.05% solution of Solway purple containing 0.5% sulfuric acid in a photographic developing dish at a temperature of 50 to 90°C. After rocking the dish for 5 min, the dye solution is poured off, and the paper is washed in several changes of warm water and finally dried. The proteins show up as violet-colored patches on the chromatograms. The front line is identified as conarachin, and the rear line as arachin.

References

1. Johns, C. O., and Jones, D. B. (1916). *J. Biol. Chem.* **28**, 77.
2. Jones, D. B., and Horn, M. J. (1930). *J. Agric. Res.* **40**, 672.
3. Irving, G. W., Fontaine, T. D., and Warner, R. C. (1945). *Arch. Biochem.* **7**, 475.
4. Karon, M. L., Adams, M. E., and Altschul, A. M. (1950). *J. Phys. Colloid Chem.* **54**, 56.
5. Thomson, R. H. K. (1952). *Biochem. J.* **51**, 118.

Recommended Reviews

1. Pace, J. (1955). *Seed proteins. In "Modern Methods of Plant Analysis."* (K. Paech, and M. V. Tracey, eds.), Vol. 4, p. 69. Springer-Verlag, New York.
2. Natarajan, K. R. (1980). Peanut protein ingredients: Preparation, properties, and food uses. *In* "Advances in Food Research," Vol. 26, p. 216. Academic Press, New York.

D. Thin-Layer Chromatography of Proteins

1. Introduction

The determination of the molecular weights of proteins by means of gel filtration on cross-linked dextran gels (Sephadex) has been described by a number of authors. Whitaker (1) has covered the molecular weight range 13,000–40,200 using Sephadex G-75, and the range 13,000–76,000 using Sephadex G-100. Wieland, Duesberg, and Determann (2) have used Sephadex G-200 for proteins of molecular weight 13,000–150,000.

The advantages of thin-layer chromatography—speed and adaptability to very small samples—make it very attractive for the fractionation of proteins on the ultramicro scale. The availability of the loosely cross-linked beadform materials—Sephadex G-100 and G-200—has extended the upper molecular weight limit for the successful chromatography of proteins to at least 180,000 (3).

2. Principle

The procedure outlined below is according to (3).

3. Apparatus

Glass plates
Plastic box
Thin-layer applicator

4. Materials

Naphthalene Black 12B   Sephadex G-100 and G-200
Nigrosin   Sodium chloride
Proteins (see procedure)   Sodium phosphate

6. Procedure

Sephadex gel powder (G-100) is distended in a suitable buffer for an appropriate time. Each gram of Sephadex G-100, together with 19 ml buffer solution (0.02 M sodium phosphate containing 0.2 M sodium chloride), is allowed to swell for 24 hr at pH 7. For each gram of G-200, 25 ml buffer is added and left for 24 to 48 hr. The gel is then spread over glass plates to a thickness of 0.5 mm with an applicator. Glass plates of dimensions 20 × 30 and 20 × 40 cm are commonly used. The coated plates are placed in a flat plastic box and leant against a buffer reservoir, which is placed on a small table of adjustable height for ready variation of the inclination of the plate. A sheet of thick filter paper (Whatman 3 MM) is applied to the upper end of the plate via a slit in the box. It ensures contact between the gel layer and the buffer, providing for a flow of liquid through the gel. It is advisable to let the liquid flow through the plates for at least 1 hr before the run. The samples are then applied with capillary pipettes as round spots 2–4 mm in diameter, corresponding to sample volumes of about 1–5 μl.

a. Development
Development is performed with the above-mentioned buffer solution (4–15 hr).

b. Detection
Upon completion of the chromatographic run, a damp sheet of Whatman No. 1 paper is deposited on the Sephadex layer so as to achieve perfect adherence between paper and gel. Under these conditions some of the chromatographed material passes from the Sephadex gel to the paper. After 30 to 40 min, the paper is carefully removed from the layer, dried, and sprayed with 1% Naphthalene Black 12B in methanol–water–glacial acetic acid, 50:40:10.

c. Recovery

The gel corresponding to the zones of the layer that, according to the detecting paper, contain the separated substances, is removed from the plate with a spatula, placed in a small centrifuge tube, and suspended in 5 ml water or some other suitable solvent. The test tubes are then centrifuged. Most of the analyzed substance is located in the supernatant. The proteins in the following table can be separated by this technique.

Proteins Separated by TLC

Protein	Molecular Weight $\times 10^{-3}$
Cytochrome C	13.0
Ribonuclease	13.6
Lysozyme	14.5
Myoglobin	16.9
α-Chymotrypsin	22.5
Trypsin	23.8
Ovomucoid	27.0
Pepsin	35.0
Ovalbumin	45.0
Hemoglobin	68.0
Bovin serum albumin	165.0
Bovin γ-globulin	180.0
Thyroglobulin	650.0
Macroglobulins	1000.0

References

1. Whitaker, J. R. (1963). *Anal. Chem.* **35**, 1950.
2. Weiland, T., Duesberg, P., and Determann, H. (1963). *Biochem. Z.* **337**, 303.
3. Morris, C. J. O. R. (1964). *J. Chromatogr.* **16**, 167.

Recommended Review

1. Fasella, P., Giartosio, A., and Turano, C. (1964). *Applications of TLC on Sephadex to the study of proteins.* In "Thin-layer Chromatography," (G. B. Marini Bettolo, ed.), p. 205. Elsevier, New York.

E. Questions

1. Define and describe a gene, a virus, and an enzyme.
2. Define the following terms: a. conjugated protein; b. derived protein; c. prosthetic group.
3. Distinguish between fibrous and globular proteins with regard to composition, structure, sources, and physical characteristics.

4. Give the name and the source of a. three proteins that are enzymes; b. one protein that is a hormone.
5. Which linkage is responsible for the crosss-linking of protein chains? Which is responsible for the helical conformation?
6. What is Edman degradation?
7. Describe three methods for determining the molecular weight of proteins.
8. Name some industrial uses of proteins.
9. What is the relationship between DNA, RNA, and protein biosynthesis?

F. Recommended Books

1. Bailey, J. L. (1962). *"Techniques in Protein Chemistry."* Elsevier, New York.
2. Fox, S. W., and Foster, J. F. (1957). *"Introduction to Protein Chemistry."* John Wiley, New York.
3. Harris, R. J. C. (ed.) (1961). *"Protein Biosynthesis."* Academic Press, New York.
4. Haurowitz, F. (1963). *"The Chemistry and Function of Proteins."* Academic Press, New York.
5. Neurath, H., and Baily, K. (eds.) (1954). *"The Proteins."* Academic Press, New York.
6. Ramachandran, G. N. (ed.) (1963). *"Aspects of Protein Structure."* Academic Press, New York.
7. Scopes, R. K. (1987). *"Protein Purification: Principles and Practice."* Springer-Verlag, New York.
8. Springall, H. D. (1954). *"The Structural Chemistry of Proteins."* Academic Press, New York.
9. Watson, J. D. (1965). *"Molecular Biology of the Gene."* Benjamin, New York.
10. Wiseman, A. (1965). *"Organization for Protein Biosynthesis."* Blackwell Scientific Publications, Oxford, England.

VI. PTERIDINES

A. Introduction

Pteridines are the pigments found in the wings of insects and in the eyes and skin of fish, amphibia, and reptiles. The pteridine system is a combination of a pyrazine ring and a pyrimidine ring.

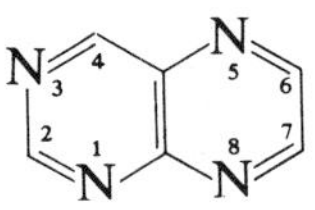

Many of the naturally occurring pteridines have a 2-amino group and a 4-hydroxy group.

Xanthopterin, isoxanthopterin, leucopterin, and erythropterin are the butterfly wing pigments; xanthopterin is also found in wasp wings, and in human urine, liver, kidneys, spleen, and pancreas.

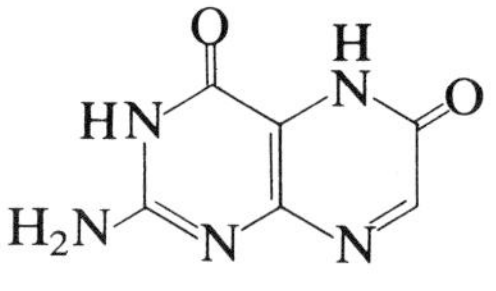

Xanthopterin

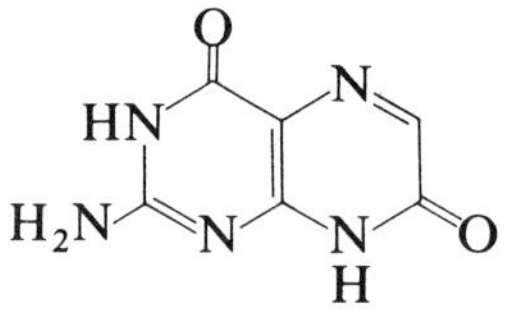

Isoxanthopterin

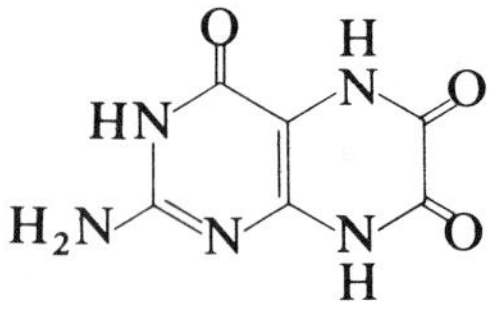

Leucopterin

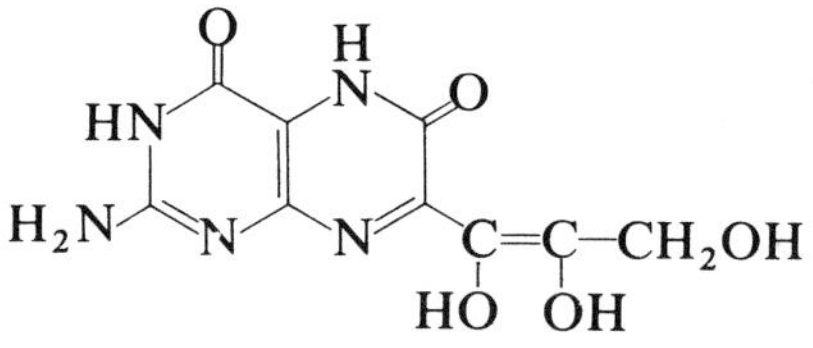

Erythropterin

Pterorhodin has been found in butterfly wings and in the eyes of *Ephestia* and *Ptychopoda*.

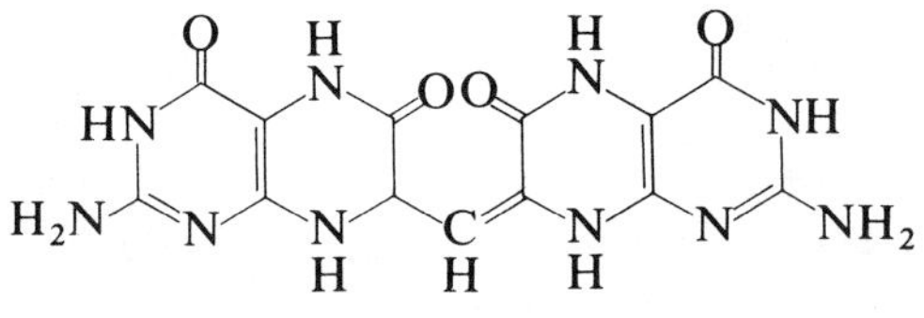

Pterorhodin

The yellow pteridines (sepiapterin and isosepiapterin) and three red pigments (drosopterin, isodrosopterin, and neodrosopterin) have been isolated from the fruit fly *Drosophila melanogaster*.

A widely occurring natural pterin is biopterin, which has been isolated from human urine, from the fruit fly, from the Mediterranean flour moth, and from the royal jelly of bees.

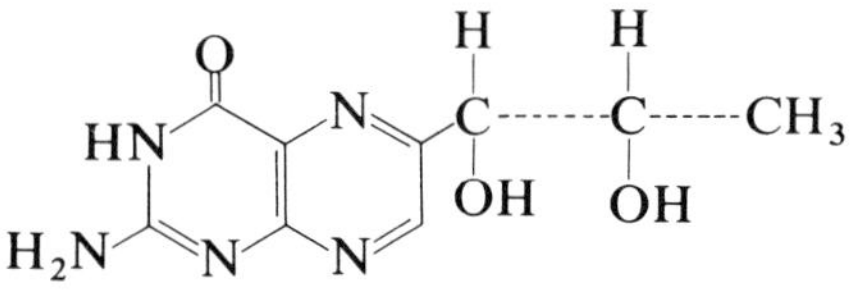

Biopterin

Folic acid, a vitamin, has a pteridine moiety in its molecule.

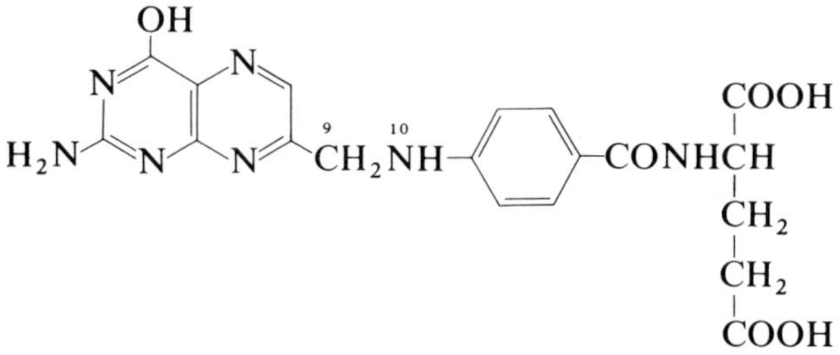

Folic acid

The two methods for the synthesis of pteridines start either from the pyrazine or the pyrimidine half of the pteridine molecule.

1. Synthesis from Pyrimidines

Since pyrimidines are relatively readily accessible, the Isay reaction, i.e., the condensation of a 4,5-diaminopyrimidine with a 1,2-dicarbonyl compound, continues to be of major importance:

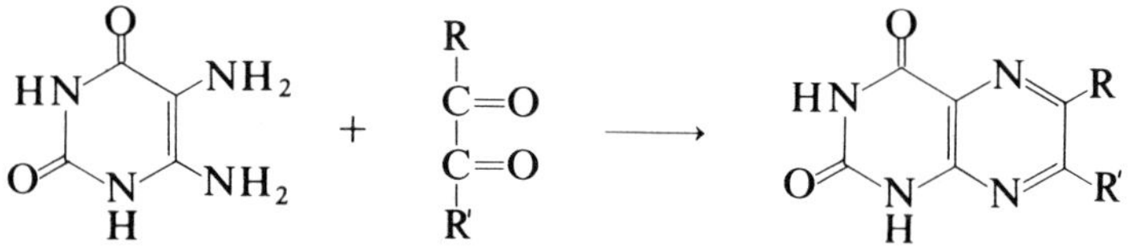

4-Amino-5-nitrosopyrimidines can be condensed not only with ketones and aldehydes, but also with nitriles and esters that have an activated methylene group adjacent to the functional group.

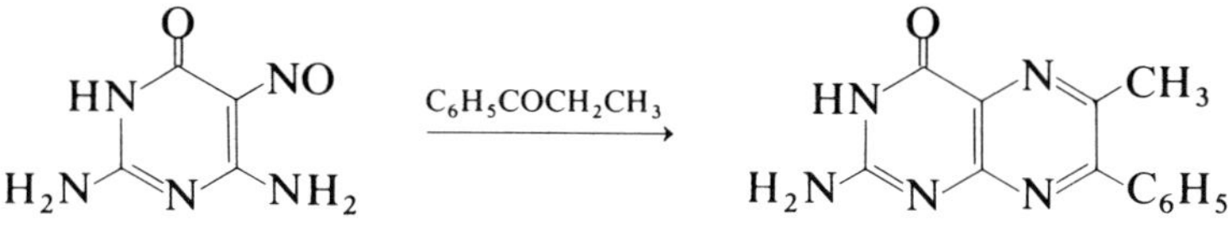

2. Synthesis from Pyrazines

Pteridine syntheses based on pyrazine derivatives have been carried out only rarely because such derivatives are not readily accessible.

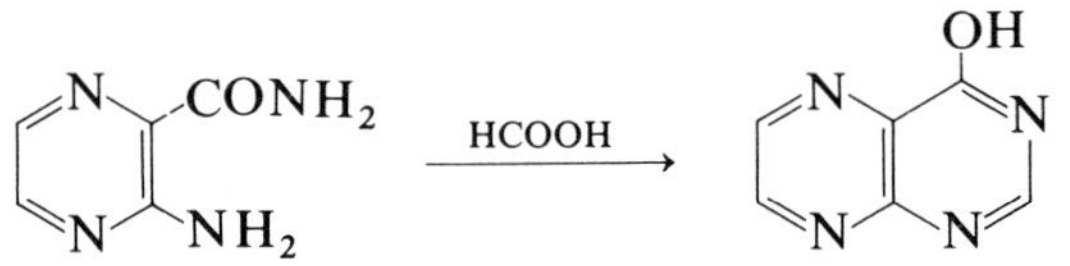

$$\underset{\text{}}{\text{pyrazine-CONH}_2\text{-NH}_2} \xrightarrow{\text{HCOOH}} \text{pteridine-OH}$$

Since the majority of pteridines are insoluble in most solvents and possess very high melting points, the most reliable physical properties reflecting homogeneity are the R_f values (obtained by paper or thin-layer chromatography) and ultraviolet spectra.

Recent chemical and microbiological research has indicated that the pteridine ring arises from a purine nucleus.

B. Isolation of Pteridines from the Fruit Fly, *Drosophila melanogaster*

1. Introduction

The yellow to red pterins from the fruit fly *Drosophila melanogaster* constitute an important group of natural pteridine derivatives.

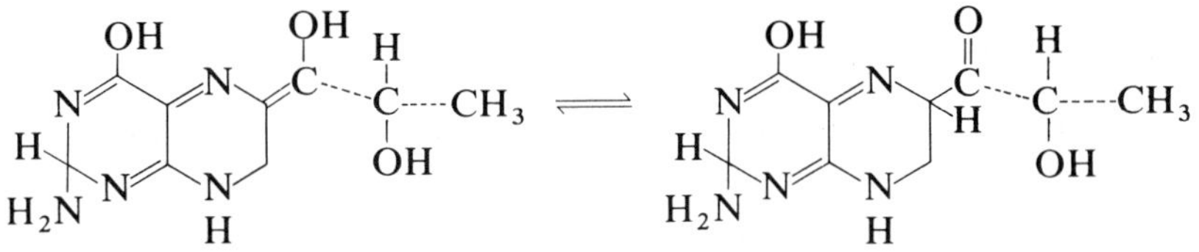

Sepiapterin

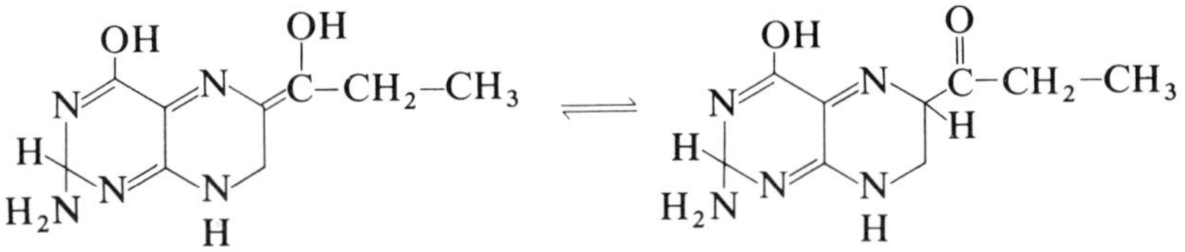

Isosepiapterin

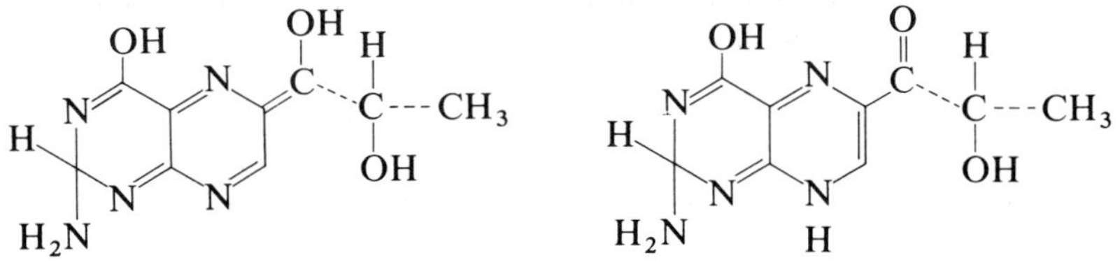

Drosopterin   Neodrosopterin

VI. Pteridines

Viscontini and his coworkers (1, 2) have shown by means of chromatography that the *Drosophila* pterins consist of the yellow compounds, sepiapterin and isosepiapterin, as well as the orange to red compounds, drosopterin, isodrosopterin, and neodrosopterin. Sepiapterin was isolated in the form of yellow crystals from *D. melanogaster*, the sepia mutant of which accumulates large concentrations of this pigment (3). It is also found in amphibian skin (4), in fish (5), and in the epidermis of the lem mutant of the silk moth *Bombyx mori* (6).

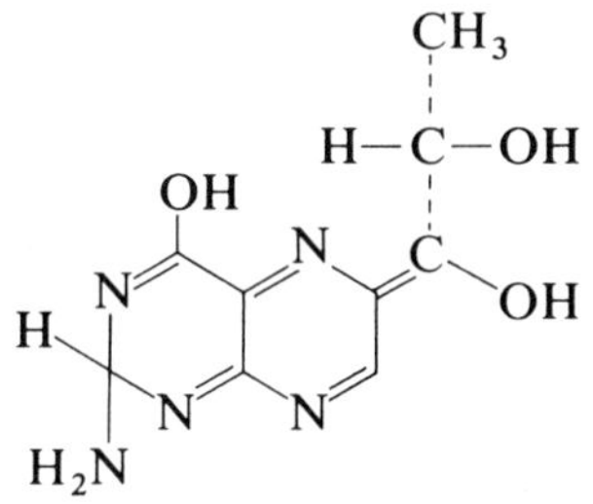

Isodrosopterin

Isosepiapterin was isolated in crystalline form for the first time from the sepia mutant of *Drosophila* (1). Larger concentrations of isosepiapterin were found in the blue-green alga *Anacystis nidulans* (7). Viscontini (2) was the first to separate the water-soluble, red eye-pigments of Drosophila into drosopterin, isodrosopterin, and neodrosopterin. These pigments also occur in the skin of some fish (8) and amphibia (9), as well as in the skin folds of various reptiles (10).

2. Principle

The procedure outlined below is based on (1) and (2).

3. Apparatus

Chromatographic column
Blender

4. Materials

Acetic acid
Ammonium acetate
Ammonium chloride
Ammodium hydroxide
Butanol
Cellulose powder
Ethyl acetate

Methanol
Propanol
Pyridine
Sepia flies (can be prepared according to Demerec, M., *The Biology of Drosophila*, John Wiley, New York, 1950).

5. Time

4–5 hours

6. Procedure

a. Isolation of the yellow pigments

Sepia flies (200 g) are macerated in a blender in 500 ml ethanol and 200 g cellulose powder. The suspension is packed into a chromatographic column (10 × 12 cm) on top of cellulose powder, and elution is immediately commenced with propanol–1% aqueous ammonium acetate, 1:1. The yellow pigment is washed from the column, and the eluates are evaporated to 30 to 50 ml under reduced pressure. This solution is applied to a column (10 × 60 cm) of cellulose packed in the same solvent. A blue pigment is eluted first, followed by a yellow zone which is eluted with butanol–acetic acid–water, 20:3:3, and contains isosepiapterin, sepiapterin, and riboflavin.

The yellow solution should be immediately neutralized to prevent decom-position of the pigments. It is then rechromatographed on a cellulose-packed column, eluting with water. The eluates are concentrated almost to dryness, leaving long, yellow needles. Paper chromatography is carried out with four solvents:

1. Butanol–acetic acid–water, 20:3:3.
2. Propanol–1% ammonium hydroxide, 2:1.
3. Ammonium chloride, 1% aqueous.
4. Propanol–2% ammonium acetate, 1:1.

Retention Characteristics of Yellow Pigments

Pteridine	R_f in Solvent			
	1	2	3	4
Sepiapterin	0.22	0.42	0.33	0.60
Isosepiapterin	0.43	0.52	0.25	0.65

b. Isolation of red pigments

Cellulose powder (400 g) and 400 g blenderized flies are thoroughly mixed with 50 ml 80% methanol. The slurry is packed into a chromatographic column (18 × 20 cm) containing 100 g cellulose powder and is eluted with 80% methanol until the red pigments are adsorbed on the upper part of the column. Elution is continued with 0.1% aqueous ammonium acetate solution until three distinct zones are formed; these are then eluted separately. Each eluate is concentrated to about 100 ml *in vacuo* and rechromatographed on 9 × 25 cm columns packed with cellulose, eluting first with distilled water to remove impurities. The pigments are rapidly eluted with 0.5% aqueous ammonium hydroxide. The solutions are concentrated under reduced pressure until turbidity, three volumes ethanol and four volumes ether are added; the mixture is left in the cold for 24 hr. The yields are 2 mg isodrosopterin, 15 mg drosopterin, and 2 mg neodrosopterin. Paper chromatography is carried out with four solvents:

1. Butanol–acetic acid–water, 20:3:7.
2. Propanol–1% ammonium hydroxide, 2:1.
3. Pyridine–ethyl acetate–water, 4:4:3.
4. Propanol–2% aqueous ammonium acetate, 1:1.

Retention Characteristics of Red Pigments

Pteridine	R_f in Solvent			
	1	2	3	4
Neodrosopterin	0.05	0.07	0.04	0.11
Drosopterin	0.10	0.14	0.10	0.19
Isodrosopterin	0.11	0.14	0.10	0.25

References

1. Viscontini, M., and Mohlamann, E., (1959). *Helv. Chim. Acta* **42**, 836.
2. Viscontini, M., Hadorn, E. and Karrer, P. (1957). *Helv. Chim. Acta* **40**, 579.
3. Forrest, H. S., and Mitchell, H. K. (1958). *J. Am. Chem. Soc.* **76**, 5656.
4. Hama, T., and Obika, M. (1960). *Nature* **187**, 326.
5. Hama, T., Matsumoto, J., and Mori, Y. (1960). *Proc. Imp. Acad. (Tokyo)* **36**, 346.
6. Nawa, S., and Taira, T. (1954). *Proc. Imp. Acad. (Tokyo)* **30**, 632.
7. Forrest, H. S., van Baalen, C., and Myers, J. (1959). *Arch. Biochem. Biophys.* **83**, 508.
8. Kauffmann, T. (1959). *Z. Naturforsch* **14b**, 358.
9. Hama, T., (1963). "Third Intern. Symp. on Pteridines." Stuttgart, 1962. Pergamon Press, London, England.
10. Ortiz, E., Throckmorton, L. H., and Williams-Ashman, G. H. (1962). *Nature* **196**, 596.

Recommended Reviews

1. Forrest, H. S. (1962). Pteridines, structure and metabolism. *In* "Comparative Biochemistry," (M. Florkin, and H. S. Mason, eds.), Vol. 4, p. 615. Academic Press, New York.
2. Viscontini, M. (1964). The structure of sepiapterin and drosopterin. *In* "Pteridine Chemistry," (W. Pfleiderer, and E. C. Taylor, eds.), Proceedings of the Third International Symposium, Stuttgart, 1962. Pergamon Press, London, England.

C. Synthesis of Pteridines

1. Introduction

Pteridines can be synthesized by two main pathways, namely, fusion of a pyrimidine ring onto a pyrazine derivative, or fusion of a pyrazine ring onto a pyrimidine derivative.

Since pyrimidines are relatively readily accessible (1), the Isay reaction, i.e., condensation of a 4,5-diaminopyrimidine with a 1,2-dicarbonyl compound, continues to be of major importance (2). The dicarbonyl compounds that have been used include a dialdehyde (glyoxal), aldehydo-ketones (phenylglyoxal), diketones (diacetyl (3), benzil (4), 13 higher homologs of dipropionyl (5), an aldehydo acid (glyoxylic acid) (6), and keto acids such as pyruvic acid (7) and dibasic acid (8). The Isay reaction can be carried out under neutral, acid, and alkaline conditions; neutral condensation usually gives the best yields. When the dicarbonyl compound in Isay's reaction is not symmetrical, two isomers can arise (9).

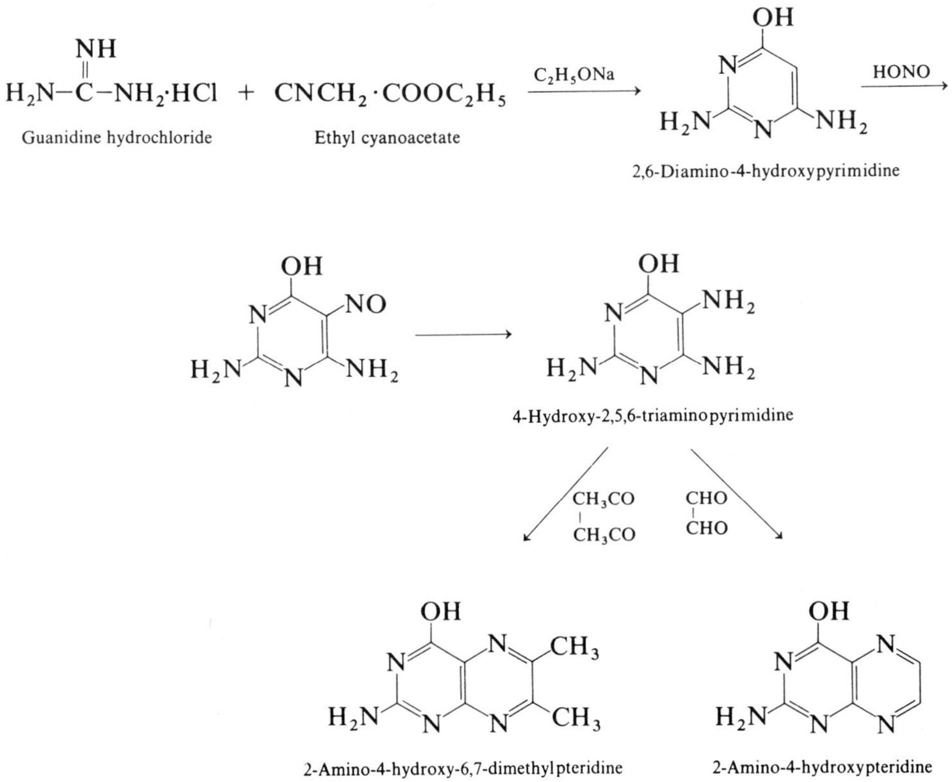

2. Principle

The following procedure (10) involves the use of 4-hydroxy-2, 5, 6-triaminopyridine, prepared by condensation of guanidine and ethyl cyanoacetate to form the 2, 6-diamino-4-hydroxypyrimidine, which is nitrosated to the 5-nitroso derivative. The nitroso group is reduced by sodium hydrosulfite to the desired substituted pyrimidine. The latter is converted into 2-amino-4-hydroxy-6, 7-dimethylpteridine by treatment with biacetyl; the reaction with glyoxal leads to the formation of 2-amino-4-hyroxypteridine.

3. Apparatus

4. Materials

Biacetyl	Hydrochloric acid
Ethanol, absolute	Sodium, metal
Ethyl cyanoacetate	Sodium bisulfite
Formic acid	Sodium hydrosulfite
Glyoxal	Sodium hydroxide
Guanidine hydrochloride	Sodium nitrite

5. Time

5–6 hours

6. Procedure

a. 4-Hydroxy-2,5,6-triaminopyrimidine bisulfite

To a solution of sodium ethoxide, prepared by dissolving 3.5 g sodium in 150 ml absolute ethanol, are added 10 g guanidine hydrochloride and 13.5 g ethyl cyanoacetate. The mixture is refluxed for 3 hr. Dilution with 100 ml water and acidification with 10 ml concentrated hydrochloric acid liberate 2,6-diamino-4-hydroxypyrimidine, which is nitrosated by slow addition, with stirring, of 10 g sodium nitrite dissolved in 30 ml water. After heating to boiling, the mixture is allowed to cool. The crystalline, bright rose 5-nitroso derivative is filtered and washed well with water. The material is then suspended in 150 ml water and, after addition of 8 ml 20% sodium hydroxide, is heated to 70 to 80°C. Reduction of the nitroso group is effected by the addition of 50 g sodium hydrosulfite over a period of 15 min, with vigorous stirring. Heating and stirring are continued for 30 min, during which period the color is almost completely discharged. The solution is then heated to the boiling point and filtered through a Buchner funnel. The light yellow product crystallizes immediately. It is filtered and washed with cold water; yield, 20 g. The product may be crystallized from two parts of water.

b. 2-Amino-4-hydroxypteridine

A solution of 15 g glyoxal sodium bisulfite in 50 ml hot water is added, with stirring, to a solution of 10 g 4-hydroxy-2,5,6-triaminopyrimidine bisulfite in 50 ml hot water. The resulting clear, yellow solution soon begins to deposit a light yellow solid, which is heated for 2 hr on a steam bath. After cooling, the solid is collected by filtration, washed with water, and dried. The yield is 4 g. The compound is recrystallized as yellow microcrystals by dissolving it in boiling formic acid and adding water to incipient precipitation.

c. UV spectrum of 2-amino-4-hydroxypteridine

$\lambda_{max}^{0,1N\,NaOH}$ 255mμ (logε 4.20); 358 (3.82)

Pteridine is a naphthalenoid, but differs from naphthalene in that it possesses some unshared electrons, namely those located on the nitrogen atoms of the nucleus. (Naphthalene main absorptions in ethanol are 275 mμ (log ε 3.93) and 310 (2.56)). 2-Amino-4-hydroxypteridine has an acidic hydroxyl group (like that of phenol) and an amino group, so that the actual constitution of the molecule depends on the pH of the medium. In alkali, the structure of the compound is that of the anion:

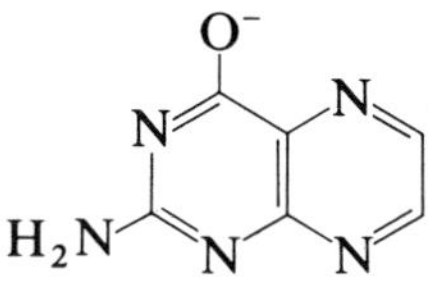

d. 2-Amino-4-hydroxy-6,7-dimethylpteridine

To a solution of 10 g 4-hydroxy-2,5,6-triaminopyrimidine bisulfite in 50 ml hot water is added 10 g biacetyl. The condensation product begins to separate immediately. In order to complete the reaction, the mixture is heated on the steam bath for 2 hr. After cooling, the solid material is collected by filtration and washed with water and then with 95% ethanol; yield, 6 g. Recrystallization from 0.5N hydrochloric acid gives yellow microcrystalline rods.

e. UV spectrum of 2-amino-4-hydroxy-6,7-dimethylpteridine

$\lambda_{max}^{0,1N\,NaOH}$ 250mμ (logε 4.34); 355 (3.94)

Note that the methyl substitution has practically no effect on the absorption.

References

1. Brown, D. J. (1952). *J. Appl. Chem.* **2**, 239.
2. Isay, O. (1906). *Ber.* **39**, 250.
3. Kuhn, R., and Cook, A. H. (1937). *Ber.* **70**, 761.
4. Cain, C. K., Taylor, E. C., and Daniel, L. J. (1949). *J. Am. Chem. Soc.* **71**, 892.

5. Campbell, N. R., Dunsmuir, J. H., and Fitzgerald, M. E. H. (1950). *J. Chem. Soc.* 2743.
6. Koschara, W. (1943). *Z. Physiol. Chem.* **277**, 159.
7. Elion, G., and Hitchings, G. H. *J. Am. Chem. Soc.* **69**, 2553.
8. Forrest, H. S., Hull, R., Rodda, H. J., and Todd, A. R. (1951). *J. Chem. Soc.* 3.
9. Elion, G., Hitchings, G. H., and Russel, P. B. (1950). *J. Am. Chem. Soc.* **72**, 78.
10. Cain, C. K., Mallete, M. F., and Taylor, E. C. (1946). *J. Am. Chem. Soc.* **68**, 1998.

Recommended Reviews

1. Albert, A. (1952). *The pteridines. Quart. Revs.* **6**, 197.
2. Elderfield, R. C., and Mehta, A. C. (1967). *The pteridines.* In "*Heterocyclic Compounds,*" (R. C. Elderfield, ed.), Vol. 9, p. 1. John Wiley, New York.
3. Pfleiderer, W. (1964). *Recent developments in the chemistry of pteridines. Angew. Chem. Intern. Edit.* **3**, 114.

D. Thin-Layer Chromatography of Pteridines

1. Introduction

Paper chromatography has been extensively applied to naturally occurring pteridines. Systematic investigations of the suitability of various solvents for the separation of natural pteridines have been carried out by Tschesche and Korte (1) with the silkworm *Bombyx mori*, by Blair and Graham (2) with green snakes, by Forrest and Mitchell (3) and Viscontini, Hadorn, and Karrer (4) with fruit flies, by Schmidt and Viscontini (5) with red ants, by Viscontini, Kuhn, and Egelhaaf (6) with the flour moth.

Recently, thin-layer chromatography was used for the separation and identification of pteridines. Nicolaus (7) used silica gel G, while Merlini and Nasini (8) and Ikan and Ishay (9) used cellulose MN–300G. This technique has the following advantages over paper chromatography: smaller amounts may be used, the spots appear smaller, and a better separation can be achieved in a shorter time.

Pteridines may be detected on the chromatoplates by observing their fluorescence under ultraviolet light (254 mμ).

2. Principle

The procedure outlined below is according to (7).

3. Materials

Ammonium hydroxide
Cellulose powder, MN–300G
Citric acid
Dimethylformamide
Pteridines (see Table 10)
Silica gel G

4. Procedure

a. Plates
Silica gel G plates, 0.25 mm thick, 20 × 20 cm, are used.

b. Development
The samples of pteridines are applied as a 1% solution in dilute ammonium hydroxide.

Development is accomplished in four solvent systems (Table 10):

1. dimethylformamide–water, 19:1
2. citric acid, aqueous, 5%
3. ammonium hydroxide, 10%
4. ammonium hydroxide, 25%

c. Detection
The spots on the chromatogram may be detected by observation under ultraviolet light, at 254 and 360 mμ.

Development of Pteridines

	R_f in Solvent System			
Pteridine	1	2	3	4
2-Amino-4-hydroxy	0.55	0.78	0.95	
2-Amino-6,7-dimethyl-4-hydroxy	0.65	0.46	0.91	
2-Amino-4-hydroxy-6-methyl	0.65	0.69	0.93	0.92
2-Amino-4-hydroxy-7-methyl	0.88	0.60	0.92	
2,4-Dihydroxy	0.00	0.88	0.92	
6-Methylisoxanthopterin				0.88
7-Methylxanthopterin				0.88
Isoxanthopterin				0.80
2,4-Dihydroxy (lumazin)				0.83

d. Cellulose plates
The suspension for five plates (20 × 20 cm) is prepared by shaking 15 g cellulose MN–300G with 90 ml distilled water for 30 sec; it is applied to a thickness of 0.5 mm with an applicator. After 30 min at room temperature, the plates are heated at 105°C for 30 min. Development may be accomplished with *n*-propanol–1% ammonia, 2:1.

References

1. Tschesche, R., and Korte, F. (1954). *Ber.* **87**, 1713.
2. Blair, J. A., and Graham, J. (1955). *Chem. Ind.* (*London*) 1158.
3. Forrest, H. S., and Mitchell, H. K. (1955). *J. Am. Chem. Soc.* **77**, 4865.

4. Viscontini, M., Hadorn, E., and Karrer, P. (1957). *Helv. Chim. Acta* **40,** 579.
5. Schmidt, G. H., and Viscontini, M. (1962). *Helv. Chim. Acta* **45,** 1571.
6. Viscontini, M., Kuhn, A., and Egelhaaf, A. (1956). *Z. Naturforsch.* **11b,** 501.
7. Nicolaus, B. J. R. (1960). *J. Chromatogr.* **4,** 384.
8. Merlini, L., and Nasini, G. (1966). *J. Insect Physiol.* **12,** 123.
9. Ikan, R., and Ishay, J. (1967). *J. Insect Physiol.* **13,** 159.

E. Questions

1. What are the main pteridines of bees and wasps?
2. How are pteridines isolated and identified?
3. Describe the occurrence, structure, and biological function of biopterin and folic acid.
4. Explain the structural and biogenetic relationship between purines and pteridines.

F. Recommended Books

1. Pfleiderer, W. (ed.) (1975). *"Chemistry and Biology of Pteridines."* W. de Gruyter, New York.
2. Pfleiderer, W., and Taylor, E. C. (eds.) (1964). *"Pteridine Chemistry."* Proceedings of the Third International Symposium, Stuttgart, 1962. Pergamon Press, London.
3. Wolstenholme, G. E.W., and Cameron, M. P. (eds.) (1954). *"Ciba Foundation Symposium on Chemistry and Biology of Pteridines."* Churchill, London.

VII. PYRAZINES

A. Introduction

Pyrazine is a six-membered ring built up of carbon and two atoms of nitrogen.

B. Isolation and Characterization of Pyrazines, the Volatile Constituents of Bell Peppers

1. Introduction

Pyrazines are important sensory constituents of foods. More than 50 different alkylpyrazines were detected in about 30 natural or manufactured, cooked, roasted, or fermented products of vegetable or animal sources. Among the 43 pyrazines so far identified in cocoa, there are not fewer than 27 alkylpyrazines. 3-Ethyl-2,5-dimethyl pyrazine, for instance, is a contributor to the odoriferous principle of cocoa aroma and one of the most potent odorants derived from potato chips. It is curious to note that the chocolate-smelling alarm pheromone of an American species *Odontromachus* (an ant species) mainly consists of alkylpyrazines.

Among the pyrazines, the acetyl derivatives occupy a particular position as flavoring agents. They have a surprisingly intense and characteristic roasted note, which is reminiscent of popcorn. Acetylpyrazine was detected for the first time in 1969 in sesame oil[1]. It was also detected in popcorn[2], roasted peanuts[3], tobacco[4], roasted coffee[5] and roasted almonds[6].

Odor thresholds of some pyrazines are summarized below:

Compound	*per 10^{12} parts of water*
2-Methoxy-3-hexylpyrazine	1
2-Methoxy-3-isobutyl pyrazine	2
2-Methoxy-3-propyl pyrazine	6
2-Methoxy-3-isopropyl pyrazine	2
2-Methoxy-3-ethyl pyrazine	400
2-Methoxy-3-methyl pyrazine	4,000
2-Methoxy-pyrazine	700,000
2,5-Dimethyl pyrazine	1,800,000
Pyrazine	175,000,000

Green bell peppers are widely accepted for use both as food and as flavoring for other foods. They differ from other members of the genus *Capsicum* in that they generally do not have the hot taste associated with the chili and the tabasco types. The hot taste has long been known to be due to the presence of the essentially nonvolatile compound capsaicin (*N*-(4-hydroxy-3-methoxy-benzyl)-8-methyl-non-*trans*-6-enamide).

The odor threshold of 2-methoxy-3-isobutylpyrazine (the main odorous principle of bell peppers) was found to be 2 parts per 10^{12} parts of water. Months after this compound was synthesized by Buttery and coworkers, his end of the building smelled of freshly chopped green bell peppers.

2. Principle

Green bell peppers are crushed and steam distilled, and the oily product is extracted with hexane.

The volatile constituents are characterized by GC/MS technique. The GC fractions are collected, and their IR and NMR specta are measured. The odorous potency can also be estimated by smelling.

VII. Pyrazines 345

3. Apparatus

   Capillary column
   GC/MS apparatus
   IR spectrometer

4. Materials

   Green bell peppers
   Hexane

5. Time

   3 hours

6. Procedure

The green bell peppers are cut open, the seeds are removed, and the peppers are macerated in a blender. The resulting puree is steam distilled, and the distillate is extracted with hexane. In a typical isolation, a 0.5 kg portion of bell pepper puree yields 1 mg of an oil.

a. Capillary GLC mass spectral analysis

A 1000-foot × 0.03 inch i.d. stainless steel capillary column coated with silicone SF 96-100 is used. The column is programmed from 70° to 170°C at 0.5°C/min. Nitrogen at a 20 psi inlet pressure is used as the carrier gas. The effluent from the end of the column is split, about 10% going to a flame ionization detector and the rest, to a silicone rubber membrane molecular separator of the Llewellyn type. The separator introduces components into the mass spectrometer. The outlet from the molecular separator to the atmosphere generally contains sufficient excess sample for informal sensory odor detection and evaluation.

b. Separation of samples for infrared and proton magnetic resonance spectra

For measurements of IR spectra of major components, samples are separated from the GLC column and collected in 150 × 1 mm borosilicate glass melting-point tubes, which are then sealed at both ends and stored at −20°C until the spectra can be measured. Larger amounts of samples are isolated using a 10-foot × 1/4 inch o.d. silicone SF 95–350 packed GLC column and collecting the samples in 150 × 3 mm borosilicate glass tubes.

   The major GLC peaks are limonene, *trans*-β-ocimene, 2-methoxy-3-isobutyl-pyrazine, and methyl salicylate. The molecular ion of 2-methoxy-3-isobutylpyrazine is 166.

   Aroma significance of bell pepper components may be estimated by smelling.

The table below summarizes the odor thresholds of some bell pepper constituents:

Odor Thresholds of Bell Pepper Constituents

Compound	Threshold in ppb
Hexanal	4.5
Furfural	3000
Hex-*trans*-2-enal	17
Heptan-2-one	140
Hex-*cis*-2-enol	70
Hept-*trans*-3-en-2-one	56
Benzaldehyde	350
2-Pentylfuran	6
Limonene	10
Non-1-en-4-one	0.4
Phenylacetaldehyde	4
Linalool	6
Nona-*trans*-2-en-4-one	0.9
1-Methoxy-3-isobutylpyrazine	0.002
Nona-*trans*,*trans*-2,5-dien-4-one	4
Methyl salicylate	40
Deca-*trans*,*trans*-2,4-dienal	0.07

The volatile oil of chili peppers also contains 2-methoxy-isobutylpyrazine in the same order of concentration as found in bell peppers.

C. Questions

1. Describe the formation pathways of alkyl and acylpyrazines.
2. What are the main pyrazines of coffee, tea, potato chips, popcorn, and peanuts?
3. Describe the organoleptic and sensory properties of some pyrazines.

D. Recommended Books

1. Maga, J. A., and Siger, C. E. (1975). Pyrazines in foods. *In* "*Fenaroli's Handbook of Flavor Ingredients.*" (T. E. Furia and N. Bellanca, eds.), Vol. 1, p. 47. CRC Press, Boca Raton, Florida.
2. Maga, J. A. (1982). Pyrazines in flavour. *In "Food Flavours, Part A."* (I. D. Morton and A. J. Macleod, eds.), p. 283. Elsevier, Amsterdam, The Netherlands.

3. Hurrell, K. F. (1982). Maillard reaction in flavour. *In "Food Flavours, Part A."* (I. A. Morton and A. J. Macleod, eds.), p. 399. Elsevier, Amsterdam, The Netherlands.

References

1. Takei, Y., Nakatani, Y., Kobayashi, A., and Yamanishi, T. (1970). *Chem. Abstr.* **72**, 35670 g.
2. Walradt, J. P., Lindsay, R. C., and Libbey, L. M. (1970). *J. Agr. Food Chem.* **18**, 926.
3. Walradt, J. P., Pittet, A. O., Kinlin, T. E., Maralidhara, R., and Sanderson, A. (1971). *J. Agr. Food Chem.* **19**, 972.
4. Demole, E., and Berthet, D. (1972). *Helv. Chim. Acta.* **55**, 1866.
5. Vitzthum, O. F., and Werkhoff, P. (1974). *Z. Lebensm. Unters, Forsch.* **156**, 300.
6. Takei, Y., and Yamanashi, T. (1974). *Agric. Biol. Chem.* **38**, 2329.
7. Buttery, R. G., Seifert, R. M., Gaudagni, D. G., and Ling, L. C. (1969). *J. Agr. Food Chem.* **17**, 1322.

Recommended Reviews

1. Maga, J. A. (1982). Pyrazines in flavour. *In* "Food Flavours, Part A." (I. D. Morton and A. J. Macleod, eds.), p. 283. Elsevier, Amsterdam, The Netherlands.
2. Ohloff, G., and Flament, I. (1976). Recent developments in the field of naturally occurring aroma components. *Prog. Chem. Org. Nat. Prod.* **35**, 431.
3. Ohloff, G., and Flament, I. (1979). The role of heteroatomic substances in the aroma compounds of foodstuffs. *Prog. Chem. Org. Nat. Prod.* **36**, 231.

BIBLIOGRAPHY

Recommended Books on Infrared Spectroscopy

Bellamy, L. J. (1958). *"Infrared Spectra of Complex Molecules,"* Methuen, London, England.

Brame, E. G. and Grasselli, J. (1976). "Infrared and Raman Spectroscopy." Marcel Dekker, New York.

Colthup, N. B., Daly, L. H., and Wiberley, S. E. (1964). *"Introduction to Infrared and Raman Spectroscopy."* Academic Press, New York.

Durig, J. R., ed. (1990). *"Applications of FT-IR Spectroscopy."* Elsevier, New York.

Ferraro, J. R., and Krishnan, K. (1989). *"Practical Fourier Transform Infrared Spectroscopy."* Academic Press, New York.

Griffith, P. R., and Haseth, J. A. (1986). "Fourier Transform Infrared Spectrometry." John Wiley, New York.

Miller, R. G. J. ed. (1965). *"Laboratory Methods in Infrared Spectroscopy."* Heyden and Son, London, England.

Nakanishi Koji (1962). *"Infrared Absorption Spectroscopy."* Holden Day, San Francisco and Nankodo Company, Tokyo.

Pouchert, C. J. (1981). *"The Aldrich Library of Infrared Spectra."* Aldrich Chemical Co. Milwaukee, Wisconsin.

Pouchert, C. J. (1989). *"The Aldrich Library of FT-IR Spectra."* Aldrich Chemical Co. Milwaukee, Wisconsin.

Rao, C. N. R. (1963). *"Chemical Applications of Infrared Spectroscopy."* Academic Press, New York.

Smith, A. L. (1979). *"Applied Infrared Spectroscopy."* John Wiley, New York.
Szymanski, H. A. (1962). *"Infrared Handbook."* Plenum Press, New York.
Wilson, N. K., and Childers, J. W. (1989). Recent advances in the matrix isolation infrared spectrometry of organic compounds. *Applied Spectroscopy Reviews,* **25** (1), 1–61.

Recommended Books on Nuclear Magnetic Resonance Spectroscopy

Bovey, F. A. (1988). *"Nuclear Magnetic Reasonance Spectroscopy."* Academic Press, New York.
Duddeck, H., and Dietrich, W. (1989). *"Structure Elucidation by Modern NMR."* Springer-Verlag, New York.
Levy, G. C., Lichter, R. L., and Nelsen, G. L. (1980). *"Carbon-13 Nuclear Magnetic Resonance Spectroscopy."* John Wiley, New York.
Linskens, H. F., and Jackson, J. E. (1988). *"Nuclear Magnetic Resonance."* Springer-Verlag, New York.
Marshall, A. G., and Verdun, F. R. (1989). *"Fourier Transforms in NMR, Optical, and Mass Spectrometry."* Elsevier, Amsterdam, The Netherlands.
Mehring, M. (1983). *"High Resolution NMR Spectroscopy of Solids."* Springer-Verlag, New York.
Pouchert, C. J. (1983). *"The Aldrich Library of NMR Spectra."* Aldrich Chemical Co., 1983.
Shaw, D. (1984). *"Fourier Transform NMR Spectroscopy."* Elsevier, Amsterdam, The Netherlands.
Wilson, M. A. (1987). *"NMR Techniques and Applications in Geochemistry and Soil Chemistry."* Pergamon Press, London, England.

Recommended Books on Ultraviolet Spectroscopy

Gillam, A. E., and Stern, E. S. (1957). *"Introduction to Electronic Absorption Spectroscopy in Organic Chemistry."* Arnold, London, England.
Jaffe, H. H., and Orchin, M. (1962). *"Theory and Applications of Ultraviolet Spectroscopy."* John Wiley, New York.
Rao, C. N. R. (1961). *"Chemical Applications of Ultraviolet and Visible Spectroscopy."* Butterworth, London, England.
Scott, A. I. (1964). *"Interpretation of the Ultraviolet Spectra of Natural Products."* Pergamon, London, England.
(See also general textbooks on spectroscopy.)

General Textbooks on Spectroscopy

Baker, A. J., and Cairns, T. (1965). *"Spectroscopic Techniques in Organic Chemistry."* Heyden and Son, London, England.
Braude, E. A., and Nachod, F. C. (1955). *"Determination of Organic Structures by Physical Methods."* Vol. 1. Academic Press, New York.
Brand, J. C. D., and Eglinton, G. (1965). *"Applications of Spectroscopy to Organic Chemistry."* Oldbourne Press, London, England.
Colthup, N. B., Daly, L. H., and Wiberley, S. E. (1990). *"Introduction to Infrared and Raman Spectroscopy."* Academic Press, New York.

Flett, M. St. C. (1962). *"Physical Aids to the Organic Chemist."* Elsevier, Amsterdam, The Netherlands.

Fretsch, E., Clere, T., Seibl, J., and Simon, W. eds. (1989). *"Table of Spectral Data for Structure Determination of Organic Compounds."* Springer-Verlag, New York.

Pinder, A. R. (1965). *"Physical Methods in Organic Chemistry."* English Universities Press, London, England.

Schwarz, J. C. P. (ed.) (1964). *"Physical Methods in Organic Chemistry."* Oliver and Boyd, Edinburgh, Scotland.

Silverstein, R. M., and Bassler, G. C. (1963). *"Spectrometric Identification of Organic Compounds."* John Wiley, New York.

Silverstein, R. M., Bassler, G. C., and Morrill, T. C. (1985). *"Spectroscopic Identification of Organic Compounds."* John Wiley, New York.

Sixma, F. L. J., and Wynberg, H. (1964). *"A Manual of Physical Methods in Organic Chemistry."* John Wiley, New York.

Recommended Books on Gas Chromatography

Burchfield, H. P., and Storrs, E. E. (1962). *"Biochemical Applications of Gas Chromatography."* Academic Press, New York.

Jennings, W. (1980). *"Gas Chromatography with Glass Capillary Columns."* Academic Press, New York.

Ioffe, B. V., and Vitenberg, A. G. (1984). *"Head-Space Analysis and Related Methods in Gas Chromatography."* John Wiley, New York.

Pery, J. A. (1981). *"Introduction to Analytical Gas Chromatography."* Marcel Dekker, New York.

Verpoorte, R., and Baerheim-Svendsen, A. (1984). *"Chromatography of Alkaloids, Part B: Gas-Liquid Chromatography and High-Performance Liquid Chromatography."* Elsevier, Amsterdam, The Netherlands.

Recommended Books on Thin-Layer Chromatography

Baerheim-Svendsen, A., and Verpoorte, R. (1983). *"Chromatography of Alkaloids, Part A: Thin-Layer Chromatography."* Elsevier, Amsterdam, The Netherlands.

Fried, B., and Sherma, J. (1982). *"Thin-Layer Chromatography."* Marcel Dekker, New York.

Kirchner, J. G. (1978). *"Thin-Layer Chromatography."* John Wiley, New York.

Stahl, E. (1965). *"Thin-Layer Chromatography."* Academic Press, New York.

Touchstone, J. C., and Dobbins, M. F. (1978). *"Practice of Thin-Layer Chromatography."* John Wiley, New York.

Wilson, I. D., and Ruane, R. J., ed. (1987). *"Prospects for Chiral TLC."* D. Stevenson and I. D. Wilson, Pleum Press, p. 135. New York.

Zlatkes, A., and Kaiser, R. E., ed. (1977). *"HPTLC-High Performance Thin-Layer Chromatography."* Elsevier, Amsterdam, The Netherlands.

Reocmmended Books on Mass Spectroscopy and Gas Chromatography–Mass Spectrometry

Gilbert, J. (ed.) (1987). *"Applications of Mass Spectrometry in Food Science."* Elsevier Applied Science, Amsterdam, The Netherlands.

Gudzinowicz, B. J., Gudzinowicz, M. J., and Martin, H. F. (1977). *"Fundamentals of Integrated GC-MS."* Marcel Dekker, New York.

Karasek, F. W., and Clement, R. E. (1988). *"Basic Gas Chromatography–Mass Spectrometry."* Elsevier, Amsterdam, The Netherlands.

Middleditch, B. S. (ed.) (1981). *"Practical Mass Spectrometry."* Plenum Press, New York.

Millard, B. J. (1979). *"Quantitative Mass Spectrometry."* Heyden.

Recommended Books on Liquid and High-Performance Liquid Chromatography

Engelhardt, H. (ed.). (1985). *"Practice of High-Performance Liquid Chromatography."* Springer-Verlag, New York.

Hamilton, R. J., and Sewell, P. A. (1977). *"Introduction to High-Performance Liquid Chromatography."* Chapman and Hall, New York.

Horvath, C., ed. (1980). *"High-Performance Liquid Chromatography."* Academic Press, New York.

Pryde, A., and Gilbert, M. T. (1979). *"Application of High-Performance Liquid Chromatography."* Chapman and Hall, New York.

Snyder, L. R., and Kirkland, J. J. (1979). *"Introduction to Modern Liquid Chromatography."* John Wiley, New York.

SUBJECT INDEX

Abietic acid
  ^{13}C-NMR of, 211
  dehydrogenation of, 209, 212
  formation of, from wood rosin, 209, 211
  ^{1}H-NMR of, 211
  MS of, 211
  occurrence of, 210
  UV spectrum of, 212
Alkoloids
  biosynthesis of, 230
  classification of, 229
  degradation of, 227
  function of, in plants, 226
  GC of, 247
  isolation of, 226
  occurrence of, 226
  physiological properties of, 226
  precipitants for, 227
Amino acids
  biosynthesis of, 257
  classification of, 254–255
  GC of, 277
  isolectric point of, 254
  properties of, 254
  resolution of, 256
  separations of, 254
  syntheses of, 255–256
  Test-tube TLC, 280
  TLC of, 273
  TLC of carbobenzoxy peptides, 282
Anthocyanins
  alkaline degradation of, 7
  classification of, 6
  colors of, 7
  color change of, at various pH values, 19–21
Aromatic herbs
  head-space GC of, 205–208

  quantitative analysis of aromatic constituents of, 208
Azelaic acid
  ^{1}H-NMR of, 31
  MS of, 32
  preparation of, 30

Betulin
  allobetulin, formation of, 215
  ^{13}C-NMR of, 214
  conversion of, to allo- and oxyallobetulin, 213
  IR spectrum of, 215
  IR spectrum of allobetulin, 215
  isolation of, from birch bark, 214
  MS of, 215
  occurrence of, 212
Bile acids
  occurrence of, 131
  structure of, 131–132
Bitter principles
  of citrus, removal of, 220–224
  of hops, 68
Brucine
  ^{13}-C NMR of, 242
  ^{1}H-NMR of, 242
  isolation of, from *Strychnos nux vomica*, 241
  occurrence of, 238
  physiological properties of, 239

Caffeine
  ^{13}C-NHR of, 233
  content of, in tea, 230
  ^{1}H-NMR of, 232
  IR spectrum of, 232
  isolation of, from tea, 230–232

Caffeine (*continued*)
  occurrence of, 230
  physiological properties of, 231
  TLC of xanthine derivatives, 232
  UV spectrum of, 232
Camphene
  formation of, 194
  ^{1}H-NMR of, 194
  mechanism of formation of, 192
Camphor
  formation of, 192–195
  ^{1}H-NMR of, 195
  UV spectrum of, 195
Capsanthin
  ^{13}C-NMR of, 112
  color reactions of, 111
  ^{1}H-NMR of, 112
  isolation of, from paprika, 110–111
  MS of, 112
  occurrence of, 110
  UV spectrum of, 113
Carbohydrates
  classification of, 70
  column chromatography of, 84
  gas-liquid chromatography of, 86
  glucosides of, 84
  HPLC of, 90
  mannoheptulose, 79
  mutarotation, 75
  oligosaccharides, 76
  thin-layer chromatography of, 89
β-Carotene
  column chromatography of, 124
  ^{1}H-NMR of, 102
  spectroscopic determination of, 125
  TLC of, 273
Carotenoids
  biosynthesis of, 109
  classification of, 105–108
  dehydrogenation of, 108
  functions of, 109
  HPLC of, 120
  hydrocarbons, 105
  occurrence of, 105
  of oranges, 117
  syntheses of, 108
  TLC of, 117
  xanthophylls, 106–108
Carvone
  ^{13}C-NMR of, 185
  enantiomers of, 187
  IR of, 185

^{1}H-NMR of, 185
MS of, 185
synthesis of, 183–184
UV of, 185
Castor oil
  formation of azelaic acid from, 30–31
  ^{1}H-NMR of, 34
  n-heptaldehyde from, 32, 33
  hydrolysis of, 30
  IR of undecenoic acid, 33
  MS of azelaic acid, 34
  pyrolysis of, 33
  ricinoleic acid from, 30
  undecenoic acid from, 33
Cerin
  IR of, 218
  isolation of, from cork, 217
Chlorophyll
  degradation of, 307
  occurrence of, 306
  structure of, 306, 308
Cholesterol
  biosynthesis of, 129
  ^{13}C-NMR of Δ^4-cholesten-3-one, 153
  Δ^4-cholesten-3-one from, 149
  Δ^5-cholesten-3-one from, 149
  IR of Δ^4-cholesten-3-one, 153
  IR of Δ^5-cholesten-3-one, 152
  UV of Δ^4-cholesten-3-one, 154
Column chromatography
  of carbohydrates, on charcoal, 84
  of carotenes, 125
  of essential oil, 198

Ergosterol
  irradiation of, 156
  TLC of, 157
Essential oil
  flash chromatography of, 198–200

Fatty acids
  classification of, 23, 24
  GC of methyl esters, 49
  isolation of, 23
  reactions of, 24, 25
  TLC of, 44
Flavanones
  chemical properties of, 5
Flavonoids
  alkaline degradation of, 3

biosynthesis of, 8
classification of, 1, 2
interconversion of, 3
occurrence of, 2
physiological properties of, 3
Friedelin
   ^{13}C-NMR of, 219
   IR of, 220
   isolation of, from cork, 219
   MS of, 220
   occurrence of, 217

Gas-liquid chromatography
   of alkaloids, 247
   of amino acids, 277
   of fatty acid esters, 49
   of steriods, 161–164
   preparation of methyl esters for, 43
Glucosamine from crustacean shells, 82
Glutamine
   ^{1}H-NMR of, 262
   IR of, 261
   isolation of, from red beet, 260–261
   occurrence of, 259
Glycylglycine
   ^{1}H-NMR of, 273
   IR of, 272
   syntheses of, 269–272

Hecogenin
   IR of, 148
   occurrence of, 146
   transformations of, 147
Hesperidin
   acidic degradation of, 12
   ^{13}C-NMR of, 11
   color tests of, 11
   isolation of, orange peel, 9
   MS of, 11
   occurrence of, 9
   physiological properties of, 9
   purification of, 10
Hesperitin
   ^{13}C-NMR of, 11
   formation of, 12
   UV of, 13
Hydroquinone
   of insects, 61
   IR of, 64
   ILC of, 63–64

Isoborneol
   acetate of, 194
   mechanism of formation of, 193
Isoflavones
   alkaline degradation of, 5
   occurrence of, 5

Leucoanthocyanidins
   occurrence of, 8
Lignans
   classification of, 48
   dehydrogenation of, 49
   formation of, 48
   isomerization of, 48
   physiological properties of, 50
Limonene
   carvacrol from, 182, 186
   carvone from, 182, 184
   ^{1}H-NMR of, 184
   occurrence of, 182
   photoprotonation of, 195–198
Lipids
   classification of, 23–24
   extraction of, 23
   phospholipids, 25
   waxes, 25
Lycopene
   ^{13}C-NMR of, 116
   color reactions of, 116
   ^{1}H-NMR of, 117
   isolation of, from tomatoes, 115
   MS of, 117
   occurrence of, 113
   UV of, 117
Lysozyme
   derivatives of, 326
   IR of, 326
   isolation of, from albumen, 324–325
   properties of, 324

Mescaline
   MS of, 246
   physiological properties of, 242–243
   syntheses of, 243–246
   UV of, 243
Methyl oleate
   azelaic semialdehyde from, 34–35
   ozonization of, 34–35
   pelargonaldehyde from, 34–35

Subject Index

Monosaccharides
  anomers, 74
  branched-chain sugars, 72
  configuration of, 71
  degradation of pentoses, 75
  deoxy-sugars, 72
  epimers, 72
  fermentation of starch, 73
  furanose form, 74
  glucosamine, from crustacean shells, 82
  heptoses, occurrence of, 71
  hydrazones of, 75
  mannoheptulose, from avocado, 79–81
  mutarotation, 75
  osazones of, 75–76
  osone of, 76
  perseitol, from avocado, 79
  pyranose form, 74
Myristic acid
  IR of, 29
  preparation of, 28
Myristicin
  dibromo derivative of, 29
  isolation of, 28
  MS of, 29
  occurrence of, 26
  physiological activity of, 26

Naringin
  action of naringinase on, 15
  bitterness of, 17
  ^{13}C-NMR of, 16
  isolation of, from grapefruit peel, 14
  naringenin from, 15
  occurrence of, 14–15
Neohesperidin
  bitterness of, 17
Neohesperidin dihydrochalcone
  synthesis of, 17
  sweetness of, 18
Nucleic acids
  classification of, 285
  deoxyribonucleic acid
    structure of, 290
  hydrolysis of, 287
  hydrolysis of nucleotides, 288
  nucleosides of, 286
  nucleotides of, 287
  purines of, 286
  pyrimidines of, 286
  ribonucleic acid
    hydrolysis of by alkali, 288
    hydrolysis of by ribonuclease, 299
    isolation of sRNA from baker's yeast, 291–292
    types of, 287
Nucleotides
  cyclic nucleotides enzymatic preparation of, 296–298
  synthesis of, 297
  uridine 2′,3′-cyclic phosphate, 296
  TLC of, 300
  thymidine 3′-phosphate, synthesis of, 295

Oligosaccharides
  classification of, 76–77
  glycosides, 76
Oxyallobetulin
  IR of, 216
  preparation of, 216
Ozonization
  of methyl oleate, 34–36
  of stigmasterol, 140–141

Peptides
  biologically active, 258
  structural determination of, 257
  structure of, 257
  synthesis of, 258
β-Phenylalanine
  azlactone synthesis of, 266–268
  ^{1}H-NMR of, 268
  IR of, 268
  MS of, 268
  syntheses of, 285
Phloroglucinols
  structure of, 66
  HPLC of, 67
  in hops, 68
Phosphate buffers, 291
α-Pinene
  camphor from, 194–195
    mechanism, 192–193
  hydrochloride of, 194
  isolation of, 194
  occurrence of, 191
Piperine
  ^{13}C-NMR of, 235
  degradation of, 236–238
  IR of, 236
  isolation of, from black pepper, 234–235
  ^{1}H-NMR of, 235
  occurrence of, 233

Subject Index   355

TLC of, 235
UV of, 236
Polysaccharides
  amylopectin, from potato starch, 97–100
  amylopectin, structure of, 95
  amylose, from potato starch, 97–100
  amylose, structure of, 94
  cellulose, 93
  chitin, 82, 96, 97
  dextrans, 96
  fermentation of starch, 94
  fructans, 95
  glycogens, 95
  heparin, 97
  hydrolysis of chitin, 82
  inulin, 95
  levan, 96
  occurrence of, 93
  pectins, 96
  plant gums, 96
  starch, structure of, 94
  wood cellulose, isolation of, 101–103
Porphyrins
  biosynthesis of, 308
  degradation of hemin, 311
  degradation of hemoglobin, 305
  function of, 304
  isolation of hemin, 309
  metalloporphyrins from petroleum, 317–320
  occurrence of, 304–305
  properties of, 304–305
  structure of, 304, 306
  TLC of plant pigments, 315
Proteins
  classification of, 321–323
  enzymes, 321
  isolation of, from groundnuts, 326
  occurrence of, 321
  structure of, 323
  TLC of, 328
Pteridines
  2-amino-4-hydroxypteridine, UV spectrum of, 340
  isolation of, from fruit fly, 334
  occurrence of, 331–332
  structure of, 332
  synthesis of, from pyrazines, 334
  synthesis of, from pyrimidines, 333, 338
  TLC of, 341
Pyrazines
  GC/MS of, 345
  isolation of, 344
  occurrence of, 343

odor threshold of, 346

Quinones
  biosynthesis of, 59
  classification of, 54–59
  ^{1}H-NMR of, 65
  IR of, 64
  MS of, 64
  occurrence of,
  in insects, 58–59, 61–62
  TLC of, 63

Rhein
  diacetate of, 61
  isolation of, 60
  occurrence of, 59
  UV of, 61
  TLC of, 60
Ricinoleic acid
  from castor oil, 30

Sesamin
  IR of, 53
  isolation of, 51
  nitro-derivative of, 52
  occurrence of, 50–51
Sesamolin, 50
  ^{1}H-NMR of, 52
  IR of, 53
Steroid hormones
  adrenocortical, 133
  androgens, 134
  estrogens, 133
  insect molting, 134
Steroids
  cardiac active, 134–136
  conformation of, 128
  GC of, 161–164
  occurrence of, 127
  sapogenins, 135
  structure of, 127–128
Sterols
  of algae, 142–146
  determination of, by digitonin, 159–161
  GC of, 161
  of marine invertebrates, 129–130
  of plants, 130
  of yeast, 131
  structures of, 129
  TLC of, 165–167

Stigmasterol
   ^{13}C-NMR of, 138
   degradation of, 139–142
   isolation of, from soybean oil, 137–138
   MS of, 139
   occurrence of, 136
   UV of, 139
Strychnine
   ^{13}C-NMR of, 240
   isolation of, from *Strychnos nux vomica*, 240
   MS of, 241
   occurrence of, 238
   physiological properties of, 239
   UV of, 241

Tannins, 8
Terpenoids
   classification of, 168
   distribution of, 168
   diterpenes,
      acyclic, 174–175
      bicyclic, 175
      monocyclic, 175
      tricyclic, 176
   essential oil, extraction of, 187
   monoterpenes
      acyclic, 169
      bicyclic, 171–172
      monocyclic, 170
   sesquiterpenes
      acyclic, 172
      azulenes, 174
      bicyclic, 173
      monocyclic, 173
   triterpenes
      acyclic, 178
      biosynthesis of, 181
      classification of, 176
      dehydrogenation of, 176
      pentacyclic, 179–180
      tetracyclic, 178
Thin-layer chromatography
   impregnation of plates for, with silver nitrate, 166
   of alkaloids, 250
   of amino acids, 273, 280
   of carbobenzoxy peptides, 282
   of carbohydrates, 89–90
   of carotenoids, 117
   of ergosterol, 157
   of fatty acids, 44
   of nucleotides, 300
   of plant pigments, 315
   of proteins, 328
   of pteridines, 341
   of steroids, 165–167
   of terpenoids, 223
   of vitamin D_2, 158
Thymidine 3-phosphate
   synthesis of, 293–296
Trimyristin
   IR of, 28
   isolation of, 27
   MS of, 28
   occurrence of, 26
   saponification of, 28
Tyrosine
   D-tyrosine, from racemic, 252
   racemization of, 262
   resolution of racemic, 264

Undecenoic acid
   ^{1}H-NMR of, 34
   IR spectrum of, 33
   MS of, 34
   preparation of, 33
Urea inclusion compounds
   of binary systems of fatty acids, 41
   of *cis-trans* isomers, 39
   dissociation temperature of, 40
   of fatty acids, 39
   of petrol, 39
   of saturated and unsaturated acids, 37
   structure of, 36–38
   of terpenes, 39
   thiourea inclusion compounds, 38

Vitamin D_2
   ^{13}C-NMR of, 158
   formation of, 155–157
   ^{1}H-NMR of, 158
   IR of, 159
   MS of, 158
   TLC of, 157
   UV of, 158

Wine
   aroma constituents, 201
   GC of aroma of, 203
   grape juice terpenes, 203

INDEX OF CHEMICAL COMPOUNDS

Abietic acid, 209
3-Acetoxy-bisnorcholanic acid, 141
3-Acetoxy-bisnorcholenic acid, 141
N-Acetyl-D-tyrosine, 264
N-Acetyl-DL-tyrosine, 263
Adenosine diphosphate, 302
Adenosine monophosphate, 302
Adenosine triphosphate, 302
Adhumulone, 69
α-Alanine, 279, 281
β-Alanine, 279
DL-Alanine
Allobetulin, 215
Allobetulin acetate, 216
Allocholanic acid, 142
Allose, 88
α-Aminobutyric acid, 279
β-Aminobutyric acid, 279
γ-Aminobutyric acid, 279
2-Amino-4-hydroxypteridine, 342
2-Amino-4-hydroxy-6-methylpteridine, 342
2-Amino-4-hydroxy-7-methylpteridine, 342
2-Amino-4-hydroxy-6,7-dimethylpteridine, 342
Amylopectin, 97
Amylose, 97
Androstan-17-one, 163
Androstan-3,17-dione, 163
Androstane, 163
Anthocyanins, 20
Arabinose, 88
Arachin, 327
Arginine, 276
Asparagine, 276
Aspartic acid, 274, 279, 281
Atropine, 249

Azelaic acid, 31
Azlactone of α-benzoylcinnamic acid, 267

Benzaldehyde, 346
Benzoquinone, 63
Benzyl alcohol, 201
Betulin, 214
Bovin γ-globulin, 330
Bovin serum albumin, 330
Brucine, 241
Butyric acid, 42

Caffeine, 231, 249
Campestanol, 166
Campesterol, 166
Camphene, 194
Camphor, 195
Capric acid, 39
Caproic acid, 39, 42
Caprylic acid, 39, 42
Capsanthin, 111
Carbobenzoxydiglycine, 284
Carbobenzoxydigly-OEt, 284
Carbobenzoxy-DL-alanine, 284
Carbobenzoxy-DL-dialanine, 284
Carbobenzoxy-DL-phenylalanine, 284
Carbobenzoxyglycine, 284
Carbobenzoxytetraglycine, 284
Carbobenzoxytriglycine, 284
Carbobenzoxytrigly-OEt, 284
Carotene, α, β, γ- 119
Carotene, β- 125
Carvacrol, 182

Index of Chemical Compounds

Carvone, 182, 189
Carvoxime, 182
Castor oil, 32, 33
Cellobiose, 88
Cellulose, 103
Cerin, 218
Chitin, 82
Chlorophyll, 316
Cholestane, 163
Cholestanol, 163, 166
Cholestan-3-one, 163
Cholestanyl acetate, 163
Cholestanyl methyl ether, 163
Δ^4-Cholesten-3-one, 152, 163
Δ^5-Cholesten-3-one, 163
Cholesterol, 151, 160, 163, 166
Cholesteryl acetate, 151
Chondrillasterol, 143
α-Chymotrypsin, 330
Cinchonidine, 248
Cinchonine, 248
Citraurin, 119
Citronellol, 201
Cocaine, 248
Codeine, 249, 251
Cohumulone, 69
Conarachin, 327
Coprostane, 163
Coprostanol, 166
Corn oil, 39
Cryptoxanthin, 119
Cyanoethylphosphate, 295
Cytidine diphosphate, 302
Cytidine triphosphate, 302
Cytochrome C, 330

5α,6β-Dibromocholestan-3-one, 151
Digitonin, 160
2,4-Dihydroxypteridine, 342
3,7-Dimethyl-octa-1-en-3,7-diol, 204
3,7-Dimethyl-octa-1,5-diene-3,7-diol, 204
Drosopterin, 337

Elaidic acid, 39, 40
Ergosterol, 143, 155, 157, 166
Essential oils, 199
Ethylbenzoquinone, 65
Friedelin, 219
Fructose, 91
cis-Furan linalool oxide, 201

trans-Furan linalool oxide, 201
Furfural, 346

Galactose, 88
Geraniol, 201
Glucosamine hydrochloride, 82
Glucose, 13, 85, 88, 89,91
Glutamic acid, 279
Glutamine, 260, 274
Glycine, 279, 281, 283
Glycosides, 203
Glycylglycine, 283
Guanosine diphosphate, 302
Guanosine monophosphate, 302
Guanosine triphosphate, 302
Gulose, 88

Hecogenin, 146
Hecogenone, 147
Hematinic acid, 314
Hemin, 313
Hemoglobin, 330
n-Heptaldehyde, 33
Heptan-2-one, 346
Hesperidin, 9, 10, 11, 12
Hesperitin, 12, 13
Hexanol, 346
Histidine, 277
Homatropine, 249
Hotrienol, 201
Humulone, 69
Hydrastine, 249
Hydroquinone, 63
Hydroxyproline, 277
4-Hydroxy-2,5,6-triaminopyrimidine, 338

Inosine diphosphate, 302
Inosine monophosphate, 302
Inosine triphosphate, 302
Isoborneol, 194
Isobornyl acetate, 194
Isodrosopterin, 336
Isofucosterol, 143
Isoleucine, 274, 279, 281
Isosepiapterin, 336
Isoxanthopterin, 342

2-Keto-bisnorcholanic acid, 142

Lactose, 88, 91
Lauric acid, 39, 40
Leucine, 279
Limonene, 162, 196, 346
Limonin, 220
Linalool, 201, 346
Linoleic acid, 39, 40, 42
Linolenic acid, 39, 40
Linseed oil, 39
Lupulone, 69
Lutein, 119
Luteoxanthin, 119
Lycopene, 114
Lysine, 277
Lysozyme, 328, 330
  bromide, 325
  carbonate, 325
  chloride, 325

Macroglobulins, 330
Maltose, 85, 88
Mannoheptulose, 81
Mannoheptulose hexacetate, 81
Mannose, 88
Melibiose, 85
Mescaline, 246
Mesoporphyrin, 302
Menth-1-ene-9-ol, 201
2-Methoxy-2-isobutylpyrazine, 340
Methyl α-arabinoside, 88
Methyl β-arabinoside, 88
Methyl benzoquinone, 65
Methyl elaidate, 39
Methylethylmaleimide, 314
Methyl α-glucoside, 88
Methyl β-glucoside, 88
6-Methylisoxanthopterin, 342
Methyl α-mannoside, 88
Methyl β-mannoside, 88
Methyl oleate, 35, 45
Methyl salicylate, 346
Methyl semiazelaaldehyde, 36
Methyl stearate, 45
Methyl-3,4,5-trimethoxybenzoate, 246
7-Methylxanthopterin, 342
Methyl α-xyloside, 88
Methyl β-xyloside, 88
Monoterpenes, 206
Morphine, 249, 251
Myoglobin, 330
Myristic acid, 28, 39, 40, 42

Myristicin, 28

Narcotine, 249, 251
Naringin, 15, 16, 17, 220
Naringin dihydrochalcone, 17, 18
Neodrosopterin, 337
Neohesperidin, 9
Nerol, 201
Nomilin, 220
Norleucine, 279
Norvaline, 279

Oleic acid, 39, 40, 42
Olive oil, 39
Ornithine, 279
Ovalbumin, 330
Ovomucoid, 330
Oxyallobetulin, 216
Oxyallobetulin acetate, 216
3-Oxy-bisnorcholanic acid, 142

Palmitic acid, 39, 40, 42
Papaverine, 249, 251
Pelargonaldehyde, 35
Pelargonic acid, 39, 40
Pentylfuran, 346
Pepsin, 330
Perseitol, 81
β-Phenylalanine, 267, 274, 279, 283
Phenylethyl alcohol, 201
Phloracetophenone
Phytofluene, 119
Pilocarpine, 289
Pinene, α- and β-, 194
Piperic acid, 237
Piperine, 234
Poriferasterol, 143
Porphyrins, 319
Proline, 274, 279
Protoporphyrin, 313

Quinidine, 248
Quinine, 248

Raffinose, 88, 91
Ribonuclease, 299, 330
Ribose, 88

Rhamnose, 13, 88, 89
Rhein, 59
Ricinoleic acid, 31
sRNA, 292
Rockogenin, 148

Scopolamine, 249
Sepiapterin, 336
Serine, 279
Sesame oil, 51
Sesamin, 52
Sesamolin, 52
β-Sitostanol, 166
β-Sitosterol, 163, 166
β-Sitosteryl, acetate, 163
Soybean oil, 39
Starch, 99
Stearic acid, 40, 42
Stigmastane, 163
Stigmasterol, 137, 138, 141, 163, 163
Stigmasteryl acetate, 141
Strychnine, 240, 249
Sucrose, 85, 88, 89, 91

Tachysterol, 157
Talose, 88
Thebaine, 249, 251
Threonine, 277, 279
Thymidine 3-phosphate, 295
Thyroglobulin, 330
Tigogenin, 148

Tridecylic acid, 39, 40
3,4,5-Trimethoxybenzoic acid, 244
3,4,5-Trimethoxybenzyl alcohol, 245
3,4,5-Trimethoxybenzyl chloride, 245
3,4,5-Trimethoxyphenyl acetonitrile, 246
Trimyristin, 27
Tritylthymidine, 295
Trypsin, 330
Tryptophan, 274
Turpentine oil, 274
Tyrosine, 263, 274, 277, 279, 281

Undecenoic acid, 33
Urea, 36, 39
Uridine 2',3'-cyclic phosphate, 297
Uridine diphosphate, 297, 302
Uridine monophosphate, 297, 302
Uridine 2'-phosphate, 298
Uridine triphosphate, 302

Valine, 274, 277
Violaxanthin, 119
Vitamin D_2, 147
Vitamin D_3, 157

Xylose, 85, 88, 89

Zeaxanthin, 119